S0-BER-596

HISTORY OF THE HOUR

HISTORY OF THE HOUR

Clocks and Modern Temporal Orders

Gerhard Dohrn-van Rossum

Translated by Thomas Dunlap

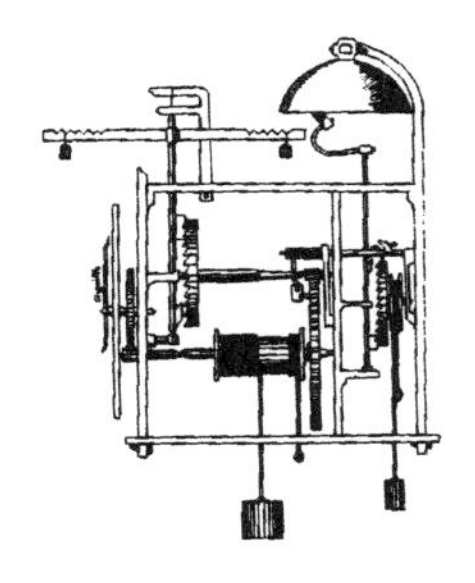

THE

UNIVERSITY

OF CHICAGO PRESS

•

CHICAGO & LONDON

Gerhard Dohrn-van Rossum teaches medieval and early modern history at the University of Bielefeld. He has served as visiting professor at the University of Chicago and the University of Zurich.

The University of Chicago Press, Chicago 60637
The University of Chicago Press, Ltd., London

Printed in the United States of America
05 04 03 01 00 99 98 97 96 1 2 3 4 5
ISBN: 0-226-15510-2 (cloth)
0-226-15511-0 (paper)

Originally published as *Die Geschichte der Stunde: Uhren und moderne Zeitordnungen,* © 1992 Carl Hanser Verlag München Wien. This edition has been supported by funds from Inter Nationes, Bonn.

Library of Congress Cataloging-in-Publication Data

Dohrn-van Rossum, Gerhard, 1947–
[Die Geschichte der Stunde. English]
History of the hour: clocks and modern temporal orders/Gerhard Dohrn-van Rossum; translated by Thomas Dunlap.
p. cm.
Translation of: Die Geschichte der Stunde.
Includes bibliographical references and index.
(pbk. :alk. paper)
1. Clocks and watches—History. 2. Time measurements—History.
I. Title.
QB107.D6413 1996
529′.7′09—dc20 95-47660
CIP

∞ The paper used in this publication meets the minimum requirements of the American National Standard for Information Sciences—Permanence of Paper for Printed Library Materials, ANSI Z39.48-1984.

Contents

FIGURES

1

Introduction

The mechanical clock and the transformation of time-consciousness

Around the year 1410, at the dawn of Europe's modern era, an anonymous author assumed the role of an English friar and described contemporary innovations in the ways people dealt with time. In his exegesis of the First Commandment, he condemned the art of the astrologers, explaining that God had created the firmament out of light and time, like a flawless clock. Light and time, the starry heavens, were meant to serve man, not the other way around. From the firmament as a great clock, he proceeded to the striking clocks, new in his day, and explained that just as the heavenly bodies did not rule the earthly creatures, clocks in cities large and small did not rule men. Rather, in cities and towns people ruled themselves by the clock.[1] With regard to the ways in which time was handled, but also to the type of time-keeping devices that were used, our preacher differentiated monastic and urban life as spheres of socially structured everyday time, quietly excluding life in the countryside.

In an inconspicuous place, a contemporary thus sketched a process which today, barely a century after its discovery, has become textbook wisdom: the transformation of time-consciousness in the late Middle Ages.

In the nineteenth century, working time became a central theme in the "social question" that divided political movements and parties. The struggles over the normal working day, and the spreading notion of working time as a factor of production that was scarce and therefore had to be put to effective use, turned many old problems of justice, distribution, and money into problems of time. Railway traffic and the telegraph had put the fact that urban, regional, and national times of day were not synchronous on the political agenda. The movements of the sky became problematic as an absolute frame of reference for scientific and technical time-measurement when astronomers, using progressively better clocks, discovered a growing number of irregularities. Old interests in chronology, the science of time-reckoning and time-measurement, were reawakened in the wake of the debates over national and international time conventions. At the same time, a host of unmeasurable, rather subjective experiences of time became topics of discussion: the intellectual experiences of the Enlightenment, the political transformations since the French Revolution, and the technical transformations brought on by the industrial revolution gave contemporaries the feeling that they were living in a era of more rapid and continually accelerating historical change.[2] A wide range of experiences with factory and office time, with an accelerated dissemination of news, and with life in the great cities was conceptualized as the experience of "fast-moving time," "scarcity of time," and "time pressure." Ever since, the temporal pattern of daily life in modern industrial society has seemed to be dominated to a high degree by the constraints and pressures of time. As contemporaries looked at their own national history as well as foreign cultures, the more recent developments struck them as the progressive loss of the possibilities of individual control of time, an alienation from its natural and humane rhythms. Scholars of European history discovered that the fundamental change in time consciousness occurred with the transition from an agrarian to an industrial society that began in the Middle Ages. Ethnologists, sociologists, and political scientists revealed foreign time-consciousness and foreign conceptions of time as fascinating problems, to be sure, but also as serious obstacles to the modernization of societies in transition. As a result, belief in the hereafter, the conception of history, expectations of the future, planning, concepts of time, time-measurement, time-control, and time-discipline have become—and rightly so—important themes of historical and anthropological research.

The development and history of mechanical clocks are topics of the utmost importance for all these discussions. The clock was and is not only the prerequisite and device for a way of handling everyday time that was at first typical of European industrial societies and then of industrial societies in general; it is also the symbol of the process of European modernization, and for describing the experience of mental differences between the old European and the modern world, between European, North American, and Japanese societies and the so-called Third World. The use of everyday time and the possession and use of clocks are indicators of modernity. The history of clocks as technical devices of time-keeping and as a symbol of the change in time-consciousness links together a series of distinct though closely related themes. As a result the history of technology and the history of consciousness or mental structures have often ended up being mixed together, and they need to be occasionally separated in order for us to connect them in a meaningful way.

The transformation of time-consciousness

The experience of time, conceptions of time, and time-consciousness are very broad concepts. Under their headings a wide spectrum of facts are dealt with, some interrelated, some in need of differentiation. The various ways in which historical change is perceived—as cyclical movement, as rise and fall, as unending progress, as accelerated or delayed change—all contain different notions about the relationship between past, present, and future. Already in Augustine's differentiation of "three times," past, present, and future are described not as external processes but as states of one's own consciousness: "Some such different times do exist in the mind . . . The present of past things is the memory; the present of present things is direct perception; and the present of future things is expectation."[3] The common ways in which memory and expectation are structured in different societies are themselves subject, in a high degree, to historical change and are shaped by religious and philosophical convictions—and this applies especially to the late Middle Ages. Doctrines of the ages of the world, the Biblical patterning of time (in the early modern period it was called "economy of time"), chronologies of rulers, the counting of the years forwards and backwards from the birth of Christ, the reckoning by centuries, modern divisions into periods such as antiquity and Enlightenment—all these things structure historical memory. In the late Middle Ages and the Renaissance we can also observe the establishment of a binding calendar with a common beginning of the year, and the spread of the dating of days according to consecutively counted days of the month. To a certain extent, expectations of imminent religious events, prophecy, and utopias imparted a temporal pattern to the future. For the individual, the recording of the date of birth made it possible to connect his or her lifetime to historical chronology. Seneca's tract "On the brevity of life," much read since the Renaissance, popularized discussions about one's own time as a precious possession to be dealt with carefully and thriftily, and under his influence Petrarch introduced notions about the "incalculable value of time" into literature. A visible sign of the continuously increasing prominence of this thematic complex was the appearance of personified time in the visual arts.[4]

Key machine of the industrial age or instrument of class struggle: The mechanical clock and the historiography of technology and modern capitalism

The history of the mechanical clock is a required topic for every history of inventions or technology. The history of technology is a more recent discipline, whereas inventions and inventors have long been the subject of a special literary genre. The tradition begins with the so-called heurematologies which, beginning in the fifth century B.C., dealt of things the gods had only gradually revealed to humankind. The most famous example in Latin literature is the end of the seventh book in Pliny the Elder's *Natural History,* where more than two hundred inventions are distributed among one hundred and forty-three inventors. During the European Middle Ages, the *Etymologies* of Isidor of Seville were the most important source and starting point for numerous texts that used a variety of arguments to compile evidence for the intellectual advances of humankind.[5] Beginning in the late Middle Ages, this classic theme underwent a conspicuous transformation. The interest in innovations ("instrumenta nova, artes novae, novitates") rose noticeably, with the artisanal-mechanical arts ("artes mechanicae") forming a focus of interest that was becoming increasingly important. The

1. Allegory of Temperantia in the fresco *Buongoverno* by Ambrogio Lorenzetti. Siena, Palazzo Pubblico. Photo: Lensini, Siena.

literally daily appearance of new inventions was described and welcomed as a permanent process.[6]

From the beginning of the fifteenth century, at the latest, the preoccupation with inventions developed a historical perspective: inventions of more recent date, the technical achievements of one's own period, were singled out, and the most important ones were arranged together into groups. A modern understanding of technology becomes visible, though at first without a separate term. The early development of this process can be grasped only iconographically in the allegorical personifications of Temperantia, one of the four late medieval cardinal virtues.[7] Probably the oldest depiction of a sandglass, a technological innovation of the late fourteenth century, is found in the figure of Temperantia on the fresco "Buongoverno" in the Palazzo Pubblico in Siena (fig. 1). The fresco was painted in 1337; this particular section is part of an area that was restored in 1355, and it can no longer be determined whether the work is that of Lorenzetti.[8] The figure of Temperantia, one of the virtues assigned to good government, is carrying a sandglass. Older personifications of Temperantia carry a large or small mixing pitcher, in keeping with the translation of "temperamentum" as measure/proper mixture/moderation, from which the word for time ("tempus") was often derived. In the depiction in Siena,

2. Cardinal William of England, with sandglass. Detail from the cycle of frescoes by Tomaso da Modena in the chapter hall of San Nicolò (1352), Treviso. Photo: G. Dohrn-van Rossum.

the mixing vessel has been replaced by a measuring vessel, the sandglass. Around the same time, Tomaso da Modena, in a cycle of frescoes for the chapter hall of the Dominican monastery in Treviso, portrayed forty eminent members of the order. In the panel devoted to Cardinal William of England, the fresco, signed and dated to 1352, shows a sandglass on a bookshelf next to the cardinal's desk (fig. 2). The sandglass is by no means an incidental prop, for in a different panel Tomaso provided the earliest depiction of another innovation that appeared around 1300 and caused great excitement: spectacles. In illuminated manuscripts of Christian de Pisan's *Épître d'Othéa,* written around 1400, "Attrempance" is shown as a female figure who intervenes in the gear mechanism of a clock to regulate it. Several examples show her to be iconographically closely related to personified "Wisdom" in the illuminated manuscripts of the *Horologium sapientiae*. This period saw the creation, presumably on the basis of a still undiscovered textual model, of a new iconographic program for the depiction of Temperantia, one that placed even greater emphasis on technological innovations. The earliest miniatures date from the middle of the fifteenth century, and the program remained in common use into the sixteenth

3. Allegory of Temperantia. From a fifteenth-century manuscript *De quattuor virtutibus cardinalibus* by Pseudo-Seneca. Dresden, Sächsische Landesbibliothek, Ms. Oc. 79, fol. 68 verso. Deutsche Fotothek. Photo: R. Richter.

century. In these depictions, the figure of Temperantia (fig. 3) carries or is surrounded by a group of technological attributes, all of which were counted among the innovations of the most recent past, or at least were attributed to the artist's contemporary, post-classical period. Temperantia has a bridle in her mouth, a clock on her head, spurs (here modern rowel spurs) on her shoes, and spectacles in her hand. At her feet is a windmill mounted on a stone tower.[9] Spurs and bridle were known in Europe before the turn of the millennium; the tower windmill, spectacles, and the striking clock were innovations of the twelfth, thirteenth, and fourteenth centuries, respectively. As the personified ideal of sensible self-control, equipped with the modern attributes of the technological relationship to the natural world, Temperantia illustrates the period's new appreciation for all technological endeavors.[10]

Technological change, or social change caused by technology, was hardly within the mental grasp of fifteenth century contemporaries. However, there are indications that various technological developments were seen as a coherent process with the power to bring about change. Shortly after 1450, Giovanni Tortelli, a widely traveled and learned humanist, completed his great work *De orthographia dictionum e Graecis tractarum*.[11] It deals with Greek terms in Latin texts and discusses the question whether one ought to limit oneself to the vocabulary of the classical authors. Tortelli's answer is a clear no, for otherwise many new inventions would have to remain unnamed. These inventions are discussed in detail in an excursus to the article "horologium."[12] Tortelli is not interested in the absolute age of an invention like the striking clock; he distinguishes merely between classical and post-classical technologies. Apart from clocks, he lists bells, portulan charts, the compass, the lateen sail, canons, stirrups, the water mill, the cembalo, the organ, cotton, spectacles, and more. Later lists of inventions arrange typical groups of three or four innovations (for example, compass, canon, clock, printing) when they praise the human gift of invention or the increasing technological competence of humankind.[13] These grouped arrangements of contemporary achievements were not only literary clichés. In Luther's writings, for example, we can see that they were at the time topics of dinner-table conversation.[14] The clock kept its prominent place in these groupings until the end of the eighteenth century. Any time the discussion revolved around technological advances in general, and frequently when evidence for the superiority of "modernity" or the Europeans was presented, the clock appeared in the list of examples.

During the industrial revolution, in the face of the flood of new and exciting inventions, the mechanical clock lost its prominent place in published writings. Instead, the importance of a rational economic use of time and the new forms of time-discipline moved to the forefront of public interest. Mechanization of production, industrial working conditions, urbanization, and the acceleration of transport and communication made people more keenly aware of the economic implications of the organization of time and working time, and of the seemingly irreversible alienation from modes of behavior and consciousness that were felt to be partly natural and partly historical (in the sense of the "good old days"). Striking accounts of this can be found in the classic works of historical sociology and economics, where we can trace the rise of modern notions about the historical change of time-consciousness in Europe.

Karl Marx described the clock as "the first automatic machine applied to practical purposes," from which "the whole theory of production of regular motion was developed."[15] Moreover, he attempted to get a theoretical handle on the problem of working time and the normal working day when he split the workday into alienated time, for the production of surplus value for the capitalistic entrepreneur, and a time for the necessary reproduction of the labor force. "Economy of time, in the end all economy is reduced to this." Time was for Marx working time, and the utopian view of control over the process of production had the double aspect of liberation from the constraints of working time and active influence on historical time: "Real saving–economy = saving of labor time = development of productive force."[16]

Together with Friedrich Engels, Marx described the conflict-ridden implementation of industrial time-discipline by using England as his primary example. The steadiness and regularity of work compelled by the use of machines, and the excessive lengthening of the working day that was made possible by the workers' inferior market position, seemed to be typical consequences of the factory system. In the writings of Marx and Engels, the measurement and control of working time is documented as a central theme of the social conflicts. Drawing on many examples from contemporary factory descriptions, Engels for the first time called attention to the factory clock and its signal, the factory bell, as the means and symbol of the industrial domination over people and of their exploitation.[17] The alienation of the worker from the possibility of an autonomous control of time had thus been discovered as a characteristic feature of modern capitalism. Through the comparison with the relatively less harsh conditions of serfdom, Marx moved the problem of working time into a historical perspective that would have important consequences. Engels painted an even clearer picture of pre-industrial work rhythms, one that was not free of idyllic overtones: "They [the workers] did not need to overwork; they did no more than they chose to do, and yet earned what they needed. They had leisure for healthful work in garden or field, work which, in itself, was recreation for them, and they could take part besides in the recreations and games of their neighbours . . ."[18]

With regard to the problem of time-consciousness, these accounts had historiographic consequences not only because they indicated an epochal change. They also made the pre-industrial ways of dealing with personal time appear to have been relatively irregular, close to nature, spontaneous (within the boundaries of natural conditions and individual perception of one's economic situation), and, above all, largely self-determined. Only in very recent times have some scholars raised objections to the cliché created at that time of an "indifference toward time" that was allegedly characteristic of the old European world.[19]

In Max Weber, discipline and method appear as essential elements in the development of rational forms of organizing work. He describes this development as part of the comprehensive process of rationalization that, along with rational science, rational technology, and rational administration, made western capitalism possible. But he does not deal with the aspects of time-organization and time-discipline in modern organizational forms. Instead, he asks about the origin of the attitudes and behaviors, the "ethos" or "mentality," that made possible rationalization and indirectly also the factory as an organizational form. He sees the monastery as the model of the rationally administrated agrarian and commercial enterprise, and the figure of the monk as the non-economically motivated example of a person who lived rationally and with a methodical division of time. The professional ethos of Puritanism, with its constant admonition to budget time, save time, and, finally, to consider time as money, was, for Weber, a secular version of the ascetic ideals of monastic life, one that played a decisive role in the success of the early modern economic bourgeoisie.[20] Despite a broad-ranging controversy and many attempts to reformulate it, Weber's "Protestant thesis" established time-consciousness as a preeminent topic in the history of mentalities.

The monastery appears also in Werner Sombart's history of modern capitalism as the nursery of a rational conduct of life. In historical terms, Sombart believes that the linking of a rational conduct of life with a systematic conduct of business, which was decisive for the development of the bourgeois type, occurred among the Italian merchants and traders of the fourteenth and fifteenth centuries, because as a group they were the bearers "of a general spirit of rationalization and mechanization" in early capitalism. Their most important achievement, according to Sombart, was the introduction of double-entry bookkeeping, which quantified the diversity of goods and services in an abstract manner. A precondition for this was the spread of arithmetic; one consequence was the beginnings of statistics. The art of land surveying in the Italian city republics and the public measurement of time were for Sombart additional indications of the "diverse causation of the modern person."[21]

In his discussion of the "public measurement of time" as one aspect of the modernizing impulses that emanated from the northern Italian city republics, Sombart could draw on a work that has become a classic for the questions we are pursuing: Gustav Bilfinger's *Die mittelalterlichen Horen und die modernen Stunden* ("The medieval Hours and the modern hours"). A Gymnasium teacher in Stuttgart, Bilfinger had produced a number of earlier studies on the division of the day, the calculation of the hour, and the time-measuring devices of ancient peoples. In 1892, not without deliberate connection to the impending introduction of Central European time, he published his "contribution to cultural history," as the subtitle of the book read. His analyses of thousands of indications of the time of day in ancient and medieval literature, originally intended only as contributions to the auxiliary science of chronology, have remained exemplary to this day. Moreover, because these analyses simultaneously took the techniques of time-keeping and the social context of time indications into account, they provided essential preliminary work for questions concerning the change in time-consciousness.

More influential for modern notions about the change of time-consciousness in the Middle Ages was Lewis Mumford's cultural history of technology, which he conceived of as a parallel work to Sombart's history of capitalism. In *Technics and Civilization,* Mumford explains that the key machine of the industrial age was not the steam engine but the clock.[22] In every phase of its development, the clock was the outstanding and most advanced device, and at the same time the typical symbol for the machine as such. According to Mumford, the first seven centuries of the machine age were characterized by profound changes in the categories of time and space. He counterposes the new time and conception of time, produced by the mechanical clock, to a lost organic time that was a sequence of elementary human experiences and events caused by man. The transition to abstract, measured hours and minutes is for Mumford, in an even broader sense than for Marx, a process of alienation from nature. On the other hand, time-keeping also made possible time-accounting and time-rationing, and in this way "eternity ceased gradually to serve as the measure and focus of human actions." Faith in "mathematically measurably sequences" promoted the application of quantitative methods in the natural sciences. He saw one source of this mechanical conception of time in the "iron discipline" of the monastic routine, which imparted to human endeavors "the regular collective beat and rhythm of the machine." The ordering and regulating of time-sequences had become second nature in the monasteries, and a technical device for striking the hour at regular intervals had been the "almost inevitable product" of the monastic need to order time. Mumford's unhappy linkage, in his description of the daily monastic routine, of modern machine metaphors with statements about the long-term change in time-consciousness in turn influenced medieval scholarship.

Long before they found an echo in Germany, Bilfinger's and Sombart's ideas were picked up by modern French social history. In his 1931 review of P. Usher's work on the history of technical inventions, Marc Bloch described the advances in time-keeping as one of the most far-reaching revolutions in the intellectual and practical life of our societies, and as one of the main events of late medieval history.[23] Shortly thereafter, Bloch added another twist to Bilfinger's "needs of urban life" when he portrayed the installation of public clocks as the work of the urban bourgeoisie. Along the way he declared that the innovation-promoting relationship between social needs and scientific-technological creativity was as difficult as it was interesting, and he called for further studies.[24]

In his 1949 work on the Italian merchants who were active in the international trade in money and goods, Yves Renouard located the transformation of time-consciousness, which previously tended to be seen as a long-term process, in a relative short-lived historical social phenomenon. He went on to describe the modernization of time measurement as a manifestation of a change in mentality carried by the class of merchants at the beginning of the fourteenth century.[25] Under pressure from the practical needs of commerce at the international fairs and markets and the requirements of bookkeeping, the merchants, according to Renouard, switched to a uniform beginning of the year for their balance sheets. Though the traditional reckoning of the hour did not bother the merchants, the recent invention of the striking clock was highly attractive to their milieu, which was shaped by logic, rationality, and a conception of culture no longer dominated by the Church. The installation of public clocks thus satisfied the civic sense of the merchant classes who held the reins of power. At the same time, abstract time became the reference point for the thinking and acting of this group. Finally, Renouard relates Mumford's discussion about the replacement of organic time and the belief in the autonomous world of the sciences to the cultural milieu of the Italian merchants, as though anything like this could in fact be detected there.

Jacques Le Goff, though fully aware of the inescapable problems involved in typifying collective notions of time, goes a step further. He describes the change in time-consciousness in the late Middle Ages, which had practical everyday as well as theological aspects, as a process born primarily by the merchants. In 1960 he coined for this process the suggestive slogan of the transition from "Church's time" to "merchant's time." According to Le Goff, the conflict between these two conceptions of time was one of the most important events in the history of mentality during these centuries, a time when the ideology of the modern world was being formed under pressure from changes in economic structures and behavior. Le Goff argues that the organization of commercial networks, the speculation in goods, the multiplication of currencies, and the expansion of exchange transactions required a more precisely measured time. Commercial documents such as account sheets, travel diaries, manuals of commercial practice, and letters of exchange made it clearer still how important measured time was for the successful conduct of business. Here, as earlier in Sombart, the measurement of time stands as a symbol for a whole cluster of processes of modernization and rationalization. Talking about the permission that was granted for a work bell to regulate the working hours in Aire-sur-la-Lys (1355), Le Goff goes on to say: "The communal clock was an instrument of economic, social, and political domination wielded by the merchants who ran the commune."[26]

The studies of Renouard and Le Goff, who saw them as suggestions for more thorough research, have today become definitive — and as such, false — formulas about the change of time-consciousness, formulas fit for handbooks and textbooks. In statements like the following, the technology of time measurement and the change in the consciousness of history are inextricably linked: "It is a fact that the sense of time and conceptions of the hereafter change, that in the cities the cyclical time of the Church is replaced by the linear, valuable time of the merchant."[27] According to a Soviet work that has attracted wide attention also in the West, the development of productive forces made time into a crucially important factor, and "the appearance of the mechanical clock was at once a completely logical result and a source of the new temporal orientation."[28] Nearly a hundred years after the appearance of Gustav Bilfinger's work, the complicated reception of his findings in the historiography on capitalism has in many cases become the received wisdom in medieval scholarship, understandable in retrospect but no longer immune to the "veto of the sources."[29] Philippe Wolff's appeal to advance the social history of time measurement through a collective research project, published in 1963 in the journal *Annales,* has not been taken up.[30] Given the importance that the more recent accounts of the period have accorded to the topic, this is striking, even more so since the preconditions for such a project are quite favorable.

My involvement with this topic began when a large television network asked me to prepare, for a wider audience, a graphic presentation in texts and images of "Church's time" and "merchant's time." This pleasant task began in Le Goff's footnotes, but it turned out to be surprisingly difficult, and fortunate finds in the archives of Valenciennes and Orvieto showed that it could not have happened the way he describes it. My attempt to assemble a gigantic puzzle of bits of information from about one thousand European cities remains indebted to his approaches and questions. It takes up Bilfinger's approach of examining the history of clocks in connection with their social use, and tries to advance along this path. In this way light is shed on a narrow but important slice of the many-layered problem of the change in time-consciousness.

The first two chapters on classical and medieval time-keeping and hour-reckoning lay the groundwork for the discussion that follows by summarizing our current knowledge. Their purpose is also to counter widespread distortions in the discussions about modernization processes in Europe. In my view it makes no sense to look at the transition from the Middle Ages to modernity in terms of ethnology as a civilizing process away from a primitive state, or in terms of developmental psychology as a maturation process from the "childhood of humankind." These approaches overlook the fact that late medieval processes of innovation and modernization were in large measure due to the traditions of older and contemporary urban civilizations. The debate over the development of the mechanical clock—a paradigm of European domination of the Old World—is being kept alive by numerous and often very scattered contributions, and it will go on, precisely because it touches, in an exemplary fashion, on the relationship of European culture to other cultures. A history of public clocks has never before been written, even though an abundance of material is provided by the history of the clock, local historiography, and the inexhaustible riches of European archives.[31] As with my earlier work on the history of clockmakers, I have tried to make a beginning that could be carried further. Processes of change in urban temporal orders, innovations in dealing with working time, and innovations in the organization of modern systems of communication reveal important steps of modernization, steps we are hardly aware of because they have become so self-evident and routine.

2

The Division of the Day and Time-Keeping in Antiquity

A clock signifies occupations and undertakings, movements and the start of transactions. For men keep their eyes on the time in all that they do. And so, if a clock falls apart or is broken, it means bad luck and death . . .

Artemidorus of Daldis, *The Interpretation of Dreams* III.66 (2nd century A.D.)[1]

The clock as a symbol for business and busyness: to our ears that sounds like a modern, if somewhat hackneyed, dream interpretation. Artemidorus treats dream symbols as an aid to prognostication, not yet as an aid to introspection, and in his interpretations we must think of successful future business undertakings as well as success in politics and the courts of law. The dream symbol presupposes a general familiarity with the use of the clock at least in the cities. For how could a clock appear in dreams if "life according to the clock" was not already a common experience in the second century A.D.?[2]

Artemidorus was by no means the only one who made this claim. A Roman comic writer has a parasite ask the Gods to confound the man who brought the sundial into the city, because people no longer ate when hungry but at fixed mealtimes. Seneca laments the unnatural innovations that make people sleep during daylight and celebrate when it is dark. Before Emperor Domitian was murdered, he was deliberately told the wrong hour, because he was known to be particularly afraid of the fifth hour. Martial jokes about the daily rounds that are completely filled up and strictly regulated according to the hours. But in late antiquity, in Cassiodorus for example, a sensible division of the hours became a precondition for a virtuous "ordo vitae."[3] The differentiation of urban life, often lamented as the abandonment of rural "simplicitas" and never welcomed as a gain, was also expressed as the experience of the social differentiation of the daily round of activities in the cities. This was a new experience, and contemporaries clearly saw that it was linked to the introduction of the clock.

Ancient authors repeatedly point out that the division of the day into hours and the use of time-measuring devices had not been customary since time immemorial, but that both practices belonged to a time of historical and datable memory. Herodotus reports that the Greeks had adopted sundials and the twelve divisions of the day from the Babylonians.[4] Prior to that, our word for hour ("hora") had described periods of various lengths, for example, a season. The Greeks divided the day into three or four segments, which were given designations like "early afternoon," or were named for mealtimes or various activities. For civil use, nighttime had no division at all; for military purposes it was broken down into three or four segments whose length varied with the seasons. It is not clear whether the calendar day began in the evening or, following popular linguistic usage, in the morning. The use of twelve division of the day, of temporal hours, and of "hora" as an hour's time is attested only from the time of Alexander the Great.

In Book VII of his *Natural History,* Pliny outlines a kind of historical anthropology in which the current level of culture is made to appear as the result of a long series of inventions. Those inventions included also conventions agreed upon among all peoples, in part tacit agreements like the use of the Ionian alphabet and the custom of shaving the beard, in part conscious and considered agreements like the observation of the hours. The Twelve Tables in Rome (451 B.C.), Pliny tells us, specified only sunrise and sunset. Sometime later noon was added, and the last hour of the day was announced by officials according to the position of the sun. The first sundial was later erected in Rome by Lucius Papirius Cursor (292 B.C.). Subsequently, according to a report by M. Varro, a sundial was brought from conquered Catania by the consul Valerius Massala (263 B.C.), though for ninety-nine years nobody noticed that its lines, because of the difference in latitude, did not agree with the hours in Rome. A properly constructed sundial was erected only later (164 B.C.), and shortly thereafter (159 B.C.) a public water clock was set up in a roofed building. In conclusion, Pliny remarks that until that time "the division of the daylight had not been marked for the Roman public."[5]

The twelve divisions of the day became possible only with the use of clocks. This deliberate division of the day, a social convention in Greek and Roman civilization, was a Babylonian legacy. The Babylonians had separated the day into daytime and nighttime, dividing each period—daylight from sunrise to sunset and nighttime—into twelve segments (hours) of equal length.[6] The duration and temporal location of these hours varied with the length of daylight, with the "sixth hour" always designating the midday point. Only twice a year, at the equinoxes, were the hours of the day and the night equal in length. These hours were called temporal hours, "horae inequales." Expressed in minutes—which were unknown at the time—the ratio of the longest to the shortest daylight hour in Upper Egypt was 67:53, in Athens 73:47, in Rome 76:44, in southern Germany 80:40, in northern England 90:30. In addition, in Rome the four divisions of the day and the division of the night into four watches continued in use, because they were sufficient for most practical purposes. In Roman cities these segments were publicly signaled by the officials. The quarters of the day were named after their last hour. As points in time, hour indications—with the exception of the first hour, "hora prima"—must always be understood in the sense of the expired hour ("hora expleta," "hora completa"). A brief example: "nona" as a space of time could designate the ninth hour of the day, more rarely also of the night, it could refer to early afternoon encompassing the seventh, eighth, and ninth hours, or as a point in time it could mean the end of the ninth hour. Instead of the uncommon "hora duodecima noctis" (or "h.d. diei"), "mane" and "vespera"/"suprema" were used to describe the natural boundaries of daylight. In addition, the cock's crow, "gallicantus," was in common use to indicate the point in time just prior to dawn.

Only the educated can be expected to have used differentiated hour indications, but even they had limited need for them. Approximate times of day and night could be described with sufficient clarity also by using terms that were oriented towards natural events and were much more highly differentiated in those days. This applies especially to transitions from light to dark, as the following qualitative breakdown of the night, which was still very common in the Middle Ages, shows in exemplary fashion: "occasus solis" (sunset), "crepusculum" (dusk), "vesperum" (appearance of the evening star), "conticinium" (silence), "intempestum" (complete cessation of all activities), "gallicinium" (cock's crow), "aurora" (waning of darkness, first flush of dawn), "diluculum" (dawn), "exortus solis" (sunrise).[7]

Hours of equal length ("horae aequinoctiales," "horae aequales") were certainly known. They were used only in the context of scientific discussions, especially astronomical and astrological ones.[8] The indication of hour fractions was rare outside of astronomy. For daily use, specifications such as "at the beginning" or "at the end" of an hour were sufficient. Half hours were common in allocating time to regulate irrigation schemes. On rare occasions fractions of hours were added to the indication of the expired hour, for example "hora quarta et quadrans," "hora quarta et triens."

Ancient sundials

Observing changes in the sun's shadow is undoubtedly the oldest method for determining the time. Suitable for such observation are the shadows of the edges of buildings or the shadow of one's own body, the length of which, expressed in "feet" or "shoes," was a very popular means of determining time in antiquity. A first reference is found in Aristophanes' comedy *The Ecclesiazusae* (392? B.C.). Here a peasant woman complains to her husband that all he does is watch his shadow until it has reached ten feet, whereupon he proceeds to his evening meal. Customary in Greece were also meetings arranged for shadows measuring six, eight, ten, and twelve feet.[9] Tablets indicating in "pedes" the length of the shadows for the hours during the different months were still very common in the Middle Ages.

Sundials with gnomons and scales had been in use since the third millennium B.C. At first they probably served only calendrical purposes,[10] the division of the day coming later. Studies of specimens found in Egypt have shown that all types of sundials from the early period gave rather inaccurate indications of time.[11] Greek astronomy subsequently developed the Babylonian models into the different types and forms described in all their variety by Vitruvius in *De architectura* (IX.8.1). Numerous archeological finds have confirmed his descriptions.[12] Greek and Roman sundials bore the network of hour lines and day curves on a variety of shapes usually carved from stone (spherical, cut spherical, conical, and cylindrical forms, vertical and horizontal faces). Monumental sundials with networks of lines over one hundred and fifty meters wide used Egyptian obelisks (taken as spoils) as the gnomon.[13]

Sundials adorned public buildings, temples, private houses, and villas. They were common throughout the entire Roman Empire. The satirist who spoke of the city filled with sundials has been confirmed by finds in Pompeii, for example.[14] Numerous inscriptions show them to have been popular donations.

Even if, as we have seen, it took some time for word to get around, the problem of constructing sundials according to geographic latitude had been solved by the third century B.C., at the latest. From that time on ancient sundials always indicated the temporal hours for a specific latitude. Localities were assigned to a "climate" in accordance with their latitudinal location; this climate could also be indicated as the ratio of the longest to the shortest day, expressed in hours of equal length. Portable hanging sundials ("horologia viatoria pensilia"), suitable for travelers, could be set for all latitudes or for a number of important localities.[15]

Since the focus of observation in ancient sundials was on the path described by the tip of the shadow, sundials could also be used to read calendric indications, such as the month, or to determine the equinoctial days, on which the extremity of the shadow traces a straight line. It is possible that this continued to be their most important function even after the appearance of the hour division.

Ancient water clocks

Time-measurement that was not dependent on weather and light was possible in antiquity only by means of so-called water clocks, clepsyd-

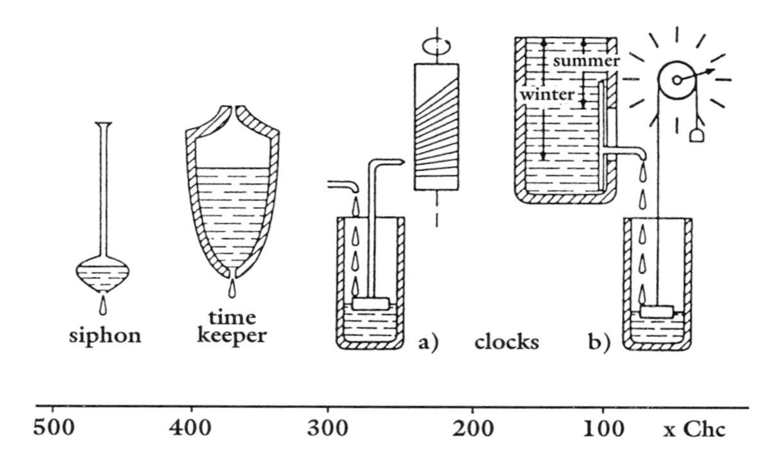

4. The development of the clepsydra from siphon to water clock. Sketch based on Volker Aschoff, *Geschichte der Nachrichtentechnik,* vol. 1, 2nd ed. (Berlin, 1989). Left to right: syphon, time-measuring device, clocks; the chronological scale runs from 500 to 100 B.C.

rae (fig. 4).[16] The term clepsydra ("water-thief") goes back to an untypical form of the device. In the sixth century B.C., various observations of pneumatic and hydraulic phenomena were made using pierced, hollow vessels with a small outflow orifice that could be sealed with the thumb; with this device one could "lift" wine, water, or oil from larger containers as with a siphon. Observation of the slow sinking of such vessels into liquid, or the possibility of delaying the outflow, led to the discovery of the clepsydra as a means for determining an interval of time.

Clepsydrae: Outflow clocks

However, the oldest and simplest way of indicating segments of time were not the wine-siphon forms, but the so-called outflow water clocks. They had no separate name and were also called "clepsydrae." Here a simple vessel was fitted with a narrow outflow spout. The outflow of a certain amount of water was then used to determine a certain interval of time, or observation of the dropping water level served the purposes of very simple time-measurement. Such vessels are mentioned in the first millennium in India and China.[17] According to Babylonian texts, operators adjusted them to the changing seasons by varying the amount of liquid they were filled with.[18] An Egyptian inscription (ca. 1550 B.C.) and a specimen found at the temple of Ammon in Karnak (ca. 1400 B.C.) reveal that the interior surfaces of the vessel, slanted in order to compensate for decreasing water pressure, were inscribed with scales that allowed the division of the night during various months.[19] Newer tests have shown that even though the physical theory behind the devices was unknown, their daily error amounted to only about fifteen minutes.[20] As far as we know, these vessels equipped with scales were used to divide the nightly watches.

It would seem that such scaled vessels appeared in Greece only at a late date.[21] Starting in the fifth century, what were at first simple outflow clepsydrae came into use in Athenian law courts and assemblies, namely, as a means of apportioning and limiting the time for speeches.[22] Since court proceedings usually had to be concluded within a day, the total speaking time in criminal trials was split three ways between the accuser, the accused, and the judges, using as a basis the approximate amount of water that flowed out of vessels on a short day (about nine hours); in civil matters the time was sometimes divided into five parts. The vessels were shut off when witnesses were heard or documents read. Aristotle subsequently reports that speaking times were also regulated depending on the amount of money involved in the litigation. The litigants were given, not a part of the day's speaking time, but a certain number of jugs (about 3.2 liters); modern calculations have shown that a jug drained in three to four minutes. The division of the day into hours was not yet customary, and thus speeches were made not "according to the time" but "according to the water," which stood in each instance for an interval of time known only very approximately. Such outflow vessels were also used for practice in schools of rhetoric. The Roman legal system adopted this technique of limiting court speeches during the republican period, the purpose being, as Tacitus tells us, to curb juridical eloquence with a bit and bridle ("frenum").[23] "Frenum" became in the Middle Ages the technical term for the escapement of the wheeled clock, and the bridle, as we have seen, appeared at the beginning of the fifteenth century in the iconographical context of the allegory of the virtue "Temperantia." Not until Cicero do we find hour-indications for court speeches, and not until Pliny the Younger do they become concrete enough that we can fix the duration of a court clepsydra at about twenty minutes.[24] In city councils, the clepsydra became a way of verifying that members of the council arrived on time; in the late Middle Ages the sandglass served the same purpose. According to a third-century B.C. inscription from Iasos in Asia Minor, on the day the council met, a clepsydra located in plain view was filled at sunrise. Those members of the council who had failed to appear by the time the water ran out forfeited their claim to the attendance fee. In Graeco-Roman times, as in Egypt, clepsydrae were also used to determine the length of the military night watches. Caesar reported that he discovered "by water measurements" that the nights were shorter in England.[25]

The use of water clocks shows that when it came to the measurement and distribution of water allowances in irrigation systems, a transition took place from vague time periods that were valid only for concrete applications and could not be generalized to abstract time periods. Clepsydrae were used for this purpose throughout the ancient world, but also in India, China, and Persia into modern times, wherever water was particularly precious.[26] Here a certain amount of water corresponded to the outflow time of a clepsydra, or to the time it took a pierced vessel floating on water to fill up and sink.[27] These water units were allocated, sold, or auctioned off. As a result the time units could also take on the name of the vessel. When clepsydrae were widely replaced by sandglasses in late-medieval Spain, where such regulating techniques had frequently been handed down from the Muslim period, the allocated half hour retained the name "pitcher" ("jarro"), in some parts of Spain to this very day, even after the spread of the mechanical clock.

The techniques of distributing water by means of time units were frequently codified in Roman law.[28] However, in imperial times people began to regulate the allocation of water also according to the hours of the day, and no longer by means of a hollow vessel. An inscription from Lamasba in Numidia (third century) lists properties, the names of owners, and the irrigation time in whole and half hours of the day.[29] This presupposes that time-measurement with sundials was already common, and this more abstract measure of time could now be used for a just distribution of water.

Clepsydrae and scientific measurements

Accounts that water clocks were used for scientific observations, and thus for early forms of the quantifying method of scientific investigation, take us to the third century B.C., the heyday of Alexandrine science.

The first account concerns the physician Herophilos of Alexandria. It is a completely isolated report, meaning that it is not confirmed by other sources and did not become part of the tradition. We are told that Herophilus used a water clock with a short outflow to determine the pulse rate in feverish patients. A comparison with the rates of healthy individuals, counted for four different age groups during the same time interval, supposedly allowed him to detect pathological changes.[30] Ancient medicine, which tended to describe the pulse as "strong," "weak," "regular," or "restless," also tended to regard adjectives such as "fast," "slow," "racing," or "faltering" as qualitative and strictly tied to the experience of the individual physician. Only in the seventeenth century did people begin to count the pulse (see chap. 8, Time-measurement and "scientia experimentalis").

More broadly reported, and still handed down in the Middle Ages, were methods—attributed to the Alexandrines—for calculating the relationship between the radius of the sun and the moon and their orbits around the earth. To do this an outflow water clock should be opened at the first appearance of the heavenly body and closed once the full "planetary disk" had passed over the horizon. At the same time a second vessel should be opened and then closed when the heavenly body reappeared the following morning. A comparison of the two volumes of water supposedly showed that the ratio of the radius of the solar disk to its orbit was 1:600 (for the moon 1:750). In the second century B.C., Ptolemy did not think such experiments were very promising. In the second century A.D., the skeptic Sextus Empiricus considered them impossible. Otto Neugebauer has said that such measurements were not feasible, and that the story was already in antiquity merely a literary cliché.[31]

Continuous time-indication and automata mechanisms

The Hellenistic period saw the development of a new type of water clock, one that had great influence on the more modern forms combining continuous time-indication with astronomical indications and with chimes and signaling mechanisms. In the so-called inflow water clocks, the water pressure is kept constant by maintaining the water in the supply or overflow tank at the same level by means of an outlet that drains off excess water. The water level in the receiving tank thus rises evenly, and a float in the tank can be used to move an indicating contrivance or some mechanism at an even rate. The only detailed source for this presumably Hellenistic innovation and the details of its construction is the account of Vitruvius on "water clocks" or "winter clocks" (*De architectura* IX.8.2–15), which is inconsistent and controversial when it comes to its technical interpretation. Vitruvius names as the inventor Ctesibius of Alexandria (third century B.C.), who is also credited with developing a hydraulic organ and a two-cylinder fire pump.

Vitruvius begins by explaining the principle elements of the contrivance: a carefully calibrated outflow spout fashioned from gold or from a pierced gem, and a float in the form of an inverted bowl. In the first variation, in which a linear movement is transformed into a circular motion, a toothed bar atop the float drives a toothed drum, which in turn causes various kinds of movements or acoustic signaling mechanisms, such as falling stones or pneumatic trumpets. Vitruvius calls these mechanisms extras. In the second variation, the float carries a pointer or pointer-figure. Two possible ways of adjusting the clock to the unequal hours are discussed. First, by changing the outflow orifice through the insertion of wedges; here it is unclear whether Vitruvius was already familiar with conical valves.[32] Second, by attaching a revolving drum marked with lines for the hours of the various months. In the third variation, an astronomical clock ("horologium anaphoricum") which, like the first arrangement, transforms linear into circular motion, a pliable chain is run from the float over an axle to a counterweight. The axle turns a dial with 365 holes that revolves behind a wire grid with the Tropics and hour lines; a pin can be inserted into the holes to indicate the position of the sun. We see here the first known version of a mechanically turnable astrolabe, which would become part of the standard equipment of astronomical clocks from the late Middle Ages on. Vitruvius concludes his account with a description of a (probably unworkable) regulating mechanism with a perforated revolving drum between the tanks.

Archeological finds have confirmed that Vitruvius was describing clockworks that actually existed. Archeologists have discovered the remains of a large, public inflow water clock in the agora of Athens, which was in operation in the third century B.C. Judging from the size of the tank, its fill-time must have been in excess of one day.[33] A "horologium" of this type was also in the Tower of the Winds erected by Andronicus of Kyros in the first century B.C. Each of the eight sides of the tower (which was about thirteen meters high) bore a sundial, with the water clock inside the tower presumably moving a variety of astronomical mechanisms.[34] And the fragment of a second-century A.D. disc with astronomical indications that was found near Salzburg was in all likelihood also part of the kind of anaphoric clock described by Vitruvius.[35]

Our current level of knowledge does not yet allow us to say how far ancient mechanics progressed in constructing for these clocks the complicated gears that are necessary for a depiction of planetary orbits. The only piece of evidence to date was a scientific sensation, but for now it remains a completely isolated piece in regard to the extant sources and current state of scholarship. This so-called Antikythera mechanism was retrieved in 1901 from a shipwreck near the island of Antikythera between the Peloponnesus and Crete. The device, about 30 × 15 centimeters in size with the thickness of a book, was used for mechanically calculating and demonstrating calendric problems on three dials. On one side one could, within the Greco-Roman calendar, set the position of the sun and the moon, the rising of bright stars, and the equinoctial dates. On the other side nineteen solar years were related to 235 synodic (lunar) months within the metonic cycle. In accordance with astronomical theory, epicyclic gears (that is, gears describing circular trajectories along a circular line) were used for this mechanism. D. Price presented this device of the first century B.C. to the public; Allan G. Bromley has suggested a reconstruction.[36] But we are still waiting for the reassessment of ancient precision mechanics and its possibilities of fashioning intricate gearing that this device calls for. For our purposes we can note the following: even if the only known example of such a mechanism was operated by hand, there can be no fundamental doubts that it was possible in antiquity to drive complicated astronomical mechanisms with water-clocks.

Classical clock technology was handed down to the European

Middle Ages in part directly from Roman late antiquity, in part indirectly by way of Byzantine and Arabic mechanics. In Constantinople, clocks of this kind were in use for the imperial palace guard. But according to John Lydos, their use for timing court speeches was no longer common in the sixth century. In the city several public buildings and churches were equipped with clocks. It is difficult to tell from the texts whether we are dealing with sundials or water clocks in specific cases. John Lydos singles out the official character of at least one of these clocks by calling it "clock of the city."[37]

The development of large-scale, water-driven automata was also continued in the Byzantine period. Procopius describes an artistic clock in Gaza (Palestine, around 530) that struck the number of the hours, counting two sequences from one to six; the striking was widely audible. The clock, erected in a large square, set in motion an elaborate mechanism of mechanical figures. The eyes in the head of the Medusa on the tympanum of the facade moved every hour. Below the Medusa head was a series of twelve doors, in front of which the figure of the sun god was moving about. Every hour a door opened, and out came Hercules with the attributes of one of his twelve mythical labors. Among the musical and figural mechanisms, Procopius singles out an eagle with beating wings that wreathed the clock. At night a light moved behind the thresholds of the doors.[38]

The *Book of Ceremonies* of Emperor Constantine VII Porphyrogenitus (tenth century) mentions a large clock in the imperial palace in Constantinople, and a clock in a building adjoining Hagia Sophia. Arabic travel accounts report that the latter, too, set figure mechanisms in motion every hour. These reports attribute the construction of this clock to Apollonius of Tyana (first century), who is not otherwise known as a mechanic.[39]

As with the monumental clocks of the European Middle Ages, the primary interest behind these mechanical marvels was in surprising effects or the visualization of heavenly phenomena. The public indication of the time of day tended to be more of a side-effect. Later, in the Islamic world, there developed an art of automata that would gradually come to exert its influence on the history of the clock in the high Middle Ages.

3

The Medieval Hours (Hora)

The Church adopted the Roman division of the day and structured the liturgy of daily prayers around it. The basic framework continued to be the double sequence of twelve temporal hours each. Prayer times were arranged in accordance with the four main divisions of the day and night. It made sense to time prayers in this way, since the New Testament, with few exceptions, used only the third, sixth, and ninth hours of the day, along with sunrise and sunset, to indicate time of day. This is especially apparent in the Passion: the high priests take counsel early in the morning; the crucifixion falls on the third hour; the ensuing darkness lasts from the sixth to the ninth hour; Jesus dies at the end of the ninth and is laid in the tomb as daylight fades. This gave rise to seven periods of prayer, reminders of the Passion: Matins, Prime, Terce, Sext, None, Vespers, and Compline. As the prescribed sequence of prayers, they were also called the canonical hours ("horae canonicae"). The approximate temporal location of the prayers corresponded to the division of daylight. Matins came at sunrise, Compline coincided roughly with sunset, Sext always designated midday. In lay use, only the Hours assigned to the quarters of the day became important: Prime, Terce, Sext, None, and Vespers. The twelve-hour division of the day was rarely used in the Middle Ages; however, it remained familiar to the educated and memory of it was kept alive, as for example in the Biblical parable of the laborers in the vineyard who are hired at the eleventh hour but are paid a full day's wage (Mt. 20.6–12).

The Hours as indicators of time: Shifting of the Hours

The linkage of the periods of prayer to the Roman division of the day (or temporal hours) was not subject to any strict canonical regulation. Though it was customary to place the prayers at the end of the three-hour segments of Terce, Sext, and None, it was not expressly forbidden to say them slightly earlier. The precise time of day for prayer was not of great concern. As a result, it was possible for the "hora quoad tempus" (hours of time) and the "hora quoad officium" (hours of prayer) to drift apart with respect to their location in time. The expression "hora prima vel quasi" ("at the time of Prime," "at early mass time") was only a very approximate designation of the first hour of the day, since the Prime prayer or the early mass could be begun or chanted before or after sunrise without violating any canonical rules. The actual temporal location changed with local customs and practical requirements, especially in the cities.

The Muslims also had a prescribed sequence of prayers, but their temporal spacing was not based on the division of the day into unequal or equal hours. The sequence took its cues strictly from the position of the sun. In the eleventh century, al-Bîruni compiled the vague and partly contradictory rules of the Koran and the Islamic schools and sects in his *Exhaustive Treatise on Shadows.*[1] In deliberate contrast to other religions, prayer was forbidden to Muslims at sunrise, sunset, and at the sun's culmination (midday). The first of the five prescribed periods of prayer began immediately after the midday culmination, "when the shadow begins to wander toward the East." It ended when the shadow became longer than the shadow-casting rod. The second afternoon prayer should be said "while the sun is still white." The third period began after the sun's disappearance behind the horizon and lasted as long as twilight. Thereafter came the time of night prayer, which had to be completed before dawn broke, before "the appearance of the white light." The period of morning prayer ended, accordingly, with sunrise. In some traditions the earliest time within each period was considered the best time for prayer. It was followed by a second best time, a third best time, a fourth time—still permitted but frowned upon because of the long delay, and finally a prohibited fifth time, which would not allow prayer to be completed before the next period began. The muezzins at the mosques, responsible for keeping the prayer times, had to observe above all the culmination point of the sun and the length of the shadows. During the Middle Ages these Muslim time experts needed and developed not only sundials, but also shadow tables, astrolabes, and water clocks.

Not only did the connection between the times of prayer and the solar day remain looser in Europe during the Middle Ages, the sequence of hours of prayer ("quoad officium") also shifted gradually with respect to the sequence of the hours of the day ("quoad tempus"). This drifting apart took place over two to three centuries and occurred at different rates in different regions. The dates and causes of the drift are a matter of conjecture, but we can describe the results. Most obvious was the shift of None to the first hour of the afternoon and then to the true midday, a shift enshrined in the English word "noon." As a result, in secular usage the Sext gradually disappeared as an indicator of time, while as a time of prayer it was moved immediately before None. Simultaneously, Vespers moved forward into the afternoon and became important in signaling pauses from work or the end of work on days preceding holidays. As a time of prayer it gradually moved, outside the monasteries, into the third hour of the afternoon.

Two thirteenth-century sundials on the Minster in Hameln allow us to observe the shift of the Hours with respect to the temporal hours (figs. 5, 6). Both dials have a half circle with identically spaced divisions. In one half-circle the vertical midday line is marked by an "M" (meridies); in the other, the somewhat later sundial, by an "N" (None).

Though the shift probably took hold in the course of the thirteenth century, even the simplest timing of hours was not obvious to everybody, and was occasionally the subject of disputes. In 1188, a duel was to take place between Gerardus of St. Obert and Robert of Beaurain before the Count of Hainault in Mons. Gerardus appeared armed around the first hour on the agreed-upon day. He waited until the bells rang for None, the end of the customary dueling period. Since his opponent had failed to appear, Gerardus then demanded that his claim in a disputed legal matter be recognized on the grounds that he had waited to the appointed time and beyond ("quod usque ad horam et ultra horam expectasset adversarium suum"). The count's counselors were uncertain about what to do. Only after observing the position of the sun and asking the clerics present for advice did they pronounce

5, 6. Two sundials for indicating the canonical hours on the minster at Hameln. Photos: Martin Neumann, Bad Pyrmont.

that None had passed.[2] Thus neither the ringing of the church bells nor the position of the sun provided undisputed indications of time. As late as the fourteenth century, the gradual shifting of the Hours led to conflicts when it came to setting the hours of work.

There is no good evidence for the reasons behind the shift of the Hours. G. Bilfinger attributed it to the spread of monastic fasting rules beyond the monastery walls among secular clergy and pious laity. Those rules stipulated that on regular fast days the main meal should be taken after None, but during the Lenten period preceding Easter not until after Vespers. And the monks were not to eat anything at all before Sext. Human weakness (i.e., hunger), Bilfinger argued, encouraged a "permitted practice" under which None and the other Hours accordingly were moved away from the temporal hours and shifted closer to the main meal, which the common people customarily ate in the morning.[3] Jacques Le Goff offers a different explanation: since None signaled the end of work on half-days of work (on Saturday, for example) pressure from urban wage workers caused the forward shift of this period of prayer, which was, after all, also an urban time signal. This hypothesis cannot be verified. The sources for the None as the end of the half-day, to which Le Goff refers, date from the years 1286–1289 and thus to a time when the None in Italy had long since moved to midday.[4] Dante, surely no friend of newfangled customs, states explicitly in his *Convivio* (1308/09) that the None should be rung at the beginning of the seventh hour of the day. To all this we can add that even within monastic communities the temporal locations of the Hours were not precisely fixed and could be moved, though prior permission was required. In 1253 the cathedral chapter in Chartres received permission from Pope Innocent IV to celebrate the night services in the early morning instead of shortly after midnight, the reason being that the monastery grounds did not offer sufficient protection at night.[5] Similarly, the chronicle of the convent of Ste. Catherine-du-Mont near Rouen reports under the year 1322 that the community, fearful of the night ("timores nocturnae"), had begun to say Matins in the morning.[6]

The temporal order of the monastic world

RECENT MISCONCEPTIONS

Ever since Max Weber and Lewis Mumford, the asceticism, discipline, and regularity of monastic life have assumed increasing importance for the historiography of modern industrial society, especially in explaining its characteristic forms of work and life. Extending Max Weber's theses about the development of western rationalism backward has led to far-reaching new interpretations of life under monastic rule. Influenced by the focus on modern mechanization and seduced by a machine metaphor that has taken on a life of its own, some authors, who tend to have a sociological outlook, have changed the picture of medieval monasticism. In many instances monasteries have been turned into prototypes of modern factories or "spiritualized and moralized megamachines" (Mumford). The mechanistic image of the monastery has given rise to partly distorted and partly erroneous theories about the monastic temporal order, the contribution of the monasteries to the development of the modern sense of time, and the history of clocks.

Max Weber's theses of the non-economically motivated, rational method that informed the way of life and the economic behavior of the monastic orders are taken a step further by Lewis Mumford: refering vaguely to Werner Sombart, he declares the Benedictines to be the founders of modern capitalism. They helped to give "human enterprise the regular collective beat and rhythm of the machine." In a later work Mumford states that in organizing their daily routine on a twenty-four hour basis, the Benedictines had anticipated a later stage of mechanization.[7] To support these theories he adds a new twist to a legendary bull of Pope Sabinianus on the ringing of church bells, turning that ringing into a continuous signal for the Hours in the monasteries. In Mumford's view, some means of striking the Hours in regular intervals was therefore a necessary product of life under the "iron discipline of the rule." There is some inconsistency between this and another statement by Mumford that "the mechanical clock did not appear until the cities of the thirteenth century demanded an orderly routine"—the statement itself is not accurate, though it does point in the right direction.[8]

H. E. Hallam, in an essay on feudal society of the High Middle Ages, also declares the "spirit of the clock" to be "wholly Benedictine." The most significant and novel trait of the Benedictine Rule, according to Hallam, is the dividing up of the day into a sort of timetable. The ideal of Benedictine life "created the need for the monastery to know the correct time and to be able to measure its lapse." The mechanical clock appeared "as services became more numerous and complex."[9]

Eviatar Zerubavel has devoted an entire essay to the Benedictine ethic as the origin of modern scheduling. The round of daily activities in keeping with the Rule, he argues, was the model of all western schedules and timetables; the Benedictine monasteries were the outstanding examples of "clockwork communities." Zerubavel, too, works on the assumption that a regular monastic time signal existed and claims that the prescribed schedule "made the standardization of the length of the hours [their absolute duration] . . . essential." The awkward situation caused by the use of the historically older canonical hours could be remedied only through the introduction of the mechanical clock. Zerubavel advances some very contradictory theories. On the one side he recognizes that clocks in the monasteries were initially used only as alarms. The standardization of the temporal location of divine services and daily activities was then followed, logically and historically, by the "rigidification of the duration" of these events. But then he turns this scenario (which is, at least, plausible) around and maintains that the Benedictines played a significant role in the development of the western, abstract conception of time because they invented a schedule that was completely based on a mechanical device for measuring time or presupposed the use of such a device.[10]

All three authors fail to distinguish between alarm devices, mechanical clocks, and striking clocks. This explains some of their misconceptions. On that basis they construct a picture of monastic daily life quite far removed from the text of the Rule, the variations in the Rules, the commentaries, the relevant literature, and monastic practice. I do not wish to play down or deny the significance of asceticism, discipline, and regularity for the development and stabilization of behavioral patterns and norms that became important to industrial society. But we ought to recall some well-known historical facts as an antidote to hasty socio-historical and socio-temporal generalizations.

THE DAILY ROUND OF MONASTIC LIFE

The Benedictine Rule, not a novel invention but the most important example of early monastic Rules and an authoritative model for all later ones, sought to be a guide for a communal life of humility and obedience in devotion to God. Renouncing the pursuit of self-interest, personal property, and personal control of one's time, the monks lived largely isolated from the surrounding world under the authority and guidance of an abbot. The purpose of this life was gradual self-perfection and preparation for life in the hereafter.

The central element in the monastic daily routine was divine worship. In keeping with Psalm 119, verses 164 ("Seven times a day I praise thee") and 62 ("At midnight I rise to give thee thanks") it was performed seven times during the day and once during the night (*Rule of St. Benedict,* chap. 16). The day was the basic unit of the monastic time system, a distinction being made between Sundays and holy days, regular working days, and fast days.[11] The next larger unit was the week. It was governed, apart from Sunday observance, by the liturgical requirement that all one hundred and fifty Psalms be chanted at least once within this period (chap. 18). Liturgically the year was split into summer and winter. With regard to the prescribed readings, the duration of the day's work, mealtimes, and the rules of fasting, the monks' year was divided into three segments. Larger units of time are not mentioned in the Rule.

The monastic day was divided into an almost unbroken sequence of divine offices, meditation, reading, work, meals, and periods of sleep. On a yearly average the main segments of the day (offices/reading, work, sleep) occupied about seven hours each.

The seasonally shifting temporal location and relative duration of the various segments of the day were determined by the divine offices. Vigils was to commence after midnight and be completed while it was still nighttime. In winter its liturgy was more extensive than in the summer, that is, it lasted longer. Matins was to be said before daybreak ("incipiente luce"). Day offices took their names and original location in time from the unequal Roman hours (Prime, Terce, Sext, None), with no particular importance being attached to a precise correspondence between the "hora quoad officium" and the "hora quoad tempus." In the monasteries, too, the canonical hours were, to a certain extent, movable with respect to the temporal hours. For example, in the rules governing work on summertime afternoons, we read that None should be chanted earlier ("agatur Nona temperius mediante octava hora"). During the winter, Terce was moved to the morning, preceding the work period ("hora secunda agatur Tertia," chap. 48). At all times Vespers should be said such that the subsequent evening meal could be completed while there was still daylight. Compline marked the boundary to complete darkness.

Despite this density of activities, the ordering of the daily monastic routine got by with remarkably few indications of time. The beginning of the offices was linked not to a particular point in time but to a signal or short sequence of signals ("signa"). The duration of the offices was determined not by a set period of time but by the prescribed liturgical elements. The remaining segments of the day were, in temporal terms, either added on behind the offices or placed in whatever gaps remained. Temporal values were pragmatic values that were not defined. This has led to the problem that modern reconstructions of the monastic day can be no more than approximations.[12] As for the duration of the elements of the day, it is often overlooked that their timing was intrinsic to them and that they were arranged sequentially.[13] Regulations governing time in the Rule were thus rarely directed towards abstract points of time or abstract periods; the same holds true for the later Rules, often many times longer than the Benedictine Rule, and for the customs (consuetudines) that took on binding force.[14] Most designations of time link the beginning of one activity or situation to the end of the previous activity. Typical formulaic expressions include: "subiungendum est; quibus dictis; quibus lectis; quam incipit cantor dicere, mox omnes surgant; post hos sequatur; post quibus lectionibus sequantur; parvo intervallo." Time indications using temporal hours and independent of the sequence of offices or Hours appear only twice in the Rule: in the nightly time of rising during winter (chap. 8), and in the rules governing work periods (chap. 48). In all other cases the time indications and offices cannot be separated, meaning that the prescribed tasks were linked to the offices, which were not precisely fixed in time.

The Rule constantly admonishes the monks to follow its temporal regulations as a group and punctually. But the implied notion of punctuality is different from its modern counterpart. In the interest of a communal performance, the handling of the rhythm of the day was elastic. If the monks rose late, the liturgy of Vigils would be shortened (chap. 11). To give latecomers a chance to catch up, the second Psalm of Vigils should be sung very slowly and with pauses (chap. 43). Though tardiness would be punished,[15] the required punctuality was not related to abstract points in time but to points in the sequence of the rhythm of collective conduct. Establishing the temporal order of a monastery was called "horas temperare" (chap. 41), an expression wholly unthinkable in schedules that are rigidly timed.

The monastery thus gave itself its own rhythm. Between post-midnight Vigils and sunset stretched a fixed sequence with times that were, to a certain extent, movable. Elements within the sequence could not be interchanged or left out. But it was unimportant whether Sext began an hour and a half or fifteen minutes before midday. This gave the various monastic orders and individual monasteries some organizational latitude.

The monastic time system was meant to set itself apart from all secular time systems, separating the monastery and the outside world. In that regard a different sense of time was virtually called for. Even if we can recognize parallels to the surrounding agrarian society in certain particulars, the result of practical necessity, the differences remain clear enough. Regularity and repetitiveness in regard to year and day and the collectively lived life produced a special rationality and required a special discipline, which becomes especially apparent in the conscious avoidance of any leisure ("otium"). Arno Borst saw here the beginning of the rationing and rationalizing of time, not, we note, of time measurement and time discipline.[16] Surely the absence of any possibility of personal control of one's time and the perspective of lifelong submission to this order produced special forms of time-consciousness, but those can hardly be described with machine metaphors.

To what extent the monastic time system had a shaping influence on its surrounding world is difficult to say. H. E. Hallam surely goes too far when he speaks of a "strict training" of medieval agrarian society at the hands of monasticism. Still, it is conceivable that important practical experiences in dealing with time were passed on. With limited but very regular applications of work in the fields, workshops, and scriptoria, the monks achieved results that may well have been regarded as exemplary, though this was only indirectly the purpose of the monastic order.[17]

Talk of "iron discipline" or the machine-like or clockwork rhythm of monastic life, even in a purely metaphorical sense, is misleading, because it suggests a time giver (a machine or clock) external to natural rhythms and the daily round of human life. In actual fact, life according to the Rule was bound in a very high degree to natural time givers, daylight and the seasons, and was by no means marked by ascetic resistance to the natural environment.[18] Of course the degree of adjustment to external time demands varied from one order to the next. Agrarian labor and pastoral care in the cities each produced different kinds of modifications to the Rules, which as a whole differed significantly from other daily routines only in that the monastic day began earlier.

Arranging the monastic time system in accordance with the Rules depended on two critical times that could not be determined without aids. From the beginning both were the topic of frequent discussions. The cycle of holy days in the ecclesiastical year was coupled to the movable date of Easter. The latter, in turn, was linked to an astronomical cycle, and until it was fixed by a calendar for the entire church, it could be determined only by observation and calculation.

The other temporal element, critical only for the monastic time system, was the timing of Vigils during the night. It was to be located as close as possible to midnight, but also in such a way that the monks could get enough sleep. In summer Vigils had to be chanted such that Matins could begin with the break of dawn. The Rule of St. Benedict stipulates that in winter the monks' rest should last slightly longer than half the night, hence it was "reasonable" that they rise at the eighth hour of the night.[19] This is the only time indication that has caused difficulties to commentators. Did Benedict mean the beginning or any point during the eighth hour? How could the eighth hour be determined at night? The great learned commentaries on the Rule, from the early modern period down to our day, have considered this passage, in particular, to be obscure, difficult, and hardly amenable to a satisfactory interpretation.[20] Right up until the end of the Late Middle Ages, monastic clocks ("horologia") are mentioned above all in the commentaries on this passage; later the passage provided grounds for presenting the division into hours as a historical fact.

The Hours as a time signal

Already in the high Middle Ages the multitude of church bell signals had become such a natural part of life that their origins were placed far back in history. Church bells were said to have been invented in the fifth century in the town of Nola in Campania. This legend was a transparent wordplay: "nola" was the name for a small bell, "campana" for a larger one. Another legend, in part repeated down to our day, attributes the public signaling of the Hours or time of day to Pope Sabinianus (604–606).[21] He is said to have ordered the churches to mark the hours of the day by ringing their bells. City chronicles during the Middle Ages regularly passed on these very widely known stories when they talked about the church bells or reported the installation of new ones.[22] In the fifth and sixth centuries, however, the spreading of bells to churches and monasteries had only just begun. The increasingly lavish outfitting with bells during the following centuries allowed the ringing of sophisticated signals for the times of prayer in the monasteries, later also from city church towers. The Hildemar commentary on the Rule of St. Benedict (ninth century) already suggested a series of signals for the sequence of Hours in the monastery (3-1-1-2-2-2-1-3).[23] Guillelmus Duranti's thirteenth-century commentary on the divine services has a sequence—apparently intended for churches—that envisaged numbered strokes for the twelve hours of the day (1-3-3-3-1(or 3)-1).[24] It should not be assumed that this was how most churches rang. Each town and each church rang in a different way to distinguish prayer times, depending on the number and type of bells available. But we can assume that in the cities of the thirteenth century the Hours were rung from many different towers more or less regularly and carefully. Countless urban time indications put this beyond doubt. An etymology circulating around 1300 falsely turned this urban phenomenon into a scoff at peasants: "Bells are called 'campane' by the peasants because they live in the fields ('in campo') and can know the hours only from these bells."[25]

Computus

Outside of the literature on the rules and practice of monastic life, information about medieval time reckoning is also found in a literary genre called "computus." A medieval computus is a handbook containing above all methods for the astronomical and calendric calculation of the movable Paschal date. Beyond that, the computus quickly became a typical forum for discussing a variety of questions concerning Christian chronology and the astronomical and mathematical knowledge needed to work on it, and for developing aids in the form of tables and mnemonic devices.[26]

Like all calendars, the Christian festal calendar is the product of compromise. It resulted from the non-synchronicity of different astronomical cycles, for example, the solar and lunar years, from their supposed and actual irregularities, and from political decisions made at various times and places. The year and day of Christ's crucifixion were and still are controversial.[27] Tradition placed it on a Friday, March 25; in the Roman solar calendar that is the day of the vernal equinox. According to the evangelists, the resurrection occurred on the day following the Sabbath of Jewish Passover. The Passover holiday followed the Jewish lunar calendar, whose festivals were movable with respect to the Roman calendar. It was celebrated on the fourteenth day (i.e. the full moon) of the first Hebrew month (Nisan). Since Easter, the day of the resurrection, was not to coincide with the Jewish festival under any circumstances, the Church, after passionate debates, placed it on the Sunday following the vernal equinox. At the council of Nicaea, the

Church, deviating from the Roman calendar, followed the observations of the Alexandrine astronomers and fixed March 21 as the Easter-limit, the date before which the holy day could not be celebrated. Papal Paschal letters and tables were intended to help establish a uniform date. But traces of competing dates for the Easter-limit can still be found as late as the turn of the millennium.[28]

The computistic writings of the Anglo-Saxon monk Bede (*De temporibus* [704], *De temporum ratione* [725]) became highly authoritative and were very widely circulated.[29] Even though the actual vernal equinox had shifted to March 17 by the eighth century, Bede, in the earlier of the two works, followed the tradition of the western Church and gave March 25 as its date. In 710, in a letter of the Irish abbot Ceolfried, possibly written by Bede himself, and then in 725 in Bede's *De temporum ratione,* we read that the date was to be fixed on March 21. Both texts justified this with repeated and very emphatic references to empirical observations with a time-measuring device ("inspectione horologica" or "consideratione horologica"). They may have been talking about observations on sundials, which were very common in monasteries at the time. It is highly unlikely, however, that these simple devices at that time already allowed observations in units on the order of only a few minutes.[30] We can hardly decide whether observations were in fact made or whether the reference to observations was intended to bolster specific arguments.[31] Bede's authority, in any case, was such that later computus authors simply adopted what he said. Voicing no reservations of any kind and invoking historical authority, these authors also took over the description of methods for measuring the diameter of planets with the help of water clocks.[32]

A different subject in the computus texts was the discussion of the largest (e.g. era) and the smallest units of time, a topic that had never been dealt with in such breadth and detail in antiquity. Here the authors used—side by side and without a uniform terminology—mathematical and astronomical divisions, Roman fractions, and weight measures (modern time values are given in parentheses). In Ptolemy the full day had been divided either into 4 quarters of 6 hours each or into 360 chronoi. The hour, accordingly, had 15 chronoi (15 minutes). In Bede and the computistic tradition the hour was divided into 4 puncti (15 minutes), or 10 minuta (6 minutes), or 15 partes (4 minutes), or 40 momenta (90 seconds), while in lunar calculations the hour was given

5 puncti (12 minutes). Bede considered the punctum the smallest unit measurable with the sundial, and the momentum the smallest perceivable unit.[33] He did not regard such divisions as natural, but as conventions agreed upon among mathematicians ("calculatores"). Bede reports on another division by which the Zodiac (full day) was broken down into 12 signs and 12 (double) hours. Each hour, accordingly, had 30 partes (4 minutes) of 12 puncti (20 seconds) each, each punctus had 40 momenta (1/2 second), each momentum had 40 ostenta (about 1/100 of a second). Bede rejected this division as an exercise of astrologers ("mathematici"), as he did the division of the hour into 22,560 atoms.[34] The high Middle Ages was also familiar with the division of the hour into 60 ostenta and, under the influence of Islamic astronomy, with the hour-minute.[35] The abundance of small and smallest divisions—the list I have given being by no means exhaustive—was frequently used in the same text side by side. However, the use of very small units of time was limited to theoretical and astronomical treatises; outside of such texts we find no trace of them. In daily life, short periods, as for instance the duration of an earth tremor, were for the most part expressed in prayer times. For example, a chronicle of Constance says about an earthquake on August 13, 1295: "around midday there came the greatest earthquake . . . and it lasted about as long as it takes someone to say a Paternoster and an Ave Maria."[36]

At times such indicated periods could be given greater precision with additional information about the tempo of the prayer. One such example comes from a time when modern hour reckoning, though not minute reckoning, had long since become customary. It is found in a letter of the Florentine Paolo Rucellai and concerns an earthquake in Naples on December 8, 1456: "It lasted the duration it would take to say Miserere quite slowly and more specifically one and a half times."[37]

Bede distinguished three kinds of time reckoning ("ratio temporum trimoda") with different degrees of binding force:

(*a*) Reckoning according to God-created nature: this included the solar year with 365 1/4 days, the lunar year with 354 days, and the nineteen-year lunar cycle.

(*b*) Reckoning according to human custom: an example of this is the thirty-day month, which corresponds neither to the solar nor the lunar cycle.

(*c*) Reckoning by authoritative decisions: on the authority of the divine law the seven-day week with the observance of Sabbath could not be changed. By the same authority agrarian labor should rest every seventh year and a jubilee should be celebrated every fifty years. *Human* authority gave rise to the four-year Olympic cycle, the fifteen-year cycle of "indictiones" (Roman tax cycle), the eight-day rhythm of weekly markets, and, finally, the division of the day into four segments ("quadrantes").[38] All Bede does is list examples, but it becomes clear that the division of time attributed to custom or recognized authority rested on human statutes and was thus changeable. In a different context, Nicholas of Cusa, in the middle of the fifteenth century, described the measuring units of time, such as year, month, and hour, as new aids which had not been present at the time of creation. Just as Ptolemy invented the astrolabe and Orpheus the lyre, the rational human soul invented these devices to distinguish and recognize the objects of the material world.[39] In medieval understanding, too, the division of time, and especially of the day, was not simply a given fact, beyond doubt and unchangeable. Instead, it was seen as determined in part by natural rhythms, in part by social convention or "political" decisions, and as subject to historical change.

4

Medieval Horologia and the Development of the Wheeled Clock

It is certain that if we knew who first invented the means of measuring time by the movement of toothed wheels, a movement controlled in such a way that the to-and-fro of the "unrest" [foliot] resists it in an alternating pattern, this person would deserve all our praise—but history teaches us nothing certain in this regard.[1]

Following these words of praise for the unknown inventor of whom "history" knew nothing "certain," Dom Alexandre, the author of one of the first modern technical handbooks on clocks (1738), outlined the questions and the period that are still the focus in the continuing discussion over the invention of the wheeled clock. His definition of the problem—"the invention of a resistance by the to-and-fro of a foliot"—is as valid today as the period he focused on, between the ninth and the fourteenth centuries. He doubted that Archbishop Pacificus of Verona, who heads the entry in du Cange's glossary, could have been the inventor, because it struck him as improbable that word of the invention, "so sorely needed for the arrangement of the nightly activities," had not reached the monasteries even a quarter of a millenium later. As evidence he cites the Rule of Cluny, where we learn that the sacristan "always had to step outside and read the time from the heavenly bodies." In Dom Alexandre's view the real inventor was Gerbert of Aurillac, something one occasionally still reads today. In

Spain, where Islamic and European-Christian civilization came into direct contact, Gerbert was said to have become so deeply versed in the "astrologer's art" and "mathematics" that he was believed to be a magician. Dom Alexandre does not pose the question why Gerbert's invention, too, did not spread more quickly to the monasteries. Instead, his historical survey leaps ahead into the fourteenth century, to the famous astronomical clock of Abbot Richard of Wallingford.

The development or invention of the mechanical clock has been more frequently discussed, and can be considered to have been more thoroughly researched, than any other aspect of the history of technology prior to the industrial revolution. Even though the evidence has been collected piece by piece for decades, it has so far proved impossible to pinpoint the place, time, and circumstances of this technological breakthrough. Most histories of clocks and technology therefore take only a cursory look at ancient or medieval water clocks, emphasize that the invention of the mechanical clock was a stroke of genius or a turning point in the history of civilization, and date the process accurately to the late thirteenth or early fourteenth century.

The appearance of the mechanical escapement did not find an echo in contemporary accounts. No entry in a chronicle, no narrative account, no description of the construction makes the invention an event we can date or locate. In this respect the history of the escapement is no different from that of tower windmills or cannons, not to mention the history of the compass or paper. Compared to these innovations the sources on the history of clocks are far more numerous. They are, however, afflicted with the problem that the transition to the mechanical clock found no reflection in language, did not cause a change in terminology. The term "horologium," a word of many meanings, was simply retained. The appearance of the escapement, today considered the decisive innovation or pathbreaking invention, does not even appear in contemporary awareness. It was at most in retrospect that it was described as significant but mysterious. In contrast, the appearance of striking clocks was registered instantly, and was felt to be technologically sensational and socially momentous. Discussions about the change in time-consciousness in the Middle Ages should take this fact into consideration.

Clearly formulated hypotheses concerning the invention of the mechanical clock are a rarity. Leaving aside the countless opinions that have been expressed, mostly in a rather casual way, hypotheses with some degree of substantiation move along two tracks which are not mutually exclusive. According to the classic view as laid out by C. du Cange, and argued most recently by J. D. North and, following his lead, also by D. Landes, the medieval monasteries are the most probable site for this technological development.[2] The overwhelming majority of European sources on medieval "horologia" refer to monastic and ecclesiastical life. Up until the middle of the thirteenth century, it is only in this sphere, which did, however, include most of contemporary astronomy, that we hear about time measurement and timekeeping devices. Even if the argument that the invention sprang from the intrinsic needs of the monastic time system is based on recent misinterpretations of the monastic Rules, it would seem reasonable, given the provenance of the sources, to assume that the development of the mechanical clock occurred in the monasteries.

But this does not yet answer the question whether we are dealing with a European development or with a technical development inspired by the reception of classical astronomy via the Islamic world. For example, the classical technology of water clocks was undoubtedly developed further by Arabic-speaking authors and mechanics. What is unclear is how much of this became known or was adopted in Europe and when this process may have begun.

The discussion was given new impulses by the publication of Joseph Needham's history of the sciences in China, and by the monograph on astronomical clocks in medieval China which he co-authored with Wang Ling and Derek J. de Solla Price.[3] Since that time, Derek Price, in particular, has repeatedly argued that the "missing link" in the history of European clocks should be sought in some inspiration from China, possibly transmitted via the Islamic world. The clock escapement was developed in Europe not because there was need for a reliable time measuring device, so the argument goes, but because Europeans were looking for the motive power of the astronomical models in China, which they knew about from hearsay. These two hypotheses, or variations that can be constructed from the two, are not mutually exclusive, because the decisive technological step, the development of the clock escapement, still eludes us in terms of place, time, and circumstances. But they do imply different models of the course history took: in the first case we have the evolution of a native technology in response to the demands of an identifiable need; in the second case the transfer of a technical vision—that of driven astronomical models—from the outside, and the gradual simplification of the complicated solutions into a usable, everyday instrument, the mechanical clock.

Interest in outside influences on the history of European science and technology has led to an intensive new scrutiny of the sources since World War II. The impulse behind this has been the effort to get away from the fixation on a Eurocentric view of history. An additional impulse has been the realization that Chinese and Islamic sources on the history of technology are, in contrast to the European political and ecclesiastical sources, far more numerous and revealing. Thus it seems reasonable to draw on non-European evidence to illustrate and explain European developments. The history of clocks is undoubtedly the most important example of this.

The clock escapement

A medieval mechanical clockwork was made up of four components: a weight drive, a mechanism for transmitting the energy through a gear train, an escapement or regulating mechanism, and an indicating mechanism in the form of a striking or pointing contrivance. Only the escapement and regulating mechanism was an innovation characteristic of the mechanical clock. The other components had a long tradition in the history of hydraulic clocks and astronomical instruments.

The verge-and-foliot escapement (fig. 7), which can be considered the normal type around 1400, is a mechanism in which the motion of a weight-driven axle is impeded or controlled in such a way that the axle's uniform rotation is suitable for use as a time standard, for example an equinoctial hour. Without such a brake the rotation of the axle would steadily increase in speed. The escapement works as follows: a crown wheel with an unequal number of teeth, mounted onto the axle or linked to it via a gear train (which has been left out in the sketch), alternately blocks and releases the verge by means of two pallets attached to the verge at a right angle to each other.

Mounted on top of the vertical verge is a scale-like beam (the foliot). By pushing on one pallet, the crown wheel moves the verge and the foliot in one direction until the other pallet stops the motion and re-

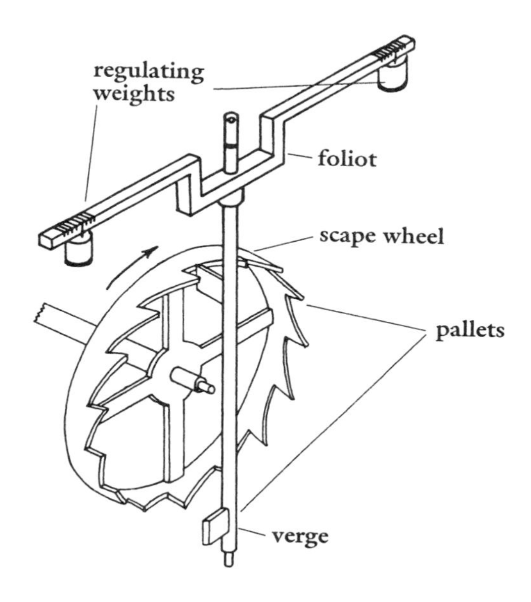

7. Verge-and foliot escapement. Drawing by G. Oestmann, Bremen.

verses the direction of the rotation. The duration of the oscillation of the inertial mass of the verge and the foliot can be adjusted by moving regulating weights on the foliot. Another way of regulating the mechanism is by changing the pull weight. This to-and-fro, oscillating movement inspired various graphic names for the device: "restlessness" (Unruhe), "foliot" (from a word describing a quivering leaf, first used by J. Froissart around 1370), more rarely also "woman's temperament."[4] The first text that gives an approximate description of this type of escapement is an operating manual which the Basel clockmaker Heinrich Halder supplied for the tower clock he built for the city of Lucerne in 1385: "And if the woman's temperament goes faster than you think it should, hang the lead blocks away from the wheel, and if it goes too slowly hang them closer to the wheel, in this way you will make it go forward and backward as you wish."[5]

A variation of this escapement appears in the older sketch of the "common clock" ("horologium comune") contained in the description of the supporting frame of Giovanni Dondi's planetary clock ("astrarium"), completed in 1365 (fig. 8). The fifteenth-century version repro-

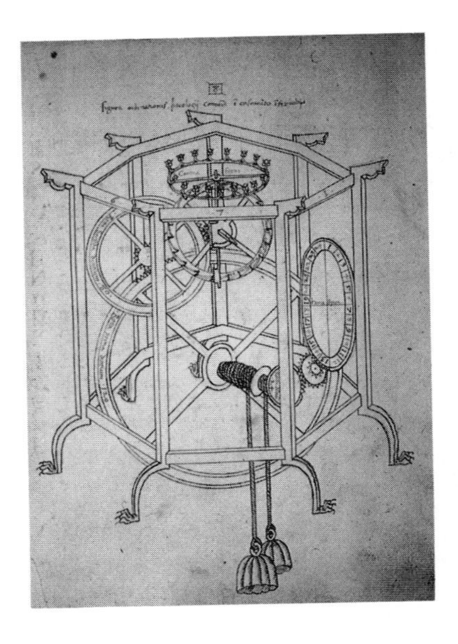

8. The astrarium of Giovanni Dondi dall'Orologio. Sketch of the supporting frame with clockwork and hour dial. From a manuscript in the library of Eton College, Windsor, Ms. 172. Reproduced with permission of the Provost and Fellows of Eton College.

duced here goes back to a nearly identical sketch in an Italian manuscript from the late fourteenth century. Dondi presupposes that the construction of the clockwork drive is well known and that various forms are possible, hence he gives only a cursory description.[6]

The crown wheel, not very clearly labeled in the sketch, is described in the text as "rota tercia, qui dicitur in horologiis communis rota freni, habens dentes ad instar serre [saw] magne lateraliter" (the third wheel, which in common clocks is called the restraining wheel ["rota freni"] and has teeth like a large saw on the side). The verge and inertial mass are called in the text "frenum cum corona." Instead of the foliot, the sketch shows a horizontally mounted wheel ("corona freni") decorated like a crown. Apparently this type of escapement can be regulated only by changing the weight or counterweight, since there is no sign of any regulating weights on the wheel. The setup is not entirely clear, since the text, in deviation from the sketch, implies two possible ways of regulating the mechanism: by changing the pull weight or by removing weights from the escapement itself.[7]

The text on the construction of the "horologium astronomicum" of Richard of Wallingford, abbot of the Benedictine abbey of St. Albans in Hertfordshire, published by J. D. North in 1976, takes us back to an

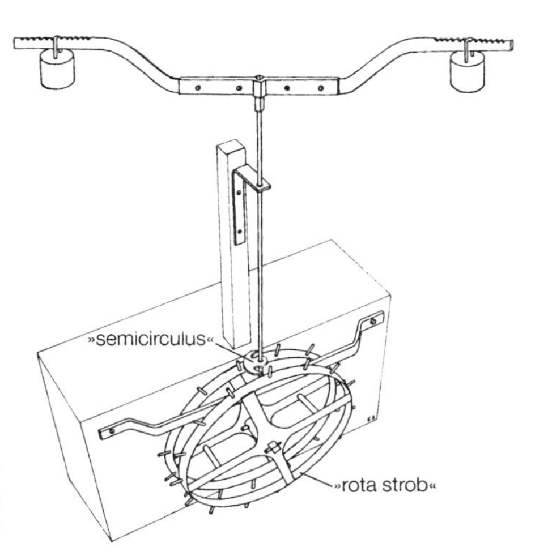

9. Reconstruction of the escapement in Richard of Walingford's astronomical clock. Drawing by G. Oestmann, Bremen.

even slightly earlier time, but also into a host of difficulties that cannot be entirely resolved. This text, too, deals primarily with the theory of the complicated gears for astronomical indicators, and only secondarily with the construction of the clockwork itself, taking a good deal for granted. The obscure references to the clock escapement appear in the sections on the striking work of the clock ("pro sonitu unius timbe" or "pro sonitu unius clok"). Our understanding is additionally hampered by a peculiar terminology which, as far as we know, was without a model and had no lasting effect. The clock, about the height of a man (fig. 13), presumably had a foliot escapement with movable lead weights; at all events, we read on one occasion of the "quadratura plumborum pro hasta strob." "Hasta strob" is apparently the vertical verge. Its oscillating movement was produced not by one but two wheels ("rotae strob"), whose teeth or pins were arranged around the edge in a radial pattern (fig. 9). Fastened to the verge itself was, instead of two pallets, a semicircular device with presumably sickle-shaped halves ("semicirculus"), whose ends alternately blocked the teeth of one of the wheels. In this way the motion of the axis shared by the two wheels was blocked in a jerking fashion and with a recoil effect.[8]

There are no conclusive texts or material remains for a reconstruction of the Wallingford escapement. But conceptual aids come from Leonardo da Vinci's sketches of clock escapements he had presumably seen somewhere or another, and from two later Italian sketches of alarms which presumably reflect technological arrangements going back to the fourteenth century.[9] A connection between the Wallingford escapement and the two later Italian sketches is plausible, but remains hypothetical. Going backwards chronologically to the oldest sources for the escapement of the mechanical clock yields the clue that the clock escapement and alarm mechanisms may have been technically related developments. But this too leaves open some important questions: Was the clock escapement of Richard of Wallingford a precursor to other escapement types which subsequently disappeared because they were not as useful or reliable? Was it a mechanically superior variation, but one that was too difficult to build or too expensive? Was it a chronologically parallel and independent special development which had no impact?

J. D. North's view that it represents a mechanically superior preform is supported only by the fact that Leonardo reports on this type of escapement and explicitly describes it as such ("tempo d'orologio"). Against it speaks the absence of concrete evidence in textual or pictorial sources. The notion of a separate development is supported by the fact that not only the escapement, but also the striking-work construction (not reproduced here) of the English abbot is without parallels in the remaining evidence. Rather, the numerous depictions of clocks for daily use show that around 1500 only the first two types of escapement mentioned above (the verge and pallets with a foliot or balance wheel) were commonly used. The extant clocks that can be dated—cautiously—to the fifteenth century usually have a foliot escapement.

Even though the notes and sketches for the astronomical clocks of Giovanni Dondi and Richard of Wallingford are the oldest reliable evidence for mechanical clock escapements, they presuppose that the existence of such escapements had long been known. As a vague terminus ante quam we can thus give for now the time around 1330.

"Clock," "mechanical clock," "wheeled clock"

Any discussion of the history of the mechanical clock runs into the problem of the broad spectrum of meanings of the word "clock"/ "horologium," which in the older scholarship repeatedly led to erroneous conclusions about the age of mechanical clocks. In the Middle Ages, "clock"/"horologium" was a generic term for all devices and aids of time-reckoning and time-indication, and occasionally also for time-ordered conduct. Regardless of how it was constructed, a water clock could be called "horologium," "horologium aquatile." The ancient word "clepsydra" was seldom used in the Middle Ages to describe clocks. A sundial, too, was a "horologium." On rare occasions the expression is more specific (for example, "horologium solarium"); more frequently we find terms where a part of the device stands for the whole: "gnomon," "quadrans," "stilus." Astronomical instruments such as astrolabes and quadrants, but also simple looking tubes, were called "horologium," since they could function also as time-measuring devices. "Horologium" was also the word for tables used to determine the time of day according to the duration of daylight or the length of shadows. "Horologium" described, moreover, the instructions for determining time by means of the length of shadows or the position of the stars in relationship to certain parts of a building. In the Eastern Churches, the daily prayer sequences that were fixed in writing were likewise called "horologium."[10] From the high Middle Ages on, bells or bell works, if they were used in some way as a time signal, were also called "horologia."

While the transition to the mechanical clock was not reflected in the language, the appearance and diffusion of clocks that struck the hours was regarded as an exciting novelty. As a result there was soon a host of differentiating descriptions for this type of clock: "horologium pulsatile," "horologium horas diei et noctis indicans," "horacudium," and so on. The words for striking clocks in the non-Romance languages make this even clearer: "clocke," "zytglocke," "urglocke," and many others.[11] The linguistic evidence therefore suggests that we carefully distinguish the development of the mechanical clock from that of the striking clock.

In modern usage, "mechanical clock" and "wheeled clock" have become the customary terms for the weight-driven clock with a mechanical escapement. I have adhered to this usage. Both terms are used uniformly and interchangeably with the same meaning. We should, however, keep in mind the lack of clarity that attaches to both. The term "mechanical clock" emphasizes the difference to simple outflow clocks (water clocks or hourglasses), to clocks with no movable components (sundials), and to the now rare clocks based on the burning of certain substances (candle clocks, oil clocks, incense-stick clocks). But this term also lacks distinctness because it excludes the complicated water clocks that certainly had mechanical components like gear trains, weights, drums, as well as elaborate automata mechanisms, and which in their day were considered "arte mechanica composita."

The antiquarian term "wheeled clock" avoids this terminological problem somewhat better and is close to the medieval usage, which frequently emphasized the iron wheels. It is, however, no less indistinct, because wheels and gears were not the specific distinguishing characteristic of the new clocks that appeared around 1300. Artificial, technical terms such as "Gewichtsräderuhr" or "escapement controlled clock" are also not able to entirely avoid these unclarities, and they pose additional problems of understanding for the layman. For all these reasons I have retained the colloquial terms "mechanical clock" and "wheeled clock."

"Water clock" is the generic term for all forms of this historical type of clock. The simple variants are also called "outflow clock" or clepsydra; the complicated variants, with an inflow that is maintained at a consistent rate or with an adjustable outflow, are called "inflow clocks" or "hydraulic clocks."

The legendary inventors

The now faded fame of Archbishop Pacificus of Verona (died 844) as the inventor of the mechanical clock was promoted above all by his epitaph on the Cathedral of Verona which Du Cange quoted in his Latin glossary. The epitaph speaks of a never-before-seen "horologium nocturnum," which, according to Du Cange, was a wheeled clock. In the same inscription, however, Pacificus is also described as the author of an "argumentum" that came with the clock; apparently the text of this "argumentum" and the illustrations that accompanied it in the medieval manuscripts were no longer known in the seventeenth century. The text and the medieval illustrations clearly show that the "horologium nocturnum" was an observation tube with a cross-shaped sighting device or scale. It was meant to be suitable for determining the night hours in monasteries "even without the crow of the cock," but like the medieval sundials it was also to be used for calendrical purposes.[12]

A similar confusion was created by the frequently cited notice attributing the construction of a clock to the scholar Gerbert of Aurillac—Abbot of Bobbio, Bishop of Reims, Ravenna, and Rome—when he occupied the papal throne as Pope Silvester II (died 1003), even though the text makes clear that "horologium" is only a functional description.[13] We are hardly in a position to determine whether the device was a water clock combined with a sighting tube for the polar star or an astrolabe.[14] When it was used as a time-measuring device it could also be called "horologium." Gerbert built a whole series of astronomical instruments for teaching and observation. He described the clepsydra as a tool for determining the equinoxes.[15] But there is no report during his lifetime that he built a clock. Only much later (around 1120) do we read in William of Malmesbury's *History of the English Kings,* which is not considered to be reliable, that the objects kept in Reims as evidence of Gerbert's learning (which was not above suspicion in religious terms) included, apart from a hydraulic organ, a "horologium arte mechanica compositum."[16] This passage, which is revealing for the medieval range of meanings of the term "mechanical clock" as well as for the reception of classical pneumatics, cannot be readily related to Gerbert as it has no connection to Gerbert's activities reported elsewhere. It is striking, however, that he is also credited—in a piece of information of unknown date and provenance—with the construction of a water clock in Ravenna ("horologium aquatile seu clepsidra figura").[17] Again, we find no mention of this during Gerbert's lifetime. Gerbert's reputation as an astronomer could be the reason why he was later also honored as a clock inventor.

The name of the Benedictine abbot Wilhelm of Hirsau appears today but seldom in the list of possible inventors. The still-extant stone sphaera he constructed (around 1260) in the monastery of St. Emmeran in Regensburg—"horologium naturale ad exemplum coelestis hemisphaeri," also "horologium nocturnalis"—was also an astronomical instrument for teaching and observation without any mechanical components.[18]

"Horologia" in medieval monasteries

The sixth-century Rule of St. Benedict places responsibility for adherence to the monastery's temporal arrangements into the hands of the abbot. Either the abbot himself or a responsible brother was to "call" the time for the Divine Offices (chap. 47). The Rule attaches no great importance to the precise temporal location of the elements prescribed for the daily routine. Critical for the monastic arrangement of time was only that the monks rise on time for the nightly offices (chap. 8).[19]

The medieval monastic Rules, the statutes of monasteries and chapters, the customs ("consuetudines"), and the commentators always stress the special importance of this particular time and the special responsibility of the person charged with determining it and announcing it to the entire monastery. At the beginning of the eleventh century, Petrus Damianus emphasized that the neglectfulness most to be avoided was that of the "significator horarum," lest the entire sequence of temporal arrangements ("ordo horarum succedentium") fall into disarray.[20] The customals for the Benedictine monasteries in Subiaco (fourteenth century) admonished the monks to be particularly fastidious about adhering to this time because the entire catalogue of liturgical duties depended on it.[21]

The Rule of St. Benedict itself does not yet mention any procedures or aids for determining this time. The Regula Magistri, written around 520 and thus still contemporaneous with the Rule, mention a "horelogium" that was to be assiduously observed day and night. This may have been a simple water clock or star and shadow tables.[22] Later versions and explanations of the Benedictine Rule frequently provide for the use of a "horologium." Here we notice right away that these "horologia" were used only at night to wake the monks for night offices or for determining their temporal distance from morning offices, which were likewise to commence while it was still dark. During the daytime we find no mention of their use.[23] Accordingly only the person charged with waking the brothers, usually the sacristan (in this function occasionally called "horoscopus"), is mentioned in connection with the supervision of the "horologia."

But not all texts of this nature, not even the majority, provide for the use of a "horologium." Other means of determining the time were at least as familiar to the monks. Frequently mentioned and often celebrated in literature was the classic waking signal of the old agrarian society, already important in the Bible: the cock's crow.[24] It has been symbolically immortalized as a wind vane on Church spires. As a movable figure it was part of the repertoire of automata of the monumental clocks in later times. For example, a mechanical figure of a cock is all that remains of the first astronomical clock in the Minster of Strasbourg.

Another method that was cultivated especially in the monasteries was the observation of the course of the stars. In the fifth century, Cassian recommended it to the watchmen of the monastery. Celestial observations were collected and passed down in the monasteries. At the end of the sixth century, Gregory of Tours composed a short tract entitled *On the Course of the Stars,* whose sole purpose was to provide aids for determining the time of nightly prayers. Methods for observing the heavens were refined over the course of time. Some monasteries put together special observation tables, which were also called "horologium." The cultivation of astronomy in the medieval monasteries found in these endeavors an important practical purpose and a compelling theoretical justification.[25] The cock's crow is not a very reliable time indicator, and celestial observations are not possible in bad weather. That is why Petrus Damianus recommended that the "significator horarum" memorize the time required to sing the various psalms and in this way use himself as the time-keeping device whenever necessary.[26]

Simple sundials were in use for gaining an approximate sense of the time during the day. They were mentioned only in early Rules, but numerous surviving pieces show that they were widely used.[27] The medieval type, a semicircle with segments marked off at regular intervals (figs. 5 and 6), is reasonably "accurate" on only two days of the year. Today these sundials are called "mass-dials" or "canonical sundials."

The burning of candles carefully calibrated as to length and weight is recommended in Rules that were influenced by the reform movement of the Benedictine abbey of Cluny as a method for determining the duration of night (fig. 10).[28] Candles, as later sandglasses, were also in use outside of monasteries into the modern era as a simple way of delimiting short periods, as, for example, at auctions or elections.[29] Measurement specifications for such candles have survived. Candle clocks as a means of dividing the entire day into equal segments are mentioned prominently in the biographies of three medieval kings. In this way their daily routines became one of the many examples of their virtues. According to a contemporary vita, King Alfred of England

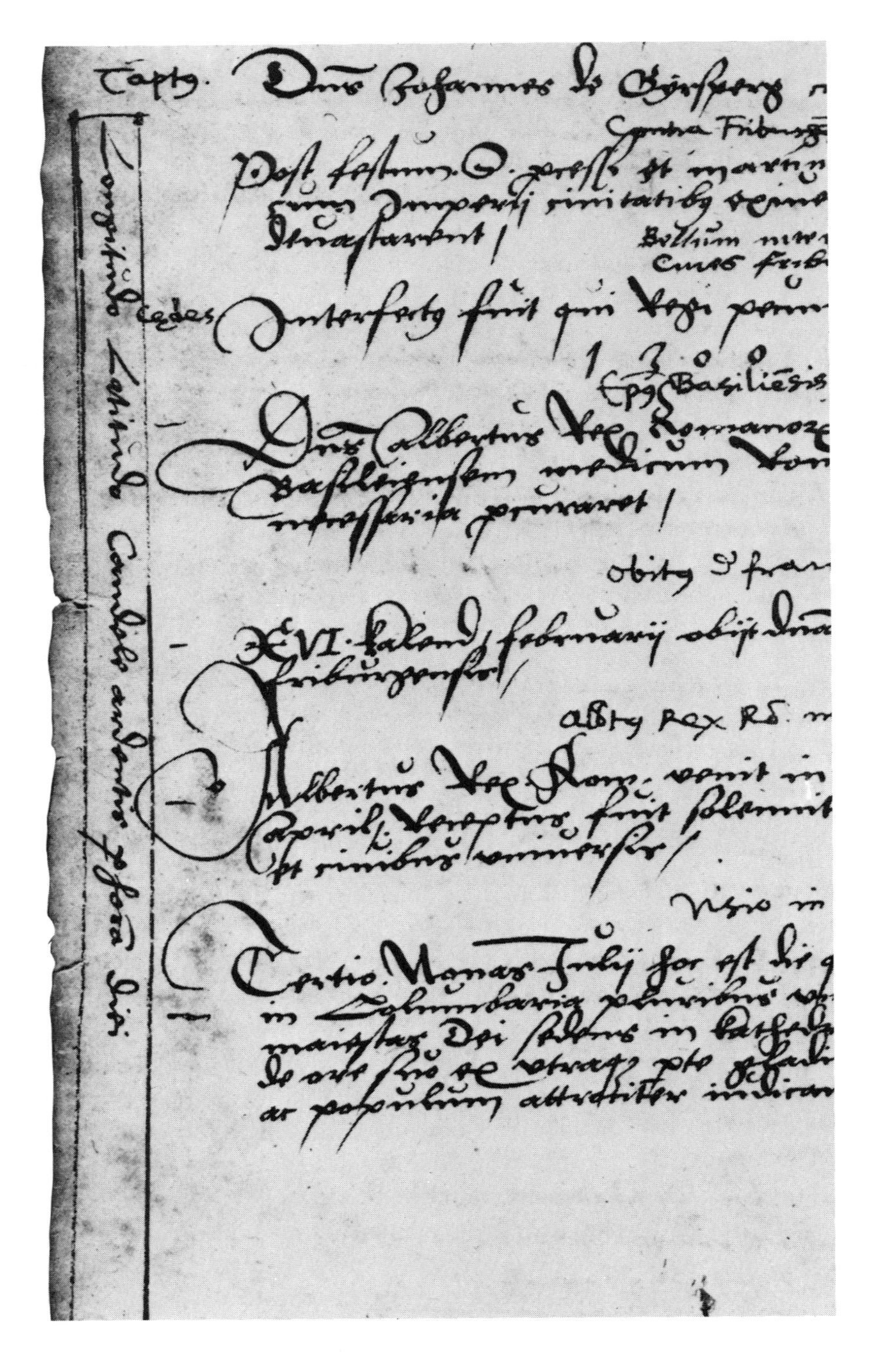

10. Sketch of a candle with a burning time of one hour in the margin of a manuscript of the *Größere Kolmarer Annalen,* Stuttgart, Württembergische Landesbibliothek, Cod. hist. 4° 145, folio 119 verso.

invented candle clocks in order to devote half of every day and night to prayer. To that end he had his workmen fashion six candles of equal weight and mark each one with twelve segments (of about twenty minutes each). The candles, protected inside a lantern on account of the bad weather, were supposedly lit at night and replaced every four hours. William of Malmesbury tells us that Alfred divided his days into three eight-hour segments, one each for study and prayer, official business, and rest.[30]

Such a setup could have worked; in the Islamic world, as well, there developed a rich tradition of graduated candle clocks. In Asia, graduated incense sticks were in use into the modern period.[31] It appears that there existed in the Middle Ages a narrative tradition about the division of daytime that had nothing in common with the monastic division of time. According to a vita of (Saint) Louis IX (died 1270), the king used candles approximately three feet in length to determine the duration of his prayers.[32] In her description of the exemplary arrangements of the daily routine of Charles V of France, Christine de Pisan reports that a twenty-four-hour candle was used for a division of the royal day into three parts.[33] Though the formal idea of an order that could be numerically conceptualized may have been what mattered most in these vitae, for our purposes what is remarkable is that the devices used allowed a division of time into segments or hours of equal length.

During the early Middle Ages, knowledge of the mature technology of classical water clocks, as Vitruvius had described it, was all but lost in the western half of the Roman empire. The scant references lead to the conclusion that only very simple forms were still in use.

From Rome Theodoric sent a sundial and a water clock to the Burgundian King Gundobad in the year 507. The accompanying letter reveals that the recipient had seen such clocks in Rome and now wished to have them in his homeland.[34] Even two centuries later, the Frankish king Pepin still received a "horologium nocturnum" from Rome.[35] In his handbook on monastic life (around 520), Cassiodorus, who in his function as imperial official had drawn up Theodoric's accompanying letters, praised the sundial and the water clock ("horologium aquatile") as highly useful inventions which the soldiers of Christ should use in fulfilling their duties. With the help of the water clock, human ingenuity compensated for the imponderables of the heavenly phenomena.[36] Apparently individual monasteries, following the example of the Roman military, had introduced outflow clocks to divide the nightly watches. The ninth-century commentary on the Rules that is attributed to Hildemar remarks, with regard to the problem of rising at nighttime, that a water clock is necessary for a sensible division of the night. This emphatic recommendation shows that while such clocks were known, they were by no means used in all monasteries.[37]

Various pieces of evidence indicate that a certain technological change had occurred by the eleventh century. To begin with, we note that in the usage of the commentators the classical term "clepsydra" for water clocks disappears from the normative texts. Used for the last time in the older consuetudines of Fleury (around 1000), in the high Middle Ages "clepsydra" became restricted again to its original meaning of "wine or water siphon." Only the dictionaries continued to drag along the word "clepsydra" with the meaning of "clock" until the fifteenth century. In addition, the qualifying words attached to "horologium"—such as "aquare," "aquatile"—disappeared from texts intended for practical use.[38]

Even though the "horologia" of the subsequent period were undoubtedly water clocks, the disappearance of "clepsydra" indicates that the simple outflow clocks originally used had been gradually replaced by more elaborate constructions.[39] In the reform statutes of Cluny (eleventh century), and in many of the texts directly influenced by them, we then hear about a clock that audibly "falls."[40] Moreover, for describing the erection and maintenance of the "horologia," a terminology came into use which can no longer be related to the observation or filling of simple outflow vessels. While we still read in the Consuetudines of Fleury "custos clepsidram previdens," in the Cluny texts the activities of the sacristan (secretarius, apocrisarius) as they relate to the clock are called "temperare," "dirigere," and "ordinare." "Temperare" remains the most common word by far; in addition "regere," "moderare," and "advigilare" also come into use.

The various developmental steps of the monastic alarm devices, which became increasingly elaborate, remain obscure. Ideas for technical experiment may have come from the reading of Vitruvius, whose work was not a rare item in monastic libraries after the eighth century, and who was often quoted and occasionally copied out.[41] At this time we cannot say whether he was only handed down as traditional learning or whether he was also used as a technical inspiration. Given that

Vitruvius's horological discussions are so difficult to understand, an inspiration is all they could have provided. We cannot prove this, however, just as we cannot provide evidence for influences from the water clocks that were imported from Byzantium and the Islamic world, a second source of possible inspiration.

Assuming that the chronology of the monastic codifications permits at least approximate conclusions about the technological development of the "horologia" that were used in the monasteries, additional developmental steps took place in the twelfth century. The sources from the French realm provide a small piece of linguistic evidence that the clocks had become, in many instances, technical ensembles of several components. First in French belles lettres, and then also in Latin statutes, the plural ("orloges/horologia") is used for a "horologium" as a linguistic mixtum compositum (as was also the case with the organ ("organa").

An important clue to a technological development is contained in a catalog of offices that was drawn up around 1100 in the monastery of Garsten or Göttweig. The document is an expanded version of the eleventh-century *Consuetudines Fructuarienses-Sanblasianae,* practical monastic texts connected to the Cluniac reform movement. We are told that in the evening after Compline, the sacristan is to set the horologium as carefully as possible. In the morning, when the clock "falls," he is to rise and, if the sky is clear, verify the time against the stars. If the time has come for the brothers to get up, he must open the door to the monastery, light candles, and set the clock by pouring water (back?) from the smaller into the larger basin, pulling the rope and the lead up (again?), and then striking the bell.[42] The horologium was evidently a water clock in which a float on the dropping surface of the water was connected to a lead counterweight. The waking signal for the sacristan was acoustic, but it was not yet directly linked to the bell. A water clock of metal is also mentioned in the animal fable *Ysengrimm,* written around 1150 in the area of Ghent.[43]

In the early-twelfth century versions of the *Liber usuum* of the Cistercians and in a series of related texts, among them also the oldest statutes of the Premonstratensians, we then encounter a type of horologium that has undergone further developments. It now produces a pleasant sound which wakes the sacristan, but which the brothers can also hear. In addition, the time of night that has passed can be determined or read from the clock.[44] The sacristan had to set the clock or the signal mechanism, which could probably also be tripped separately, such that it rang at certain times. The context and the choice of words—"pulsare" and "signum vel horologium"—suggest that the "horologium" was already connected to a bell. Further evidence is also the shutdown of the clock in the period before Easter that is provided for in the statutes of the Cistercian monastery of Wettingen (thirteenth century).[45] Moreover, on festive occasions the "horologium" became part of the bell ensemble. According to the regulations enacted in 1250 in the Benedictine abbey of St. Albans, the newly elected abbot stepped up to the altar accompanied by much ringing of the bells—"pulsato classico [name of a bell], sonantibus burdones [name of a bell] cum horologio."[46] From the twelfth century on, a bell that was linked to a clock became a fixed part of the increasingly elaborate bell ensemble in churches and monasteries.

As early as 1166 the commentary on the offices of Johannes Belethus mentioned the clock bell in the list of the six bell types for church use. In this list, which is arranged according to use—"squilla in refectorio, cymbalum in claustro, nola in choro, nolula vel duplula in horologio, campana in capanili, signa in turribus"—the clock bell assumes a middle position in terms of size. This terminological suggestion is far removed from contemporary usage. At the most "squilla," "cymbalum," and—missing in Belethus—"tintinnabulum" describe small clock bells, "campana" larger ones. But "nolula" or "duplula" appear nowhere in connection with a "horologium." Nevertheless, the far more influential commentary on the offices of Guillelmus Duranti (1284) handed down this terminology.[47]

Up until the middle of the thirteenth century, normative texts are nearly the only usable sources on the monastic use of clocks. These texts were spread throughout Europe in thousands of monasteries. But does this permit us to assume also a corresponding diffusion of monastic clocks, even though no remains have been preserved? There are two conclusions we can draw from the frequent, though by no means universal, mention of clocks in the monastic codifications. When the honored and widely copied or imitated founding statutes of the reformed Benedictines (under the influence of Cluny), the Cistercians, the Premonstratensians, the canons regular, and many other orders mention a "horologium," we can assume either that the existence of such "horologia" in the affiliated monasteries is a matter of course, or that the monasteries will make an effort to obtain such a device. In this way the mother house may have been the impulse behind the spread of the clock. The other possibility is that while the venerable texts were copied everywhere, the "horologia cadentia" or "sonantia" were not always taken over along with them. The picture we get from the sources is not entirely clear. "Horologia" were surely not common in Hildemar's time (ninth century). Petrus Damianus (eleventh century) does not mention them. From the end of the eleventh century they are mentioned with increasing frequency in important texts. However, it is notable that the Consuetudines of the Vallumbrasian Order (before 1193), for example, mention a "horologium sonans" in discussing the waking time in winter, while later redactions return to the text of the Rule of St. Benedict ("iuxta considerationem rationis").[48] How do we explain this? Reverence for the historical text? "Horologium" as such a self-evident object it did not need to be mentioned? Bad experiences with the clock's reliability? The Cluniac texts quite unequivocally do not mention the clock as the sole means of determining the time at night. Should the clock fail, we read there, the sacristan is to use candles or observe the night sky.[49] And still at the end of the thirteenth century, the Dominican Humbertus de Romanis pointed out that while every monastery should have a clock, in a pinch the sacristan had to be able to wake the brothers on time without one.[50]

A notable reference appears in the vita of Saint Hermann Joseph of Cologne. He had entered the Premonstratensian monastery of Steinfeld in the Eifel region as a twelve-year-old handicapped boy, and had risen to the post of sacristan. According to the extant manuscripts, his vita, published in his canonization documents, was undoubtedly written shortly after his death in 1230. We read there that Hermann Joseph knew how to build a clock ("horologium instrumentum"), and that he was frequently asked by other monasteries to build one if they did not have one ("nova [horologia] ubi non erant"), or to renovate and reset ("distemperata iterum temperare") existing ones.[51] This piece of information about the first medieval clock builder known by name shows that even at the beginning of the thirteenth century not all monasteries possessed the technically more demanding and sophisticated alarm devices.

At the same time we can see the beginnings of a new situation. Possessing a clock became obligatory, and maintenance of the clock was more rigorously inspected. In 1198, the archbishop of Sens obligated the servant performing weekly duties at the cathedral to set the clock accurately, on penalty of a fine.[52] A visitation record for the Benedictine monastery in Redon in the Bretagne took positive note in 1232 of the fact that a reliable clock was at hand. The cathedral chapter in York was rebuked for failing to properly maintain the clock (1290–1294).[53]

Clocks had thus become objects of daily use in churches and monasteries of the thirteenth century, and their maintenance was a fixed part of the expenses of ecclesiastical administration. Like books, relics, incense, bells, vestments, and chalices, they were part of a church's ornamentation, the "ornamenta." This emerges for the first time in a letter from Abelard to Heloise in 1136, in which he makes suggestions concerning the Rule and furnishings of her small oratory of the Paraclet.[54] Because such ornamentation was not superfluous but necessary, in the spirit of the Cistercian reform it should be as modest as possible. But unlike chalices or bells, clocks never became objects that were liturgically distinguished through special consecrations.[55]

On the construction of medieval water clocks

While we have numerous references to monastic water clocks in normative texts, there is very little information about the construction and appearance of these clocks. To date only a single text has become known that can shed light on the details. It is found as a fragment only in one codex from the Benedictine monastery of S. Maria in Ripoll (Catalonia). The codex contains various tracts that are mostly astronomical in content and represent for the most part translations or adaptations of Arabic sources, as well as excerpts from Bede and a few texts on the astrolabe that are attributed to Gerbert of Aurillac. It is not entirely clear where the codex is from. At present it is dated to the eleventh century and attributed to a scriptorium in Lotharingia.[56] What is undisputed, however, is that the untitled fragment on the water clock is not based on Arab models.

The beginning part, missing in the codex, may have contained the instructions on how to construct the (water)clockwork, the extant final part discusses the ringing mechanism and how to make the weights, set the clock, and position it properly on the floor of the church.[57] What MS Ripoll 225 describes is a construction that is slightly taller than a man and wherein small bells ("schillae" or "tintinnabula schilarum") produce a continuous ringing for a short period at a time that could be set at will. This was accomplished by means of a horizontal shaft which was turned by a weight running through a rope pulley that was attached to a beam. Mounted at the ends of the shaft were discs with pins that alternated in striking the bells. The text leaves many questions unanswered. The number and possible tuning of the bells remain undetermined. Unclear are the shape and arrangement of the frame ("domus orologi") and of the clock case ("scrinium") that apparently also existed. What is mentioned are water vessels and the problems that could be caused by a shortage of water, polluted water, and insufficient oil lubrication; we remain in the dark, however, about the construction of the outflow and its regulation. The vocabulary and style are completely independent of Vitruvius's descriptions. Interpreting and situating this description historically is made difficult also by the fact that we lack any possibility of comparison with other medieval European texts.[58] Only the alarm mechanism bears some resemblance to the sketch of a bell work that was recently discovered in a fourteenth-century Catalan manuscript (fig. 11).[59] Particularly noteworthy is the mention of two setting dials, one containing the division of the full day ("hore diei ac noctis") into quarter-days, and the other a—otherwise unknown—division into hundredths. Noteworthy also is that on the setting indicators, the hours of the full day were numbered consecutively (VI, XII, XVIII). Functionally this clock, too, was an alarm, though an elaborate one. Technically it was a water-driven striking mechanism with cylinders and weights. The regulation of the clock seems to have involved the use of an unconventional and not otherwise known form of dividing the hours.

The sources offer little on the development during the subsequent (twelfth) century. The diffusion of large water clocks is confirmed by an episode in the chronicle of the monastery of Bury Saint Edmunds (Suffolk). We are told that one winter night in the year 1198, shortly after the clock had "fallen," a fire broke out in the church. Stricken with fear and panic, the brothers came running to fight the raging flames. Some fetched water from the well, others took it from the apparently sizable reservoir of the clock.[60] The moral of the story: Saint

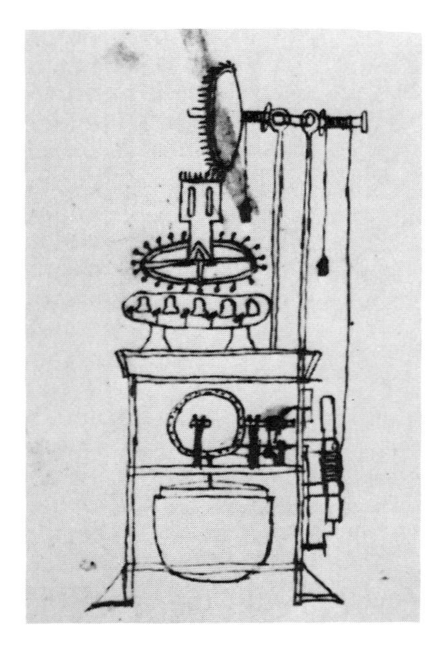

11. Sketch of a water-clock-driven carillon on the verso of the last folio of a legislative compilation drawn up during the reign of King Jaume II of Catalonia (1291–1327). Barcelona, Archive of the Crown of Aragon, Canc., Caixes de leg. no. 2, II. Photo after E. Farré-Olivé in *Antiquarian Horology* 18 (1989).

Edmund punished the brothers if they did not carefully guard his relics in the church, but he also helped them by waking them with the clock and providing water to put out the flames.

The often complicated regulating of the clock by means of mathematical methods is also reported in the Consuetudines of the monastery of St. Victor in Paris (after 1140).[61] Here the registrar is charged with adjusting ("temperare") the clock during the second half of the year in such a way that the portion of the night preceding Matins would be initially as long as the portion following it. Gradually the night (that is to say, the sleeping period) preceding Matins should be extended during the individual months, until at the winter solstice the sleeping period was twice as long as nighttime after Matins (that is, after the time Matins began). Based on the not entirely accurate premise of a uniform increase and decrease in the length of the day, the text formulated a rule of thumb that was independent of what was provided for by nature, and used only the halves and thirds of the longest and shortest nights as the starting units for its calculation.

Even more complicated regulating instructions and a new variant of the division of the hour are contained on slate tables that were found in 1894 in the ruins of the Cistercian Abbey of Villers-la-Ville in Brabant. Preparatory notes concerning the maintenance of a water clock were discovered on three of five pantile-like tablets; the notes were made between the late summer of 1267 and early spring of 1268, presumably concurrent with the building of a clock in the choir of the monastic church.[62] According to the notes, this was a relatively primitive type of water clock which, in keeping with the text of the Rule, served primarily as an alarm. The falling of the clock ("cadere") is explicitly mentioned. In order to compensate for the influence of atmospheric fluctuations and the decreasing water pressure, the outflow vessel was refilled with a measured amount of water twice a day at different times. In nice weather the sunlight falling through the choir windows served as an aid in setting the clock.[63] Only the twenty-four letters of the alphabet appear as time indicators, never hours or fractions of hours. According to the reconstruction, the clock had a dial with a pointer that was driven by a float and a drum. The dial itself was divided into four fields numbered with Roman numerals, each field sporting twenty-four letter fields along the edge. Even though the day—which began at night, around six o'clock—was divided into three segments of eight hours each, one complete rotation of the pointer corresponded to a period of thirty-two hours. By comparing the sunrises indicated on the tablets for certain feast days (e.g., "Exaltatio sancte Crucis diescit super N") with the calculated sunrises for these days, the time value of each letter field can be calculated on average to a third part of an hour (twenty minutes). Apart from being rung for waking, the bells were also sounded for the various offices. In addition, the text gives the duration of some of the offices, for example for Matins "in die Sancti Stephani tenent VII litteras" (= 2 hours, 50 minutes). The sacristan's sleeping time is also carefully noted; for example, for the Christmas season it reads "dormi decem litteras" (= 3 hours, 20 minutes).[64]

Two methods of observing the time of day were combined in Villers. Tracking the sunlight that fell through the windows of the church—which may have been constructed with this in mind—corresponded to a customary practice. Earlier, in the itinerary of Benjamin of Tudela, we read that the great mosque in Damascus had a wall with many glass openings that corresponded to the number of days in a year, and that one could read the hours of the day from a scale.[65] For calendric purposes, such as the spring equinoxes, windows were appropriately placed also in European cathedrals of the Middle Ages. In the Minster of Strasbourg, for example, a green beam of light still today indicates this event over a period of six days, weather permitting. In Villiers, and surely not only there, the beams of sunlight flooding in through the windows were used for the daily regulating of the water clock. The puzzling thing about this clock is the otherwise uncommon division of the day into thirds and hour-thirds. Even more remarkable is the fact that the daily routine of the monks, which varied with the duration of daylight, was regulated in this case by means of a device that could indicate only intervals of equal length.

The division into hundredths on the setting dial in the manuscript from Ripoll, and the clock reconstructed on the basis of the slate tablets from Villers, suggest that the methods of regulating water clocks according to the unequal hours by changing the outflow opening or by means of interchangeable dials, methods described in Vitruvius, were either unknown or had proved to be impractical. It would appear that during the Middle Ages a change was thus made in some clocks to regulating the passage of the unequal hours of the day by means of clocks that were guided, if not by hours of equal length, at least by artificial intervals of equal length. There is no way of knowing how widespread these types of clocks were, but other clues point in a similar direction. At the end of the thirteenth century, the late-antique tables that indicated the length of the day in terms of shadow measurements or the changing number of equal hours came once again into wider use. Bernhard Ayglerius, abbot of Monte Cassino, used them in his commentary on the eighth chapter of the Rule of St. Benedict, which deals with the time for rising during winter.[66]

Depictions of medieval "horologia"

Information about the external appearance (the cases) of the large medieval water clocks comes to us from two illustrations. Around 1235, the architect Villard de Honnecourt, a native of Picardy, drew in his sketchbook a "portrait" of a "maizon d'une ierloge" (fig. 12) he claimed to have seen in person—which means it would have been somewhere in the region of northern France. The probably wooden Gothic case with four floors and a gable permits no inferences of any kind about

12. Sketch of a case for a clock in the sketchbook of Villard de Honnecourt (around 1235). Paris, Bibliothèque Nationale, Ms. fr. 19093, folio 6 verso. Based on H. R. Hahnloser, *Villard de Honnecourt* (1972).

13. Abbot Richard of Wallingford in front of his astronomical clock. Miniature in a manuscript from the end of the fifteenth century. London, British Library, M. Cotton Nero D. 7, folio 20.

the technology of the clockwork, indicators, or bell works. Hence the comment by the editor of the sketchbook that this is the "first example of a wheeled clock" remains pure speculation.[67] Comparable Gothic clock cases can be found later in the miniatures of a Brussels Suso manuscript (fig. 24) and in an English manuscript from the late fifteenth century that shows Abbot Richard of Wallingford in front of his astronomical clock (fig. 13).

Richer in detail, though difficult to interpret, is the only medieval European depiction of a water clock in a miniature of a moralized Bible that was produced around 1250 for the French court (fig. 14).[68] The illustration is for a story handed down in 2 Kings 20.11 and in Isaiah 38.3. In it the Lord, through the prophet Isaiah, promises Hezekiah, the mortally ill King of Judah, that he will extend his life by fifteen years. Hezekiah asks for a sign, and Isaiah announces that the

14. Water clock in a miniature from a French *Bible moralisée* (around 1250). Oxford, Bodleian Library, Ms. Bodly 270b, folio 183 verso.

Lord will bring back by ten steps the shadow on the steps of the palace of Ahaz, Hezekiah's father. As he had predicted, so it happened. It is unclear whether those handing down the story were already referring to the sundial mentioned in most of the older Bible editions, or whether the long shadow was only a symbol for the evening of life.[69] The miniaturist has put a water clock in the place of the "horologium Ahaz."

Ever since C. B. Drover called attention to this miniature in 1954, its "technical" interpretation has been a matter of controversy.[70] Lynn White, Jr., points to the resemblance between the clock case and the drawing in Villard's sketchbook and speculates that the clock stood not in a church or a monastery but in a royal palace. He interprets the central component as a wheel composed of fifteen metal cones connected with each other by small openings, though the holes visible in the sketch "may well be schematic rather than visually naturalistic." The compartments were fifteen in number because "the *hora equalis* corresponds to fifteen degrees of the equinoctial cycle," and the wheel thus probably made one complete rotation each hour. The driving power appears to come from a weight hanging from a cord visible on the left, while the breaking action is produced by the slow trickling of the water from one chamber into the next. "Behind the braking wheel, and mounted either on the same axle or on one collared on to it, is a large toothed wheel which appears to control the striking train for the bells. At the left of the clock is a turbine-like wheel of vanes which is probably a fan-escapement to slow the action of the chime, at the striking of the hours, by friction with the air." Finally, he points out the similarity between this cone-wheel and the illustration of the mercury clock in the approximately contemporaneous *Libros del Saber*, and concludes by suggesting that this is the first instance of a weight drive, the present evidence indicating that it was a Western innovation.

J. D. North disagrees, arguing that the fifteen chambers in the miniature are probably a reference to the fifteen years of life mentioned in the biblical text. The so-called fan-escapement wheel is in reality a symbol of the sun, which is frequently depicted in the other miniatures of this manuscript. Finally, North maintains, White's interpretation ignores the clearly visible animal head spout below the large wheel. The resemblance to the Alfonsine mercury clock is entirely superficial. What appear to be twenty-four teeth would indicate that the clockwork turned only once a day.

There will be further attempts to interpret the illustration. For example, one could ask about the significance of the single bell that is visible at the lower left of the toothed wheel and complements the bell work in the upper part. The miniaturist was not a dreamer but a keen observer of contemporary technology. Still, one should not stretch the technical interpretation of a miniature from the middle of the thirteenth century too far. By intent and purpose it was not a realistic depiction. Hence all we shall note for the moment is that the combination of bell striking mechanisms with large water clocks was known around 1250, and that its setup was possibly based on a division of the day into twenty-four hours.

Water clocks in the Islamic tradition

The rise of Islam in the seventh century and its subsequent expansion created a cultural sphere that comprised the Mediterranean and the Near East between Spain and Afghanistan. Islam was the unifying religion, Arabic the predominant language of science. The heyday of Islam (eighth–fourteenth centuries) was characterized, even more so than had been the case with the Hellenistic period, by the reception and assimilation of the cultural traditions within its entire sphere of influence.

There is no need to recall the role that Arabic writers played in the transmission of the Hellenistic sciences, especially of astronomy. As far as the theory and technology of time-measurement are concerned, we must additionally take into account not only the process whereby the classical stock of knowledge was assimilated and developed further, but also the direct continuation from and development of the late antique traditions of skilled technical craftsmanship. What is more, consideration must be given to the traditions of the Syrians and Persians, and here scholarship runs into even greater difficulties.[71] Also largely unclear is the contribution of Arab writers to the transmission of science and technology from India and China to Europe.[72]

It is easier to trace the Arab clockmakers who continued in a direct line from Byzantine mechanics. The first report of independent Arab clockmaking is also evidence for the relative technological backwardness of Western Europe. An entry in Einhard's Royal Frankish Annals, frequently copied in the Middle Ages, speaks of a gift from Sultan Harun-al-Rashid to Charlemagne. In the year 807, the Eastern ruler sent to the Western emperor a mechanical marvel, "a brass clock, a marvelous mechanical contraption." The striking similarity of the figurative automata program with that of the late-antique clock in Gaza immediately catches our eye.[73] Of course the Gorgon's head and the twelve labors of Hercules from classical mythology have disappeared. Now a horseman emerges each hour from one of the twelve windows. At the same time a small bell ("cimbalum") is struck. Variations and enlargements upon this figurative program, involving, among other things, singing birds, musicians, female slaves, and execution scenes, are found with many large and small Islamic clocks in the subsequent period. In nearly every instance we hear of balls that fall into a basin as an hour signal and of a moving or revolving writing or pointing figure.

Descriptions and building instructions for such clocks are known from the tenth century on. The most important are the Andalusian treatise on automata called *The Book of Secrets about the Results of Thoughts* (eleventh century) and *The Book of Knowledge of Ingenious Mechanical Devices* by al-Jazari (1204–1206).[74] Various Arabic tracts mention Archimedes (third century B.C.) explicitly as a precursor. A work by Archimedes on the construction of water clocks is, however, not otherwise known. Perhaps the mention of Archimedes was merely intended as a generic name that honored the Greek authorities as a whole. Today scholars speak of Pseudo-Archimedes.[75] Heron of Alexandria is also cited by Arab authors. Hellenistic theory and practice was thus clearly absorbed to a considerable degree. Among the components that were perhaps newly developed by Arab clockmakers, or possibly taken over from East Asia, we should point out the use of water wheels—with experiments conducted also with sand as a flowing medium—and of scales on which the outflow vessels were mounted, a practice otherwise known only from China. Astronomical simulations with these clocks are reported also in European sources. In 1232, Sultan al-Ashraf of Damascus presented to Emperor Frederick II an extraordinarily precious "artificial sky" on which the course of the stars and the hours of the day and night could be read. One account calls this contraption "a marvelously fashioned tent, in which the images of the sun and the moon, in accord with their positions, run their courses and indicate unerringly the hours of the day and night." Another account speaks of a "golden, gem-studded astronomical heaven that contains a mechanical course of the planets."[76] However, a textual variation that is occasionally cited in books on clocks and that emphasizes more strongly the mechanical aspect ("ponderibus et rotis incitatae") appears only in a later redaction that already dates to the time of the mechanical clocks.[77]

In medieval Europe, such automata from the Islamic cultural sphere were known only by hearsay. Often the accounts have a certain plausibility about them, as for example the tale—in a story told by Scheherazade on the 357th night—of a peacock that flapped its wings and cried out each hour, or they are full of fantastic embellishments.[78] The seventh-century palace of the Persian King Chosroes II with its many—and probably for the most part fabled—automata provided the model for the literary form of the Grail. In *Titurel,* attributed to Albrecht of Scharfenberg, we find a description of an artful heavenly mechanism of the Grail which, driven by a hidden "oroloei," showed the movement of the stars against an artificial vault and sounded the seven periods of the day (that is, the Hours) on golden cymbals.[79] The author has combined the fabled oriental automata with the concept of the automatic ringing of the Hours, which, as far as we know, never existed.

Two other aspects stand out in a perusal of the sources on the Islamic tradition of water clocks. Most of these clock automata, costly and difficult to maintain, were toys used for entertainment at the courts and in wealthy homes and to amaze visitors. Compared to the far simpler versions, whose explicit purpose was to determine prayer times, they were not widespread.[80] But in addition there also existed a host of large public clocks, and thus possibly also a tradition of public time-indication that went back all the way to Byzantine times. Leaving aside the clock in Gaza, the oldest reference comes from a Chinese travel account about a golden water clock in the city gate of Antioch, which could have dated back to Byzantine times. It was shaped like a scale from which a ball dropped every hour with a ringing sound. As the account goes on to say: "This served to indicate the parts of the day without the least error."[81] In Damascus there is mention of a public clock on the great mosque between the tenth and fourteenth centuries; according to Ridwan, it also served to indicate the times of prayer.[82] Remains of a public clock of the fourteenth century have survived in Fez (Morocco).[83] According to a Greek-Latin-Arabic inscription, the Norman King Roger II of Sicily also erected a device for indicating the hours ("opus horologii") in Palermo in the year 1142.[84]

What makes these references to public time indicators peculiar is also the fact that after the appearance of mechanical tower clocks in Europe, Islamic rulers balked at their introduction in their own territories. Our information about the nature of the hour signal in these Arabic clocks is also poor. The accounts of the clocks in Gaza and Damascus emphasize that they were intended for the hours of the day and the night. At least during the day, the clock in Gaza twice struck the numerical sequence 1–6; in Antioch a signal is said to have been set off every hour. The automata theater of the so-called "palace clock of the unequal hours" of al-Jazari was to be set into motion during the day at the sixth, the ninth, and the twelfth hour, and at night at the sixth and twelfth hour (fig. 15). The astronomical model presented to Frederick II supposedly indicated "the hours of the day and night." This wording is noteworthy, since it later became the specific attribute of mechanical clocks. But in none of the mentioned forms of indication can we detect a connection with the Muslim times of prayer.

We are also struck by the fact that the Arabic horological tracts for water clocks and candle clocks describe constructions for indicating temporal as well as equinoctial hours without revealing a clear preference for one or the other. Equinoctial hours are also here parts of the full day or of daylight during the equinoxes, and are not yet defined by smaller units of time, for example, minutes.[85] Though E. Wiedemann translates the Arabic terms for temporal hours ("crooked or temporal hours") and for equinoctial hours ("uniform hours"), he does not explain which reckoning of the hours was in use when and for what purposes.

A use of equal time segments that was due solely to the technical exigencies of clock construction, as is likely for the clock described in MS Ripoll 225 and for the clock reconstructed from the slate tables in Villers-la-Ville, can be ruled out in this case, since the simple float-scales in outflow vessels also show both ways of counting.[86] We can thus surmise that in the Islamic sphere, unlike in Western Europe, both forms of counting the hours were in use throughout the entire Middle Ages, and that in individual cases equal hours may have also been publicly indicated. It must remain open, however, whether and to what extent this simultaneous use of two ways of counting the hours was in use also outside of learned circles.

Astrolabe and time-measurement

While we can only conjecture about the significance of Islamic impulses for the development of European water clocks, the key role that Islamic authors and builders played in the introduction of the astrolabe—the most important innovation for medieval astronomy and time-measurement—to central Europe can be traced much better and has been more thoroughly studied.[87]

15. Viewing side of a water clock of al-Jazari according to a thirteenth-century manuscript. Istanbul, Library of the Topkapi Serail. Based on A. J. Turner, *Waterclocks* (1984).

The astrolabe, the "star grasper," is a combination calculation, demonstration, and observation instrument. Its construction is based on the stereographic projection of points and lines of the imagined celestial sphere onto a disc. A circular backplate ("mater"), usually of brass, with a sighting device on the back carries a plate ("tympanon") with engraved coordinates and base circles (horizon, meridians) of the visible celestial sphere, taking into account the geographic latitude of the place from where the observation is made. A movable, openwork skeletal plate ("rete") that is concentrically attached over the plates indicates the position of prominent stars. The altitude of a star is determined with the help of the sighting device and the scales on the back. Then the rete is rotated over the plate with the co-ordinates until the observed position of the star matches the corresponding markings. A host of different astronomical calculations can then be carried out for any point in time. By means of hour lines one can also read the time of day of the observation according to unequal hours or convert unequal into equal hours.

Stereographic projection has been known since Hellenistic times; in the second century A.D., Ptolemy used the adjective "astrolabic," explained the theory of the projectors for practical astronomy, and combined them with a time-measuring device that made possible the reckoning of the hours of the night with the help of the position of fixed stars. Prior to that the use of an astrolabe dial to simulate astronomical constellations on a water clock had been described by Vitruvius. Astrolabes and instructions for using them were known in Byzantium and Alexandria in the sixth century, whence knowledge of the instruments spread quickly throughout the Islamic sphere and thence all the way to India. Arabic authors were the first to praise Ptolemy as the inventor of the astrolabe, which was used for horoscopes, for the calendric reckoning of the month of fasting (which depended on the position of the moon), and for the times of prayer (which depended on the sun). In the Orient the astrolabe became an attribute of wisdom and power; according to a ninth-century Arab author, scholars and kings used water clocks at night and sundials and astrolabes during the day. The "muwaqqit" (calculators of the hours of prayer) employed at the great mosques used astrolabes as late as the twentieth century.[88]

In connection with the constructional advancement of the astrolabe in the Islamic sphere, we should call attention to a sideline of the development whose connection with the classical mechanical simulations has not been entirely clarified, but whose importance for the history of astronomical clocks is obvious. Around 1000, the Arab constructor al-Bîruni speaks of a gear with toothed wheels that is to be mounted on the back of the astrolabe dial and serves to depict the orbit of the sun and the moon as well as the phases of the moon that are so critical for the Islamic calendar. The Science Museum in London has an astrolabe on whose back is a similar, somewhat simpler "calendar computer" (D. J. Price) with toothed wheels; the instrument was built in Isfahan in 1221/1222.[89] Within the tradition of astronomical mechanization, which reaches back via Islam into Hellenistic times, another element of the possible pre-history of mechanical clockworks has thus been documented.

Knowledge of the astrolabe reached Western Europe through Arab authors and through the numerous instruments that were imported throughout the Middle Ages. All of the oldest extant specimens come from the eastern Islamic realms. The Arabic texts, on the other hand, made their way to Europe via translations made in Spain. Here the astrolabe is occasionally called "horologium regis Ptolomei" after its Greek origins — Ptolemy having been promoted to the status of a king of Egypt — or "horologium secundum Alcoran" to reflect the Islamic provenance of both the texts and the instruments. In a foreword to a late-tenth-century Latin tract on astrolabes, astrology is defended and emphasis placed on its importance for a performance of the mass that is conscientious and exact in regard to time. Every cleric, we read, should learn to calculate ("computare") past and future times (of the date of Easter, for example), and the hour-watcher ("horoscopus") needed sundials ("horologia gnomonis") and astronomical instruments to perform calendric calculations and determine the time of day. The reader is advised to study also the books of Ptolemy and the clocks of Vitruvius, to the extent permitted by the censorship of the Holy Church. This foreword, which has survived in several collections of older tracts on the astrolabe, including MS Ripoll 225, is to my knowledge the only instance where Vitruvius's descriptions of clocks are explicitly referred to in the Middle Ages as instructions for building clocks.[90] How did the Christian translator come to know these texts?

A tract on the usefulness of the astrolabe, produced in the circle of Gerbert of Aurillac's students, describes it as an instrument of astronomy, which one could also call the art of measuring time ("horologica disciplina"). With its help one could construct accurate sundials for any latitude, and determine the equal and unequal hours ("horae naturales vel artificiales"). Both things were important for mass, for spiritual exercises, and for the elimination of false clocks ("pseudohorologiae").[91] The first instructions on how to construct and use an astrolabe that were usable and important for the spread of the instrument in Europe were produced in 1045 by Hermann the Lame, a monk from Reichenau. Among other possible uses, he also suggests the astrolabe for the construction of portable sundials ("horologia viatorum"). His vita therefore praises him as an instrument builder and clock constructor.[92]

The spread of astrolabe tracts that we can observe since the end of the eleventh century reveals the importance of this instrument as a scientific and practical innovation in astronomy. All astrolabe tracts have sections on the conversion of equal into unequal hours and vice versa. The astrolabe did not, however, become an everyday time-measuring device. Theologians could not ignore its proximity to astrology; they considered it a tool of the devil and frequently also evidence of unseemly theoretical curiosity. Though it was used in monasteries, it never got to the point where it was mentioned in the texts of the Rules. Thomas Aquinas did not approve of its ecclesiastical use: "The Church does not strive for restriction through clever study of time. One does not need to use an astrolabe to know when it is time to eat."[93] Since the use of the astrolabe required only a little astronomical knowledge, an increasing number of scholars were able to determine or convert points in time or time periods. A miniature in a psalter for a high-ranking lady, later identified as the French Queen Bianca of Castille, shows around 1220 a noble user with an astrolabe and a looking tube (fig. 16). One helper is assisting with a work in Arabic letters, presumably a star chart, while a second one enters the results into a Latin book.[94] The infrequently documented use of this time-measuring device remained limited to astronomical and physical events (eclipses, earthquakes, and the like), and in these instances, too, one must not overestimate the extent of empirical observations and the accuracy attained.[95] Among these rare references are two accounts by Walcher, prior of the Benedictine Abbey of Great Malvern in Worcestershire. Walcher reports observing the lunar eclipse on October 30, 1091, in

16. Miniature in a psalter for a noble lady (beginning of the thirteenth century). Paris, Bibliothèque de l'Arsenal, Ms. 1186, folio 1 verso. An astronomer with a sighting tube in his hand is taking the bearings of a star with the alidade of the astrolabe. One assistant is holding a book with Arabic ciphers (a star table?), another is writing the results into a Latin book.

Italy, expressing his regret that he had no "horologium" available for accurately determining the time. The following year in England, he fixed the lunar eclipse on October 18 with the help of an astrolabe.[96] The astrolabe as the earliest analog-computing device in the history of European science and as a new kind of timekeeper was gradually displaced in the thirteenth century by the quadrant, which made time-reckoning easier but hardly permitted any astronomical calculations,[97] and from the fourteenth century on by the mechanical clock.

The "mercury clock" in the Libros del Saber de Astronomia *(1276)*

The Arabic, Jewish, and Christian scholars, physicians, astronomers, and technicians whom King Alfonso X (the Wise) of Castille and Leon gathered together at his court in Toledo produced a series of works that were of great importance for various branches of the history of European science. The king not only had Ptolemy's major astronomical work, the *Almagest,* newly translated; he was also the impulse behind the publication of the *Alphonsine Tables,* which were, as far as their subsequent influence is concerned, the most important astronomical charts of the Late Middle Ages.[98] Of interest to us is that these tables consistently use the sexagesimal division, by which the degrees of arcs, days, and hours were divided into minutes, seconds, and "terciae." The most important division are the "minutae diei," each of which, according to an attached conversion table, corresponded to "2 minutae 30 secundae horarum."[99] An equal hour thus corresponded to "24 minutae diei." Despite the ambiguous terminology, the rapid and broad diffusion of the tables must have promoted knowledge of the hour-minutes and hour-seconds in learned circles.

The *Libros del Saber de Astronomia* are a collection of sixteen tracts on astronomy and related fields that King Alfonso commissioned in 1276. The tracts are in part translations from Arabic, in part newly written works on the basis of Arabic models.[100] The aim of the collection was to make the treasures of Islamic science accessible. The fourth part of the collection contains five books with descriptions of various astronomical instruments and clocks. It can be seen as a kind of summary of Islamic horology, which was to be made known in Christian Europe as exemplary and superior.

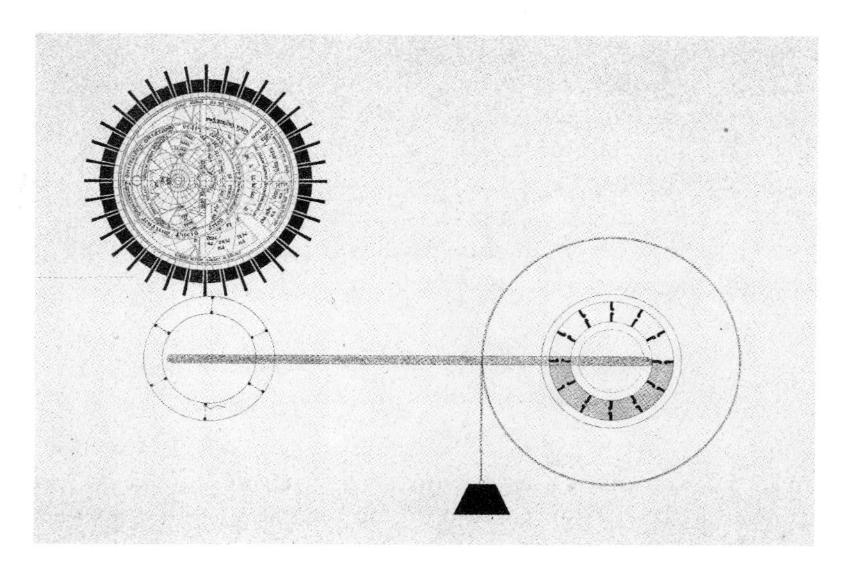

17. Drawing of the "mercury clock" in the *Libros del Saber de Astronomia* (1276). After M. Rico y Sinobas, *Libros del saber de Stronomia del Rey Don Alfonso X de Castilla* (Madrid, 1863–66).

The sections of the collection dealing with sundials, water clocks, and candle clocks remain within the tradition of Islamic clockmaking, some interesting details notwithstanding, among them the "Palace of the Hours," a sundial in the form of a building. Only in the description of a mercury clock, "relogio del argent vivo," do we encounter a remarkable attempt to solve the problem of a uniform drive. For the design of the clock, the author, Rabbi Isaac ibn Sid, refers to principles from a (lost) work of Heron of Alexandria (third century B.C.) on how to lift heavy objects. The clockwork consists of an annulus of twelve chambers separated by partitions, with the chambers connected through small openings in the partitions (fig. 17). Half of the chambers should be filled with mercury and the wheel mounted on a shaft. The large wheel is turned by a weight attached to a chain, whereby the descent of the weight is slowed by the inertia of the heavy mercury as it percolates slowly from one chamber to the next. According to the author of the treatise, the task of breaking the descent of the weight would be solved optimally if the contraption could be balanced such that the large wheel makes "a complete revolution in a day and a night—no more and no less—just as the ninth heaven makes to produce day and night." If the wheel turned more quickly, the movement could be slowed as required by a lantern pinion.

Concerning the construction of the clock, the text refers to experiences with the building of Arabic water wheels and mill-wheels. The use of mercury, whose viscosity—unlike that of water—hardly varies with the ambient temperature, had been mentioned by Chinese, Indian, and Islamic constructors, and it also appeared in medieval projects for a perpetuum mobile.[101] But in spite of the similarity of the compartmented annulus to the central element of the water clock in the French moralized Bible (fig. 14), it is quite unlikely that such a mercury clock was ever built. A modern attempt at reconstructing it has shown that, quite apart from problems with the material and corrosion, a sufficiently accurate and reliable operation of such an assembly is not possible. We are thus dealing merely with a suggestion of how one might gradually approach the solution to the problem of a continuous drive. It apparently made no impact at all. The text has survived in very few manuscripts and was translated only once, in 1341 into Italian in Florence.[102] We have no knowledge, however, of the actual existence of such contraptions during the Middle Ages. Only Leonardo da Vinci mentions mercury again as a flowing medium. The concept of the "compartmented cylindrical clepsydra" was not picked up again until the sixteenth century, and water clocks were built on this principle until the end of the eighteenth century.[103]

The works of the mercury clock were intended to move an astrolabe dial that was to rotate "as though by itself," and from which one could read the position of the starry sky and the unequal hours of the day and night. An auxiliary mechanism (not described) with twenty-four pins was to set off a striking mechanism with small bells. It would have necessarily rung the equal hours, but a continuous series of signals was apparently not intended in the first place, for the author remarks that one could ring the bells at whatever hours or segments of the day one chose.[104]

Once again we observe a contradictory juxtaposition of technological developments and technical designs.[105] Assuming a uniform speed of the clockwork, it is entirely possible to read the unequal hours by designing the astrolabe discs or the scales correspondingly. It is also possible to sound an acoustic signal at a specific time. By contrast, the continuous mechanical-acoustical indication of the unequal hours is hardly possible with the envisaged clockwork designs.

The "Heavenly Clockwork"

When Jesuit missionaries at the end of the sixteenth century brought clocks as gifts to Japan (after 1551) and China (after 1584), these "self-sounding bells" were marveled at in those countries as convincing proof of European inventive ingenuity and scientific-technological superiority. The Jesuits made use of this and in subsequent years used complicated clocks and trained clockmakers at the Asian courts as tools for the "propagatio fidei per scientias."[106]

The European image of China, characterized by a low opinion of the Chinese sciences and a view of Chinese technology as barely capable of development, has its origins in these reports of the Jesuits; its after-effects carried over into our time.[107] These reports also spoke of the Chinese technology of time-measurement, which at the time supposedly knew only poorly functioning water clocks apart from antiquated sundials. The Jesuits could hardly have arrived at a different judgment, since by this time the important tradition of Chinese water-clock technology had been practically forgotten, that is to say, it could be accessed only in obscure books.

In 1954, in the first volume of his monumental history of Chinese science, Joseph Needham still called the mechanical clock the last specifically European invention that was imported into China.[108] A few years later, after the rediscovery of a host of medieval accounts and their examination—jointly with Wang Ling and D. J. Price—had led to a complete revision of the history of Chinese timekeeping, he retracted that assertion. Since then the now recognizable tradition of complicated water clocks in China has given rise to a renewed, intensive scrutiny of Islamic and European sources and to a no less intensive discussion.[109]

Needham and Price believed they had found the missing link in a chain of evolutionary technical development stretching from ancient water clocks to the mechanical clocks of the European Middle Ages. They concluded that the mechanical clock as a timekeeper was a partially misused, primitive offspring of the drives of the great astronomical clocks of medieval China, in the words of Price, a "fallen angel from the world of astronomy."[110] The main source for this hypothesis is the description of a large astronomical clock that was completed for the Chinese court in the year 1092 A.D. A tall wooden tower, more than ten meters high, housed an entire ensemble of automata. Set in motion were, among other things, an armillary sphere (an instrument for depicting certain planetary orbits on a globe made of rings), a celestial globe, and a drum with a procession of figures and automata that gave aural and visual indications of calendar dates and the time of day according to hours and quarter-hours.

The driving force for this heavy machinery came from a large water wheel (about three meters in diameter) with thirty-six buckets. The buckets were attached to the spokes of the wheel by means of short, pivoted arms which acted as counterweights. From a reservoir, maintained at a constant level, water poured at a uniform rate into one bucket at a time. Once the bucket was filled, it was pulled down by its weight and tripped a scale-like stop mechanism which blocked one spoke at a time during the filling processes and prevented the recoil of the wheel. The wheel then rotated slightly "clockwise" and was again blocked. The functional principle behind this escapement, which was called the "heavenly scale," is shown in a sketch by D. Penney (fig. 18).[111] Additional regulating mechanisms, on the water outflow, for example, were not necessary.

Two systems of equal "hours" had been in use in China from as early as the second century B.C. The only disadvantage was that it was impossible to "reconcile" them. The full day was divided either into one hundred "quarters" (each being equivalent to 14 minutes, 24 seconds), or into twelve equal (double)-hours, which were linked with the astronomical succession of the signs of the zodiac. Sometimes the double-hours were divided into a first and second half. This second system may have been adopted from the Babylonians, whereas the first is considered to be authentically Chinese. In addition, the night was divided into five night-watches of variable length.[112] In Japan, by contrast, the unequal hours were kept. They were abolished only on January 1, 1873, in conjunction with the introduction of the European calendar.[113] The time it took to fill a scoop and trip the mechanism on the great astronomical clock probably corresponded to the length of a Chinese quarter-day, which means that over the course of a full day the clockwork was noisily set into motion one hundred times.[114]

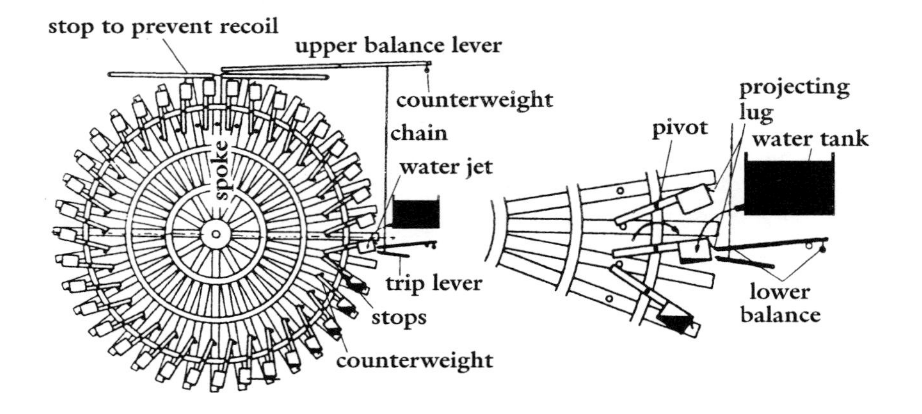

18. Reconstruction of the escapement mechanism of the "Heavenly Clockwork." Based on A. J. Turner, *Waterclocks* (1984).

This astronomical clock was the highpoint and presumably also the end of centuries of "collective developmental work" on water-driven "heavenly machines" in the Chinese empire, work that reached into the time of the Sung dynasty, whose rule has been compared with the European Renaissance. A few comments on this development: simple water clocks had already been known in China—as in Egypt—in archaic times. Perhaps they were imported from Mesopotamia. From the second century B.C. on, inflow clocks with floats and scales were in use, and beginning in the seventh century A.D. we hear about models with foliots and weights. In smaller models for shorter intervals, experiments were also conducted using mercury. From this time, at the latest, public devices for indicating time were maintained.[115] An Arab traveler reported in the ninth century on water clocks with scales in city bell-towers.[116] Following the Chinese model, bell-towers with water clocks existed also in the large cities in Japan. In Korea in the fifteenth century, elaborate water clocks were constructed for astronomical-calendrical observations.[117]

In China the tradition of water-driven astronomical models went back to the time of Christ's birth. Needham thinks Greek and Byzantine influences are a possibility, though they cannot be proved. As Needham and his collaborators have plausibly demonstrated, the escapement type I have described for the "Heavenly clockwork" was developed in the eighth century, and from that time on a whole series of large automatic globes and armillary spheres were built in China. The word "development" seems appropriate here because the descriptive texts refer to each other, discuss known construction flaws, try to make improvements at the next attempt, and so on. The "heavenly clockwork," completed in 1092, is said to have been in operation for a few decades. In the subsequent period similar but increasingly simpler automata works without difficult astronomical simulations were built. Needham and Price have interpreted this as a process of devolution, in which the water-clockworks more and more seldom drove astronomical automata, and in which the measurement and indication of the time of day was left as an atrophied remnant of an important technological tradition. The image of the "fallen angel" transfers this devolution model to European history. Though Needham and his collaborators emphasize the political and social significance of astronomy and calendric sciences for the Chinese imperial dynasties, they did not investigate the social function of everyday timekeepers. In this regard their thesis of time-measurement as a derivative, atrophied remnant of an ingenious technology remains untested.

The great waterwheel connected with the "heavenly clockwork" was surely a functionally appropriate device, whose main advantage was that it drove and regulated a complete ensemble weighing several tons. The result of a long series of preliminary tests, it probably worked reliably as long as the requirement of continuous and careful maintenance—by no means a given—was met. In terms of the mechanical details, one can see the solution as an independent contribution to the escapement problem; the escapement can be described as "mechanical" and the clock as an "escapement clock." However, it is not evident why this clock should be the "missing link" in the history of European horology. The principle of the Chinese escapement is pivoting balance levers that stabilized a stop-and-go motion. The principle of the European escapement, which employs the centrifugal force of an oscillating inert mass, does not resemble it in any way whatsoever. For that reason David Landes has called the Chinese development "a magnificent dead end."[118]

Apparently Europe had no knowledge of the Chinese escapement. The faint external resemblance between the Chinese spoked wheel and the central, circular component in the French miniature of the moralized Bible is the only weak argument to the contrary. The obvious differences and incomparabilities have led J. Needham and D. J. Price to formulate a kind of ersatz hypothesis: reports about Chinese astronomical clocks with usable escapements, possibly transmitted via the Islamic world, provided a stimulus ("stimulus diffusion") to European clockmakers who, driven by an imitative instinct while simultaneously possessing insufficient technical information, set to work on the escapement problem themselves and eventually came up with their own solution. This attractive thesis has two disadvantages: first, it shifts the burden of proof into the area of intercultural exchange processes in the Old World, an area that is diffuse and will never be fully elucidated. It cannot be proved, but neither can it be refuted. As a substitute J. Needham offers colorful but not very helpful speculations: stories may have traveled via caravan, in which case the quality of the information would have undoubtedly suffered; or this counterfactual scenario: a Chinese scientific delegation around 1250 spared the constructors in Europe "a good deal of trouble," and so on.[119] Moreover, Donald Hill is surely right in saying that we haven't even begun to understand the direction and chronology of possible "stimulus diffusions" between the Islamic, the Indian, and the Chinese spheres.[120] Second, this thesis fails to accomplish precisely what it is supposed to for "scientific universalism." It does not situate the seemingly sudden and surprising appearance of a superior technological variation—in the figurative sense: the invention—within the concept of an evolutionary process, in which scientific-technological progress appears as the product of common efforts among various cultures, possibly noncontemporaneous but all aiming in the same direction. Quite apart from the problem of "stimulus diffusion," there are strong clues that the problem of suitable drives and regulators for astronomical simulations or clocks was being worked on in Europe at the end of the thirteenth century. European water clocks were apparently not suitable for this purpose. The only question, then, is whether we should in fact look for the development of the mechanical escapement in the realm of astronomical mechanization.

"An impending invention"

The era of the water clock came to an end in Central Europe in the second half of the thirteenth century. While complicated water clocks continued to be built during the fourteenth century in the Islamic sphere (all the way to Spain) and in China, the slate tablets from Villers-le-Ville (1266–1267) are the last evidence for a European monastery clock that was unquestionably a water clock. To be sure, scholars and engineers continued to take an interest in the old water-clock texts, but we hear nothing about practical results.[121] The replacement of hydraulic by mechanical clockwork drives, that is, the technical innovation brought about by the mechanical escapement during these decades, has left no traces and can thus be dated only very approximately.

Around 1240, William of Auvergne, Bishop of Paris, compared natural and artificial bodies with respect to the independence of their parts and their movements. As an example for "artificia" he gave the water clocks with weights that were used by the astrologers.[122] Even if astrology and astronomy were at this time concepts that were hard to separate, William is surely not talking about the monastic use of clocks. Though the bishop's comparison is only didactic in purpose, he did not think much of these clocks. They ran only for a short time and required frequent interventions to regulate and repair them. Not too much later we find comparable comments. In a letter about the magnet, the military engineer Peter de Maricourt (Petrus Peregrinus) reported in 1269 that work was being done in many places on the problem of the perpetuum mobile, a technical vision that had come from India and was much discussed in the thirteenth century.[123] His suggested solution to this problem was to mount a magnetic globe—similar to an armillary sphere—parallel to the celestial axis. If it were possible to mount it without friction, it would turn precisely in accordance with the celestial rotation. One would then no longer need a clock, because one could read the astronomical dates at any given hour, just as the astronomers/astrologers wanted.[124]

In 1941, Lynn Thorndike published a brief text that is highly noteworthy for the history of mechanical clocks.[125] It is a commentary insertion by Robertus Anglicus in a standard astronomical textbook for courses on astronomy at universities, Johannes de Sacrobosco's *De*

Sphera. In discussing how cumbersome the reckoning with equal and unequal hours was, Robertus remarked that to this day (1271) there was no clock that met the needs of astronomy. So far, clockmakers had been trying in vain to construct a wheel that turned precisely in accordance with the equinoctial circle (that is to say, the movement of the equinoctial point around the earth during a full day). Should their experiments succeed, one would have a correct and good clock, one more suitable for determining the equal hours than the astrolabe or other astronomical instruments.

Robertus then contributes his own suggestion: a wheel of uniform weight throughout should be turned by a suspended lead weight such that it performed exactly one rotation from one sunrise to the next. The shifting of this point in time by one degree every day (that is to say, the difference between sidereal time and solar mean time)[126] should be regulated with an astrolabe. This text states explicitly that there were efforts under way among clockmakers or instrument constructors—who are for the first time mentioned as a group—to equip astronomical instructional and observational devices with an uniform drive, something water clocks evidently could not provide. Like Petrus de Maricourt earlier, Robertus Anglicus contributed a suggestion that is interesting in theory but of no practical use.

According to Thorndike, this brief text is a terminus ante quem non for the invention of the mechanical clock. For if a suitable regulating mechanism had been known, Robertus would not have made his suggestion; the authors of the *Libros del Saber* would undoubtedly have mentioned it. It would appear from this evidence that the invention was around the corner, and Thorndike speaks of the "impending invention." But we cannot fail to note that the champions of astronomy were not thinking in the least of a mechanical escapement. Their "dreams" went into an entirely different direction as they pondered the possibilities of the technical realization of a perfect rotation that simulated the movements of the starry sky. Peter de Maricourt wanted to use the rotation of the earth, the magnetic stone (relatively new in Europe), and frictionless mounting. This suggestion, oriented on the idea of a floating compass needle, can be reconstructed, but it is not suitable as a drive. Robertus Anglicus sought to balance a weight drive perfectly; but even without any knowledge of the laws of falling bodies and of inertia, he could have easily ascertained that the rotation of a drum would inevitably accelerate. These designs were utopian armchair technology, as far removed from the possibilities and experiences of European water-clock technology as they were from the solutions for an escapement devised in China two centuries earlier. Nevertheless, it is possible that Robertus Anglicus was right in what he said about the fruitless attempts of clockmakers or constructors. Perhaps they were experimenting around 1270 with the problem of drives for astronomical simulations. Such simulations did exist, even if we know little about them. Let us recall the costly present to Frederick II (1236), the astrolabes with gear drives, the armillary spheres mentioned by Petrus of Maricourt, and finally an extravagant planetary gearing that showed the equal hours (though we know nothing about its drive)—the "Opus quorumdam rotarum mirabilium quibus sciuntur vera loca omnium planetarum et etiam hore dierum ac noctium" that was described around 1300.[127]

Even though the influence of Islamic science on European astronomy reached its zenith in the thirteenth century, when it came to employing or adopting hydraulic drives for such instruments Europeans hesitated or had to struggle with problems. Perhaps these drives were not reliable if put to a different use. Perhaps the European instruments were too complicated and too heavy. One modern argument that is often heard in this context is that the hydraulic drives were unsuitable because of the freezing temperatures in the northern latitudes. The argument is attractive by virtue of its simplicity, but it is not convincing. For how could there have been a three-hundred-year tradition of water clocks in the monasteries—which used these clocks precisely during winter nights—if the instruments were unsuitable? The reason why the tradition of hydraulic drives came to an end can only be connected with rising demands as to reliability—as would have been demanded by the astronomers—or mechanical efficiency.

If one tries to push the above-mentioned terminus ante quem (ca. 1330) for the appearance of the mechanical escapement back chronologically, the sources concerning technical questions become somewhat clearer, albeit not clear enough for a dated development of the escapement.

An inventory of Charles V of France, drawn up in 1380, mentions a clock that supposedly belonged to Philip IV the Fair (died 1314). Fashioned entirely of silver without any iron, it is described as having two silver weights that were filled with lead. We believe we can recognize a mechanical clock and learn that such clocks were normally made of iron. The Chronicle of the Dominican monastery of St. Eustorgio in Milan also emphasized, under the year 1306, the procurement of an iron clock.[128] Larger quantities of iron, metal, and charcoal were delivered to the weapons smith Gilebert de Louvre—he lived next to the palace—in the summer of 1301 at the expense of the royal treasury. He was to use them to make a clock ("opus quorundam horologiorum") which the king wanted to give to the new abbey in Poissy.[129] Blacksmith's work ("un pieche de fer as orloges") was also paid for in 1300 at the Castle of Hesdin, where the Counts of Artois were setting up a kind of medieval amusement park with numerous musical automata, trees that rained water, and other machines ("engins d'esbattement"), the sort of things known previously only from oriental accounts.[130]

References to new technological developments are also found in literary sources. In a well-known passage in the portion of the *Roman de la Rose* that was composed by Jean de Meun between 1275 and 1280, the hero tries to bring a statue to life by means of various musical instruments, one of which is called "orloges."[131] The reference to artfully arranged wheels probably concerns a bell work; their ceaseless motion can be understood as a reference to a mechanical drive, though this conclusion does not emerge conclusively from the text. A fifteenth-century manuscript of the *Roman de la Rose* places one of the fairly new chamber clocks among a collection of musical instruments (fig. 19).

Classic evidence for the appearance of mechanical clocks is two extended metaphors in Dante's *Divine Comedy,* which he completed between 1315 and 1321. In the horological literature they are frequently regarded as the first descriptions of mechanical clockworks.[132] The first scene is a typical monastic waking situation, elevated through the motif of the heavenly wedding. Benvenuto da Imola's great Dante commentary, which he finished in 1380, imparts a worldly nuance to the scene by relating it also to the scholars' pre-dawn period of study. Dante describes an alarm mechanism in which we can recognize movements in opposite directions ("tira e urge"). The waking signal ("tin tin") is an onomatopoeic description of the repetitive striking of a small bell.[133] A second passage describes the wheelworks ("cerchi") of a small clock ("oriuolo"), and what mattered to Dante here were the

19. *Roman de la Rose,* miniature in a manuscript written around 1420. Valencia, Biblioteca de la Universidad, Ms. 387, fol. 141.

very different rotational speeds ("il primo . . . quieto pare, e l'ultimo che voli"). Though these verses also describe a clock, they can be related much better to the action of the mechanical striking mechanism that was connected to the clockwork. Nonsensical is the notion, common in modern Dante commentaries, that the clockwork was driven by a sequence of interlocking wheels, the fastest of which was regulated by a wheel of wind vanes.[134] Such an arrangement was never found in clockworks, but always in striking works. The scene and the choice of words also rule out the interpretation by Landes, according to whom Dante was reporting a technological sensation with pride and excitement. On the other hand, there is no reason to share Lynn White Jr.'s skepticism and doubt that Dante was already thinking of a mechanical clock for the simple reason that while we do read about gears, there is no mention of an escapement.[135] Dante is speaking, in a conventional scene, of a monastic alarm, which was undoubtedly common by his time. He apparently did not notice that this machine differed from its precursors in one important technical detail.

The first indications for the adoption of the equal hours outside of astronomy, that is to say, for the waning of the binding force of the Hours also in the sphere of the Church, come likewise from the beginning of the fourteenth century. Dante briefly describes the equal hours in his work *Convivio,* but he still says that they are common among the "astrologi" (III.6 and IV.23); he also does not connect them to clocks. In contrast, the title and form of the best-known Book of Hours from the fourteenth century shows the gradual penetration of the new hour-reckoning made possible by the new technology. The *Horologium Sapientiae* (around 1334) by Heinrich Seuse, a Dominican friar from Constance, divides the didactic material into 24 "materiae" and briefly talks about the title in the prologue:

> So the mercy of the Savior deigned to reveal this present little book in a vision, when it was shown as a most beautiful clock, decorated with the loveliest roses and a variety of well-sounding cymbals, which produce a sweet and heavenly sound, and summon the hearts of all men up above.[136]

The text is framed as a dialogue between Sapientia (divine wisdom) or Temperantia (the virtue of moderation) and their servant, the author. In the fifteenth century it gave rise to the most important illustrations of clocks in medieval miniature paintings (figs. 24 and 31). Even if the "rosis speciosissimis" in Seuse's prologue cannot be replaced by the obvious variant reading "rotis speciosissimis,"[137] the question arises what sort of clocks Seuse had in mind. They evidently had nothing in common with the clocks astronomers were demanding at the end of the thirteenth century.

From the end of the thirteenth century there is a noticeable increase in the number of sources on clocks and clock-use, and this holds true even if we adjust for the spread of written accounting. All over Europe new clocks were being built or bought. During this transitional period—that is to say, between the presumed termini post and ante quos, framed by the dates ca. 1270–ca. 1330—it is hardly possible to determine in any given instance whether we are still dealing with a hydraulic clock, already with a mechanical clock, or with a bell that was part of a clock.

A conceivable scenario is a boom in water clock constructions, in the course of which the new escapement was found and substituted in a few places. This scenario has frequently been given plausibility with reference to a street of clockmakers in Cologne. It is also conceivable that the development of mechanical clocks made possible this surge. As a result various clocks of the transitional period were and are retroactively declared to have been mechanical clocks. But apart from the above-mentioned references to the use of iron and weights, the large number of other sources does not permit us to make a decision either way that is technically grounded; the context of the sources reveals no change in customary time-measurement.

Contrary to what was the case in the High Middle Ages, but much like in the communes in the later fourteenth century, the procurement of new clocks was immortalized in monastic chronicles. One example is found in the Annals of the Priory of Dunstable under the year 1283: "we built a clock which is located on the rood-screen"; a later one appears in the chronicle of the founding house of the Dominicans in Orvieto (1305).[138] The sheer size of a clock and its cost became topics of discussion. In 1279, the Dominican nuns of the Johannis Convent in Colmar, whose prioress Adelheid of Rheinfelden had nighttime visions before the sound of the clock, recorded the purchase of a clock, which would hardly have been the first, "for six marks." Storm and fire had ravaged the church of the Augustinians in Barnwell in 1287, and among the damage—"which only God the Almighty could assess"—they listed in a memory book paintings, lead (for the roofs), the clock, glass windows, and bells. The record of the accomplishments of a prior of the cathedral chapter in Canterbury, who had served in his post with distinction, notes a "large new clock in the church" for the considerable price of thirty pounds (1292). The commemorations of donors or the use of donated funds and bequests for clocks show that these instruments were expensive acquisitions. The memorial book of St. Vincent in Mâcon in Burgundy (thirteenth century) immortalizes a deacon for having given his church an "arrologium." Endowment funds were used in 1305 at the cathedral in Augsburg to purchase a well-functioning clock ("orologium bonum et bene instructum"). The French king was not alone with his gift of a clock to the Abbey of Poissy, which he had founded. In 1329, the pope in Avignon had a clock built at his expense for the Carthusians in Chorges (Provence).[139]

The donation charter for a new clock in the cathedral of Lincoln (1324) makes it clear that churches and monasteries competed with one another in the expenditures for clocks. The charter plainly points out that a well-functioning clock was not only a necessity (the clock from the thirteenth century having become feeble with age), the cathedral had also fallen behind most other churches and monasteries in the acquisition of a new one.[140] Finally, the costs of maintaining and repairing these clocks also reveal the rising expenditures. The new profession of the clockmaker or clock guardian ("orologiarius") becomes visible from about 1270 on.[141] The surge in clockmaking at the end of the thirteenth century remains distinctly perceptible even if we drop the one piece of evidence most frequently cited in the newer literature: a guild of clockmakers or a street of clockmakers in Cologne.

A guild of clockmakers in Cologne in 1183? A street of clockmakers in Cologne in the thirteenth century?

Important modern works on the history of technology and horology speak of a guild of clockmakers attested in Cologne since 1183 or of a street of clockmakers in Cologne in the thirteenth century, as though this information—in many ways sensational in terms of the history of European trades—were secure historical knowledge.[142] Zinner and White base themselves on two short works by E. Volckmann on Cologne as the oldest seat of German clockmaking (1918) and on the old German trade streets (1921).[143] Volckmann himself does not speak of a guild of clockmakers, but of clockmaker's trade ("Gewerbe"), industry, or art. The misunderstanding is created by the translation of the German word "Gewerbe" by the English word "guild," which connotes an organized association. The misunderstanding is perplexing, since this information would turn the historiographic literature on guilds completely upside down. Nowhere in thirteenth-century Europe were such highly specialized trades being organized into guilds. Nowhere else are guilds of clockmakers attested prior to the sixteenth century. Perhaps there were clockmakers in Cologne outside of the monasteries, but we don't know of a single one. That they should have been numerous enough to organize themselves is utterly implausible.

More difficult is the interpretation of the street name. Volckmann's source is the great work on the topography of Cologne by H. Keussen, who suspects that a resident called "Urloug" gave the street its name (the German word for clock is "Uhr").[144] A chronological arrangement of the old name forms collected by Keussen shows that a connection with clocks existed only in one case, and here the link was perhaps coincidental: 1170–1183 "Urluge" as the name of the owner; "platea urlugen" 1230–1232; "Urlogesgassen" 1251; (between 1297 and about 1450 also "Urlus-," "Urloch-," "Urloigis-," "Urlis-" and the like); "Horlogesgazen" 1266–1271; "Ourliges-" 1264; "Orloges-", "Orliges-" and the like 1271–1336; "Oyrle-" 1385; "Oirloich-," "Ortloess-," "Ortlich-," "Ortlinx-," "Ortlieff-" and the like in the fifteenth century.

Are we dealing with references to clocks or clockmakers? Linguistic usage in Cologne from a later period, when clocks and clockmakers certainly existed, would argue against it. In 1374 a sundial or a dial were called "uurtafel," clocks were called "ure" or "uyre," clockmakers were "u(y)rklockenmecher" (1395–1418).[145] The conventional interpretation is also contradicted by the fact that the name of the street in the fifteenth century bears increasingly less resemblance to old words for clocks. Finally, in opposition to the view of Volkmann and Zinner, we should point out that according to Keussen's notes this street did not have any of the typical features of a street where smiths lived. Very few metalworkers lived in the general area, and not a single one in this particular street.[146] K. Maurice has argued that the street could have been named after a house whose distinguishing feature was a clock (a sundial?).[147] This interpretation, in turn, is contradicted, on the one hand, by the linguistic usage in Cologne, and, on the other hand, by the fact that this would be another instance of a practice that could not be attested anywhere else in medieval Europe. J. Leclerq has offered an alternative explanation, one that looks plausible only at first glance.[148] In Middle High German, a clock was called "urlei," "or(o)-lei," urleige," "urlei," and "ore." But there is a similar word for war and battle and everything related to them: "urliuge," "urlei," "orleige," "urloge," "orloge,"[149] and likewise in Middle Dutch, "(o)orloge," "oirloge," "orlage," and "ourlage."[150] Is it therefore not possible that weapons or armor were manufactured in this particular street in Cologne? This explanation might make sense in terms of the history of trades. Cologne was an important center of metalworking. However, it clearly contradicts Keussen's study, according to which the Spinnmühlengasse ("Spinning-wheel Street") was not a street with such a trade.

A guild of clockmakers is thus not mentioned in the Middle Ages. The name of the street is contradicted by the linguistic usage in Cologne and by the broader European historical context. Until a better explanation appears, one should thus take the "Horlogesgazen" (1266–1271) in Cologne to be a mistake in writing or hearing, and no longer refer to either a guild of clockmakers or a street of clockmakers. David Landes, who did not even investigate this particular problem, instinctively came up with the right question: "after all, why Cologne?"[151]

A noticeable rise in expenditures for clocks can be observed during this period, not only in the churches and monasteries, but also—and this is new for central Europe—at the great courts. The Counts of Artois had a clock set up in their amusement part in Hesdin. Philip the Fair, the king of France, commissioned work not only on a clock for the monastery of Poissy, but also on an elaborate construction for his own court.[152] With the clocks of this transitional period, the rising expenditures were caused by more expensive materials and more elaborate supplementary contrivances. The latter were in most instances not astronomical indications of some kind or another. Costs had gone up for something else.

Clock bells

As early as 1270, Humbertus de Romanis, visitor of the Dominican Order, criticized the increasingly costly outfitting of monastic clockworks. He demanded that clocks in the monastery be reliable ("verax et certum") but not attract attention with their costliness or superfluous frills ("curiositates").[153] What did he mean by superfluous frills? It cannot be ruled out that he wanted to ban from the monastery, among other things, astronomical mechanisms attached to the clocks. What is more likely, however, is that he was talking about acoustic-musical gadgets. It is known that clock bells, which were alarm bells in functional terms, had separate names already in the twelfth century and could be used on special occasions to augment the bell ensemble. In the thirteenth century one apparently began to increase the melodiousness of the wake-up signal provided for in the statutes. We have clear evidence to this effect. The tract on music by Hieronymus of Moravia (before 1304) has the usual discussion of the art of tuning bells through changes in weight or the relationship between the weight of the bell and its pitch. On this occasion Hieronymus, as shortly afterwards also Engelbert of Admont, speaks explicitly about the artistry of attaching small, tuned bells to clocks for the purpose of producing music.[154]

The connection between bells, clocks, and organs becomes visible from the thirteenth century on also as a field in which mechanics were active. In Exeter, the bishop engaged Walther de Ropford in 1284 to maintain the bells, the organ, and the clock of the cathedral and repair them if necessary. In the park of Hesdin a clock was connected to an organ. The large clock of St. Paul's in London was built by Walther the Orgoner (1344). The engineer Giovanni degli Organi from Modena, who was in the employ of the Visconti, built the tower clock for Genoa in 1354. He apparently inherited the honorable surname "degli Organi," and was thus descended from a renowned organ-builder. In 1381, Bodo of Hardessen cast the clock bell for the city of Blankenburg in the Harz and immortalized himself on it as "organista et orlogista." And in the Genoese colony of Caffa on the Crimea, one Jacobus de Novaria, "magister orologii et organorum," was active in 1410.[155]

Sequences of tuned bells had long been in use. The inventory of Prüm mentions a bell-wheel in front of the altar in the ninth century. In the monastery of Abingdon (Berkshire), a bell-wheel ("quadam rota tintinnabulis plena") was built in the tenth century and set in motion on feast days to promote the spirit of devotion.[156] Slats with tuned bells were also used as musical instruments that could be played (figs. 20, 21).[157] New was that such bell-wheels were connected to clocks. The reason for this could have been in the function of both as wake-up signals, regardless of technical details. Late, though clear, evidence for this comes from the Chronicle of Windesheim. We read that a lay brother, Heinrich of Laer in Westphalia, sacristan and able mechanic, built for the monastery a wheel with seven bells, hammers, and iron wheels, and mounted it between the dormitory and the cell of the custos.[158] In the fifteenth century, bell-wheels were considered instruments typical of monastic clocks.[159]

Leaving aside simple alarm-bell mechanisms, we can trace the tech-

20, 21. Initials "E" for Psalm 81 in two breviaries of Philip the Good of Burgundy. Brussels, Bibliothèque Royale, Ms. 9511, fol. 307; Ms. 9026, fol. 163 verso.

22. Tubal-Kain, a legendary inventor of musical instruments, at a mechanical bell-wheel. Relief on the façade of the Cathedral in Orvieto (around 1320). Photo: G. Dohrn-van Rossum.

nical development more or less from the pictorial sources; however, this means that the dating of this development will necessarily remain vague. In the middle of the thirteenth century, the water clock in the moralized Bible (fig. 14) shows a slat with bells. A bell-wheel which Tubalkain, one of the legendary inventors of musical instruments, is striking on a relief on the facade of the cathedral in Orvieto (around 1320), is turned by a pinwheel and a drive (fig. 22).[160] Some kind of

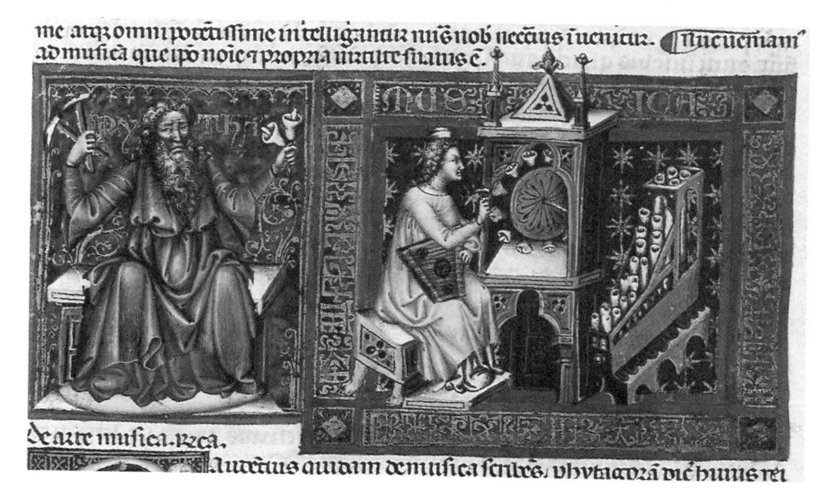

23. *De arte musica:* Miniature in an Italian collection of classic texts from the library of Petrarch (beginning of the fourteenth century). Paris, Bibliothèque Nationale, Ms. lat 8500, folio 39 verso.

mechanization is apparent; the drive remains hidden. The same holds true for an Italian miniature to *De arte musica* (around 1340) (fig. 23). Perhaps it contains, if read from left to right, a reference to the technical progress of the most recent past. In the left field is seated David (or Pythagoras?) with handbells and hammers. The figure in the right field, seated next to a pneumatic organ, is striking—still by hand—an apparently mechanically operated bell-wheel (a new technological achievement?).

That these early forms of the playable carillon were admired innovations and may also have become fashionable at the beginning of the thirteenth century is revealed, on the one hand, by the title and motto of Heinrich Seuse's *Horologium* and, on the other, by a notice in the chronicle of the convent of Benedictine nuns in Rouen. We read in the latter that in 1321 a clock was installed whose hymn, *Conditor alme siderum,* was still audible five kilometers away.[161]

The miniature of the Brussels Seuse manuscript (fig. 24) shows, as in a collection or display, eight different devices for keeping and indicating time that were known in the middle of the fifteenth century. On the left is a large clockwork with a dial (twice twelve hours) which, as

de leurs bonnes euures de leur vie par
faitte z de leur saincte conuersacion.
Ceulx ci queroient dieu en grant
bonte z en parfaitte simplesse de cuer.
Car toute leur entente z leur estude
estoit les occupacions mondaines ou
blier. leurs pechiez gemir z plourer.
les choses diuines assauourer. Eulx
dedens eulx contenir. et garder le corps
et a seruitude ramener. Et seulement
a la diuine sapience acordre. et leur [illegible]
occuper. Mais las dolent au iourdui
le monde est enuieilli en malice et lamo[z]
diuine est es cuers de plusieurs tellement
refroidee quelle est peu sen fault estainte
Car on en treuue pou qui estudient
en deuocion ne qui aient diligence de
renouueller de la grace de dieu pour
eulx eschauffer ne qui par desplaisance
de leurs faultes aient souuent la ler

me a loeil ne qui desirent la presence
la visitacion la doulce alloqucion
et plement de la grace diuine. Ains
se occupent z estudient en vanitez en
narracions en genealogies et hystoires
vaines z temporelles et en delices cor
porelles esquelles ilz sont endormis
par vng grief z pesant sommeil.
Et pour tant la diuine sapience
qui tousiours est songneuse du sau
uement de toute humaine creature
en desirant damender la vie de ses
esleus et en voulant oster z anientir
tous vices de leurs consciences veult
et entent en ce liure principalment
enlumer les estains z enflamer les
les refroidiez les pecheurs resmouuoir
les mal deuots a deuocion rappeller et
promouuoir. Et les endormis par negligence
a lestude des euures vertueuses esueiller

24. Miniature in a manuscript of a French translation of Heinrich Seuse's *Horologium Sapientiae* (around 1450). Brussels, Bibliothèque Royale, Ms. IV, iii, folio 13 verso.

the full page reveals, trips an hour-striking bell in a distant belfry via guide beams. At the base we see an astrolabe, which was presumably used to regulate the clock.[162] In the middle is a bell-striking mechanism, placed next to the clock undoubtedly for reasons that had to do with the spatial arrangement of the miniature. We should imagine that it was located on top of the clock and was set in motion by the latter via levers and pulleys.[163] Because of this pictorial arrangement the weight drive also remains hidden. Hanging from the table are a small tower sundial and a quarter-circle equatorium, an instrument derived from the astrolabe. On top of the table on the right side stand the most modern instruments that were developed only in the fifteenth century: a table-clock with a case that has been deliberately left open and in which we can see the spring tension and fusee, and a modern sundial with equator ring and a polar staff parallel to the axis of the earth. The bells of the striking works of the frame in the center are rigidly mounted on a ring. Each bell is assigned a hammer with a lever that is tripped by a horizontal wheel with studs or teeth. This is a variation of the monastic alarm bell mechanisms, here arranged as a sequence.

The development of the mechanical escapement

So what does the development of increasingly elaborate bell-striking mechanisms—which, no matter how elaborate, were essentially more sophisticated alarms—have to do with the development of the mechanical clock escapement? When Dante was describing the sound of a monastic alarm ("tin tin sonando"), he had a very simple arrangement in mind. We can get an idea of what it might have looked like from the simple monastic alarms and tower warden clocks of the subsequent period. The few extant specimens cannot be dated, and the common museum caption "fifteenth century" is nothing more than a makeshift solution.[164] There is no reason why the simple clocks at the beginning of the fourteenth century should have looked or functioned any differently. Behind the going train sits a striking mechanism that is tripped by a lever (fig. 25). A drum with a weight moves the repetitive (clanging) hammer stroke on a bell, which lasts until the weight has run down. What is noticeable is that the escapement of the going train and the repeating mechanism have one essential component in common: a vertically mounted scape wheel with sawlike teeth and a verge

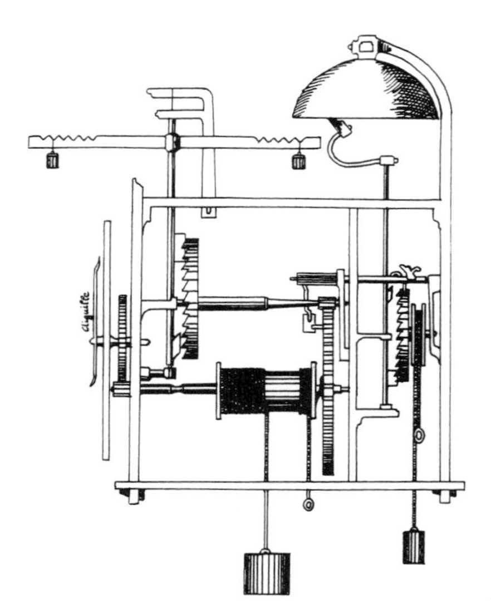

25. Tower warden clock with alarm bell mechanism. Drawing by G. Oestmann, after C. Sandon, *Les Horloges et les Maîtres Horlogers à Besançon* (1905).

with two palettes that alternately block the teeth. In the striking mechanism, the verge, which is fitted with a hammer lever, is moved a short distance and is then blocked and thrown back. The same thing occurs with the escapement, only much slower, that is to say over a longer interval of time. At the same time the rhythm of the oscillation is regulated by the inertia of the foliot which can be adjusted by small weights. This observation first led J. Drummond Robertson (1931) to the conjecture that the clock escapement could have been developed out of older mechanisms for the repetitive striking of a bell. More recently, J. D. North has made this once again the probable line of development through his discovery of another double escapement mechanism in the Wallingford clock.[165]

Thus the development of the mechanical clock escapement could be described something like this: for some time, monasteries had been using contrivances for repeatedly striking the waking signal on a single bell, on several bells, or on a bell wheel. At the end of the thirteenth century it was discovered—apparently in several places—that such contraptions were also suitable as a regulator if one slowed down the oscillating of the bell-hammer lever by increasing its weight in relationship to the drive weight and made it regulatable by changing the swinging mass. This evolution of the repeating mechanism into the clock escapement presupposes numerous practical experiments that were prompted by the work of mechanics and smiths on the bell works in churches and monasteries. We cannot say when such attempts began to be undertaken. Since the astronomers were looking in different directions entirely, and since the results did not meet their stated needs, the generally accepted terminus post quem (1270) is problematic. On the other hand, the abundance of newly acquired clocks seems sufficient reason to date the success of this developmental work to the second half or the last third of the thirteenth century.

This evolutionary history remains a hypothesis, though one that is secure against the "veto of the sources." It cannot be proved that repeating striking mechanisms with crown wheel and verge and pallets, or, as in the Wallingford clock, with a double pinwheel appeared historically before the clock escapement. All evidence for this comes from a later period. The sequence is plausible, however, since mechanical striking works, even very heavy ones like that described in the MS Ripoll 225, could initially very well have been tripped by water clocks. The mechanical clock escapement was in all likelihood an independent European development, since neither in China nor in the Islamic sphere can we observe a comparable development toward a more elaborate bell technology.

The "angelic clockwork" of Villard de Honnecourt

Apart from the technically unrevealing drawing of the "maizo(n) d'une hierloge" (fig. 12), the sketchbook of Villard de Honnecourt also contains a depiction of a weight-driven revolving frame for an angel whose finger is supposed to point at the sun at all times (fig. 26).[166] The first editor of the sketchbook, J. B. A. Lassus, surmised in 1858 that the figure of the angel was supposed to complete one rotation during twenty-four hours. After the expert C. Frémont subsequently had declared the revolving frame a precursor of the mechanical escapement, this notion made its way into the handbooks on the history of technology whence it was widely disseminated.[167] Frémont tried hard to explain how the contraption might have worked: the rope was supposedly slowed through friction against the spokes of the wheel which oscillated back and forth. This "oscillating friction-escapement" was regulated by means of the two weights. It didn't bother any of the interpreters that Villard or the contemporary authors did not in any way describe this

26. Mechanism for a pivoting angel figure, the so-called "angelic clockwork," from the sketchbook of Villard de Honnecourt (around 1235). Paris, Bibliothèque Nationale, Ms. fr. 19093, folio 22 verso, after H. R. Hahnloser, *Villard de Honnecourt* (1972).

contraption as a clock or even as performing a regulated movement. Thus the editor of the facsimile edition of the sketchbook, invoking the verdict of experts even in the new editions, could continue to write about the "earliest clockwork with wheels," in this way putting art historians, too, on the wrong track.

Following Lynn White Jr.'s skeptical comments on the "angelic clockwork" ("ambition rather than achievement"), Needham and Maurice, among others, have pointed out that the contraption simply could not have functioned the way Frémont had conjectured, and that we were dealing at most with a turning movement that was set off from a certain distance.[168] Still, works that continue to report this invention keep appearing. M. Daumas has tried hard to clear up these misunderstandings, which were caused in part by faulty tracing of the sketches in the first edition. His practical experiments have demonstrated that this was definitely not an early form of a clock escapement. As for the "angelic clockwork" as a precursor to the clock escapement, we can thus join in the opinion of P. Mesnage, who said as long ago as 1965 that this notion should once and for all disappear from the history of technology.[169]

Automata displays

The term "regulator" (German "Zeitnormal," lit., "time standard") is surely too ambitious for the early and undoubtedly crude clock escapements, since contemporaries did not even take notice of them as improvements in the technology of time-measurement; in fact they didn't even mention them. From this quarter we thus cannot expect any consequences for the conception of time or changes in time-consciousness at the end of the thirteenth and the beginning of the fourteenth century. But mechanical clockworks now offered the possibility of driving large and heavy automata and keeping them operating more reliably than had previously been the case with hydraulic drives. And from the beginning of the fourteenth century, such automata were being built all over Europe. They drove carillons and became the focal point for various mechanisms of the kind that were known from classical and Islamic tradition.[170] In the cathedral of Norwich, a large clock was constructed between 1322 and 1325 with a heavy dial (eighty-seven pounds), a carillon, and a procession of monk figures. During the same period the Chronicle of Glastonbury extols an abbot who had enriched the church building with a number of "spectacular" furnishings, among them a clock with automata and figures and an organ of marvelous size.[171]

In the accounts for the clock in the hospital of St. Jacques in Paris, where thousands of pilgrims stopped off each year on their way to Santiago, we hear for the first time — as far as I know — of the procession of the Three Magi (1326), which had now taken the place of Heracles or the wild Arabic horsemen of the Islamic tradition of water clocks. In 1334, the goldsmith Mondinus, from the city of Cremona, fashioned in Venice an artful clock worth eighty guilders for the king of Cyprus. In St. Paul's Cathedral in London, a mechanism turned a dial and an angel (1344). The astronomical clock in Cambrai displayed moving pictures and a calendar (1349). The chronicles report the construction of the Strasbourg clock (1352–1354) and its procession of the Three Magi. All that is left of this first clock is a mechanical cock with flapping wings, but the description of a copy, the clock in the Minster of Villingen (1401), gives us fairly precise information about its appearance in the Middle Ages.[172] Only a few years later the town of Frankenberg in Hessen also installed such a clock.[173] Today these clocks are called monumental astronomical clocks. At the time most of them tended to be automatic displays and carillons. Astronomical and calendrical dials were rotated, kings moved past the Virgin Mary and bowed down, cocks flapped their wings and crowed, moveable skeletons appeared as reminders of death. As the protocol of the cathedral chapter in Chartres from the year 1407 clearly bears out, the stated purpose of these contraptions was, right up into the nineteenth century (interrupted by some iconoclastic phases), to lure people into church, to astound them, and thus strengthen the authority of the Church.[174]

The replicas of the starry heavens, which were moved at an unimaginably slow and even pace, surely did not fail to have an effect on the viewers' conception of time. But this "mechanization of the liturgy" (K. Maurice) was of no importance in terms of timekeeping or time-indicating. When people looked at these mechanisms, they were not "looking at the clock."[175]

The development of such automata-clocks led individual astronomers to concern themselves with the problem of sophisticated mechanizations. But the achievement of Richard of Wallingford or Giovanni Dondi lay in the development of special gears for displaying the positions of the planets. They were not interested in the drive, since clock drives were sufficient for such automata from a didactic and demonstrational perspective, while being much too inaccurate for observational purposes. This did not change until the end of the fifteenth century. Understandably enough, astronomers thus did not cry out "eureka" when the mechanical clocks appeared.

Hour-striking clocks: A technological sensation in the fourteenth century

A chronicle of the city of Milan reported, under the year 1336, the installation of a clock in the tower of San Gottardo. Its manner of striking the hours made it admirable and useful for all social classes.

> In the spire are many bells, and there is a clock which is admirable because it is a very large bell (tyntinabulum) that strikes a bell (campana) twenty-four times according to the number of the twenty-four hours of the day and night, in such a manner that it gives one sound in the first hour, two strokes in the second, three in the third, four in the fourth, in this way distinguishing the various hours. This is exceedingly necessary for people of all estates.[176]

The chronicler Galvano Fiamma is certainly suspect as a biased reporter who exaggerated the significance of the contraption and the merits its builder deserved. As the chaplain and scribe in the service of the Visconti, the Dominican monk naturally attributed the flowering of Milan—and around 1300 a large number of various bells was a common indicator of thriving prosperity—to the princes of the city, his employers. The purpose of his eulogy to the rulers was for their city to outshine all others in Italy also when it came to such novelties. Clocks caught Galvano Fiamma's attention. In the chronicle of his monastery he reported the construction of an iron clock. Among the splendid furnishings of the princely palace, which he did not want to list in their entirety, he emphasized another clock. In the list of innovations ("novitates") that were made in Milan under the Visconti, he enumerated, next to powerful warships, the use of large clockworks to drive mills. With the help of these marvels with wheels and weights, a boy could easily grind four bushels of grain into flour of the highest quality.[177] In technical terms Galvano Fiamma's description of the new clock in the tower of San Gottardo is meager. He calls it a large bell that strikes another bell. But he is not really concerned with such details. The novelty is not the clockwork but the indication of the hours as they are counted, the striking work, which made it possible to distinguish the sequence of the hours as they were struck from one evening to the next.

Three years later we find in another Milanese chronicle, on the occasion of a civic event of the first order, a reference to the forms of the division of the day as they had been altered by the new clock. We are told that the man who had commissioned the striking clock of San Gottardo, Azzo Visconti, died on August 14, 1339, in the twentieth hour,[178] that is to say, around three in the afternoon, the first modern hour-indication in an urban-civic context.

Galvano Fiamma was not exaggerating. Clocks in urban towers or municipal hour signals did exist earlier, but they can be clearly attested only in Italy, and we know nothing about how they indicated the hours. There is, therefore, no reason to doubt that Galvano Fiamma's report was newsworthy. The information from the subsequent period confirms that the diffusion of an entirely new technology had begun. A few years later, in 1344, Paduan chronicles report the construction, in the tower of the palace of the city's rulers, of a twenty-four-hour clock with automatic striking of the hours; numerous other accounts refer again and again to the same situation.[179]

The appearance of the striking clocks was for many the first encoun-

27. The monk Lazar shows Archduke Vasilij Dimitrievich and two of his vassals the clock in the Kremlin. St. Petersburg, Library of the Academy of Sciences, Ostermanovskij manuscript (sixteenth century). From Edward V. Williams, *The Bells of Russia* (Princeton, NJ, 1985).

ter with a mechanical clock, and they were thus hardly able to distinguish the going work from the striking work. But only the invention of the striking mechanism made possible the spread of the clocks, and only this invention popularized the notion of clocks as fascinating automata.[180] This still comes out very clearly in the description of the first tower clock in Moscow (1404) (fig. 27):

> This hour-marker is called an hour-measure; each hour a hammer strikes the bell, measuring and counting the hours of the night and of the day . . . No man strikes it, but it is somehow wondrous strangely fashioned to look like a man and sound and move of itself, by man's cunning, with great invention and cleverness.[181]

In Giovanni Tortelli's retrospective list of inventors, too, the automaton which, seemingly animate, did the work for humans appeared as the true innovation and a socially significant event:

> [For not only does it show and register the hour to our eyes,] but also its bell announces it to the ears of those who are far away or staying at home. This bell is placed on top to distinguish the number. Hence in a way it seems to be alive, since it moves of its own accord, and does its work on behalf of man night and day, and nothing could be more useful or more pleasant than that. However, it is a new invention.[182]

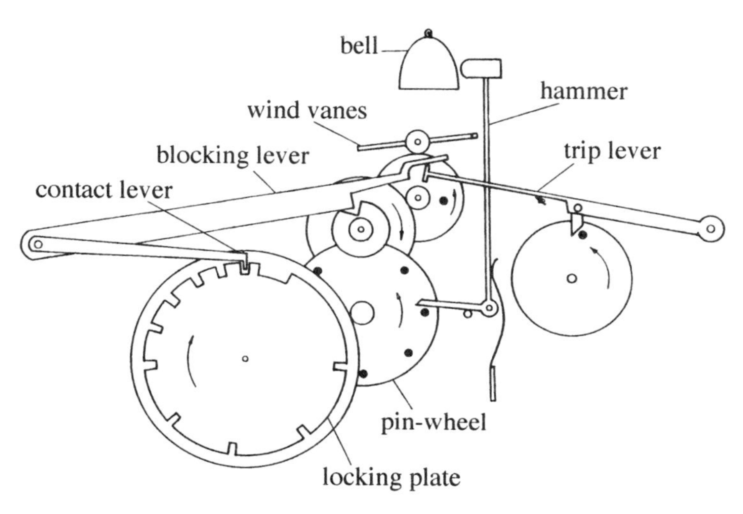

28. Locking-plate striking mechanism. Drawing by G. Oestmann.

What was this new invention? The mechanism for striking the hours is, even "if one knows how it works," a far more complicated mechanism than the movement of a clock. It arose—as the clock may also have—from modifications to a repeating arrangement which, unlike the repeating works with a crown wheel and a verge with pallets, was also suitable for delivering the stroke of the hammer to heavy bells and to bells mounted some distance away. Since this mechanism, so momentous to the history of late medieval society and mentality, is rarely explained, a brief description follows (fig. 28).

By means of pins on a rapidly turning wheel (pinwheel) the bell hammer is briefly activated in regular intervals via a wire-pull. The rotation of the pinwheel is transmitted and is steadied by means of a shaft with wind vanes. The ringing mechanism is tripped by lifting a blocking-lever from a notch of the starting-wheel. In order to limit the duration of the ringing, that is, in order to prevent the entire length of rope on the weight-driven wheel from unwinding, when the blocking-lever is released it is simultaneously lifted from a second notch of the concurrently rotating "locking-plate" (also called "rota serandi") and is held up until it can fall into the next notch of the locking-plate. The duration of the ringing, and with it the number of strokes, is thus determined by the distance between the notches on the radius of the locking-plate. We can make out a striking work with a locking-plate that is notched on the inside in the rear half of the large clockwork of the miniature in the Brussels Seuse manuscript (fig. 24). The ringing program can be varied by notching the locking-plate in the appropriate way. Any number of strokes can be arranged on the "program wheel" in any desired sequence. It is therefore possible that the locking-plate was already developed earlier for the ringing of the Hours in churches, since the locking-plate striking work can also be set off manually at any given time.[183] A development independent of the clockwork is thus conceivable. But moving the locking-plate back in time remains a speculative step.

In the context of various arrangements for repeated striking, the locking-plate striking work was presumably a new invention, the product of efforts to limit or differentiate the sequence of signals.[184] Even if it arose from attempts to automate stroke-sequences for variable segments of the day, such efforts were technically feasible only for portions of the day or hours that remained constant. There was, from the medieval perspective, no other option. In his 1537 edition of Bede, Johannes Noviomagus explicitly emphasized the connection between the use of equal hours and the striking clocks.[185] A ringing of the Hours is technically possible, though very complicated and awkward: one could, as in the Japanese clocks of the seventeenth century, provide for two foliots and exchange them in the morning and evening.[186] In addition, the running of the clock would have to be adjusted continuously, or at the very least every fourteen days, to keep it more or less in sync with the changes in daylight. Nothing of the sort is mentioned anywhere. Speculations that the sequence of Hours was automatically rung in the Middle Ages do not appear until the modern period.[187]

The notice in the Milanese chronicle is a plausible terminus ante quem for the development of the striking clocks, the name one can give to mechanical clocks with an automatic striking of the hours, even if Seuse's title may argue for a slightly earlier date. Several reasons suggest why we should surmise that the development took place in Italy, perhaps in an Italian city. G. Bilfinger had already concluded as much from what were in his day meager references to striking clocks, and especially from the early use of the modern hours in Italy.[188] New sources have reinforced this impression. Public clocks or public hour-signals appear from the beginning of the fourteenth century in Italian cities.[189] The only exception is the clock signal mentioned in Valenciennes. These were probably tower warden clocks or tower warden alarms by means of which some of the hours of the day, but by no means the entire sequence, were signaled. All secure reports about striking clocks, usually large tower clocks, from the first half of the fourteenth century refer to Italian cities.

Various forms of modern hour-reckoning

If the public indication of the hours according to their sequential numbering was technically feasible only with the use of equal hours, this did not yet settle the question of how the hours would be counted, and models were not known.[190] The Hours had never been rung according to their sequential number. There were thus no venerable customs that one could have picked up on. The arguments that were exchanged in this open situation—the future arrangement of the time of day was up for debate—would be far more revealing for an analysis of late medieval conceptions of time than all technical details combined. Even though such discussions must have occurred in various places, we know nothing about them. The only known argument was that of the astronomers, who had emphasized the advantages of equal hours for the observation of heavenly phenomena.[191] But this is never mentioned in connection with the introduction of clocks, and astronomers could use astrolabes to determine the time.

Different solutions were found in different places—clear evidence that a decision had to be made. The solutions did not pick up on the kind of artificial divisions of the day that were known from the regulating of some water clocks, nor directly on the astronomical divisions. Instead, all solutions were attempts to avoid a complete break with the accustomed system of unequal hours. Three characteristics marked the old system: day and night were separated, two series of twelve hours each were counted, and the beginning and ending points were shifted along with daylight.

The three most important forms of modern hour-reckoning which became customary around the middle of the fourteenth century tried in various ways to link up with the old forms.[192] This linkage was of a formal nature; the Hours as a sequence of prayers no longer played a role here. In this respect the transition to modern hour-reckoning was in fact a reflection of a broad process of the laicization of the way of life of townsmen and city-dwellers.[193]

In what is presumably the oldest Italian form of hour-reckoning, unequivocally attested for the first time by Galvano Fiamma, the twenty-four hours were counted through from one evening—more precisely: one half hour after sundown—to the evening of the following day. The twenty-fourth hour was the last hour of daytime. The only linkage to daylight has to do with timing the point at which the counting begins in the evening. Reckoning the day from evening to evening was customary in the Islamic sphere. But more important for its introduction into Europe was probably its agreement with the Judeo-Christian notion of the feast day. The influence of the latter on shaping the new system was, however, limited: the change of date usually occurred in the morning, that is to say, events that took place at night were assigned to the date of the previous day. The "Italian clock" (also called "Ganze Uhr" or "Ganzer Zeiger") spread via Bohemia even into Silesia, and for a time it was also used in Poland. The drawback to this method of counting—often lamented—was that the striking clocks had to be continuously moved ahead by a certain amount during the first half of the year, and moved back during the second half. To avoid this, the clocks were adjusted only after changes in the length of the day had reached a certain point, and they were then set with the help of tables.[194] In the diary of his journey through Italy, Goethe recorded with bemusement how cumbersome it was to convert this method of counting, a method that was "intrinsically" connected to the nature of the Italian people and should not be replaced by the "German clock hand."[195] The Italian way of counting was in use—with many exceptions—in the geographic border regions in Bohemia and Silesia until the seventeenth century, and in Italy itself up to the wars that followed in the wake of the French Revolution. The drawback to this method in terms of the technology of the striking work was the fact that the three hundred strokes required each day wore out the material and also needed a lot of rope, that is to say, the weights had to be frequently drawn up. For that reason some localities in Italy soon switched—with or without changes in the method of counting—to ringing sequences of 2 × 12 or 4 × 6.[196]

The "Nuremberg Clock" (also called "Great Clock") followed the old tables of the length of the days; it kept the division into day-hours and night-hours as well as the movable beginnings of both sequences, but it dropped the sequences of twelve. According to this method, attested for Nuremberg since 1374, and two years later also for Regensburg, in December the day had eight and the night, sixteen hours.[197] Until June the day was lengthened by one hour approximately every three weeks and the night shortened accordingly; after the mid-point of the year the procedure was reversed. The advantage of having the hour-reckoning conform to daylight was counterbalanced by the disadvantage of having to use tables whose accuracy was only approximate. All public civic activities that were fixed by the hours, and work time in particular, had to take this continual adjustment into consideration.[198] An additional, serious drawback was the fact that the clock and ringing tables became increasingly more complicated as efforts were made to take into account the varying rates in the increase and decrease of daylight. According to the setup of 1489, corrected by the astronomers, the period of eight-hour days lasted fifty-two days, that of thirteen-hour days, seventeen days, that of fifteen-hour days, twenty-two days, and that of sixteen-hour days, fifty-seven days. A few southern German cities followed this practice, which died out at the end of the eighteenth century. The Great Clock was never able to establish itself as the exclusive method even in Nuremberg, and it remained limited to the cities.

The method of hour-reckoning familiar to us was called "small" or "half clock" in the Old World, from an Italian perspective "orologio oltramontano." All it retained of the old system was the dual sequence of twelve hours. Daylight was no longer of any consequence. It was the most abstract form of division with respect to the conditions of the natural environment, though it did divide the day into two, a practice alien to everyday life. Prior to this the middle of the night had been customary as the dateline only among Roman jurists. Ludovici Guiccardini, in the account of his travels through the countries of northwestern Europe (1507), points to what was possibly the decisive advantage of this method of counting the hours: technically it is the most elegant solution, because in practice the clocks had to be set only once a day (in theory never) without any tables, using only a primitive sundial.[199]

Of limited local importance was the "Basel Clock." As travelers often noted in astonishment, the clock in Basel struck one at midday instead

29. Fight over the "Basel clock" (1779). Colored etching by L. de Rütte. Staatsarchiv Basel, Bildersammlung Falk A 500.

of twelve. The origins of this striking sequence, which once more sheds light on how fluid the situation was at the beginning of the fourteenth century, are unclear. As the humorous drawing depicted in figure 29 shows, at the end of the eighteenth century, scholars and merchants tried to adjust the striking of the hours in Basel to European customs, against the opposition of the artisans, but Basel's special custom did not come to an end until the French Revolution.[200]

The method common today, that of counting twenty-four continuous hours starting at midnight, appeared in the Middle Ages very occasionally in astronomical time indications.[201] Prior to the development of modern transportation systems it played virtually no role at all. Because all forms of communication were so slow, the co-existence of various methods of counting the hours were not a major annoyance—only travelers noted it with puzzlement. However, the lack of synchronicity in the methods of reckoning the hours in the cities did attract the attention of Emperor Charles V. For the eastern part of his empire, "on which the sun never set," he had a clock built that "showed the hours according to the Flemish and Italian methods."[202]

"Hour-glasses": Sandglasses and modern hour-reckoning

Another new time-measuring device appeared in Europe simultaneously with the wheeled clock: the sandglass. Because the sandglass has a certain similarity with the ancient clepsydra, scholars long believed that it was very old. Even today one can frequently read that there were sandglasses in the High Middle Ages.[203]

Since sandglasses were in use until the beginning of the nineteenth century to regulate ship-watches and, together with the log line, to calculate a ship's speed, scholars have conjectured that the device originated in the realm of seafaring. Such a view would be supported by the earliest references to their use and the linguistic usage in the fourteenth century. The oldest reference to their use is not entirely clear semantically. In the *Documenti d'Amore* (verse 1313), Francesco Barberino describes the good helmsman who minds the course and speed of the ship by using compass and clock ("arlogio") without losing a single hour. At the beginning of the Hundred Years' War (1345–1346), "glass clocks" ("orologii vitrei") were purchased for the English ships in the port of Sluys in Zeeland. The first description of a sandglass in an inventory of Charles V of France (1380) likewise shows that contemporaries located the origins of this time-keeper in the world of seafaring.[204] However, at this time we cannot prove its nautical origins, and it is noteworthy that sandglasses appear regularly in ships' inventories only in the fifteenth century. By contrast, they were already used in the fourteenth century for regulations relating to the organization of time without any connection to nautical problems. Despite the exceedingly meager textual sources, these early accounts do attest a certain familiarity with the sandglass already in the first half of the fourteenth century. It follows from this, too, that sandglasses—no matter where they had been developed—spread through Europe at about the same time as wheeled clocks. It was only with the introduction of equal hours that one could find use for sandglasses ("hourglasses" or "sand-hours").

Sandglasses were cheap, quiet, and reliable companions and competitors to wheeled clocks. In numerous tower warden rooms they were used to check the going work of public clocks, and in Spain the hours were for a time also struck according to sandglasses. When the large striking work of the old tower clock in Perpignan broke down once again in 1387, the king ordered his procurator to find two artisans who could work at their craft in the tower. In two shifts they were to strike the public time signal on the large bell according to a sandglass. The king added a copy of a similar arrangement for the cathedral in Perpignan.[205] The same procedure is described in detail and codified in the ordinances for Zaragoza, promulgated by Ferdinand I of Castille in 1414.[206]

Sandglasses appear here as replacement or auxiliary clocks in the early stages of a tower-clock technology that had not yet been perfected. But their importance was far greater. When it came to the introduction of the social use of the new hour-reckoning, and to the testing of new techniques of temporal organization, sandglasses played a role that was at least comparable to that of wheeled clocks and extended far beyond the Late Middle Ages.

The story of the success of the wheeled clock has largely driven the story of the use of the sandglass from our awareness. Through that story, however, we can trace how people learned—along interconnected and yet independent paths—to deal with the sequence of equal hours and with abstract periods of time.

The first small clocks

The technology of the going train and the striking train in the large tower clocks seems to have changed very little during the next two centuries. We can assume, however, that there were improvements to

30. Hannibal dines in Capua. Miniature with a wall clock, from *Histoires romaines de Jean Mansel* (fifteenth century). Paris, Bibliothèque de l'Arsenal, Ms. 5087, folio 221 verso.

and variations in their design, the result of increasing experience in craftsmanship. Technologically closely related are the house and chamber clocks, whose use became customary in wealthy households from the beginning of the fifteenth century, at the latest. In pictures we find them from this time on in some numbers as wall clocks or set on pedestals of stone or wood (fig. 30). Among the oldest examples is a miniature with a golden background in a French manuscript of Heinrich Seuse's *Horloge de Sapience* that was dedicated to Marie de Berry in 1406 (fig. 31). The advances in the construction of astronomical drives were of no consequence for the development of public time indication.[207]

A "portable clock" that was purchased in 1365 for Pope Urban V in Avignon is discernible as an innovation only in terminology; it is the same kind of clock which the French king had his court clockmaker build in 1377.[208] These were transportable clocks, apparently sought-after rarities in those centuries. In 1385, the Duke of Burgundy had such a clock overhauled prior to a trip to Avignon. An order by the infante of Spain in the same year reveals that these clocks had a pedestal and that one expected their mechanism to be such that one could travel

31. Miniature in a manuscript of Heinrich Seuse's *Horologium Sapientiae* (around 1406) dedicated to Marie de Berry. Paris, Bibliothèque Nationale, Ms. 926, folio 113.

with them and stop anywhere. The French king is said to have employed a "clock-carrier" on his travels in 1387–1389.[209]

About a century after the beginning of the spread of hour-striking clocks that could be operated only in one place and were for the most part publicly accessible, a new clockwork drive was developed, one that made possible the construction not only of transportable but of truly portable clocks.[210] The first references, still vague, take us to the end of the fourteenth and the beginning of the fifteenth century. A biography of the builder of the cathedral cupola in Florence, Filippo Brunelleschi (1377–1446), reports that Brunelleschi, a trained goldsmith, had built clocks and alarms in his youth. In the process he had also experimented with all kinds of mainsprings, "varie e diverse generazione di mole."[211] This is the first time that springs are mentioned in connection with the construction of clockworks. Compared to the drive with weights and rope, a wound-up steel spring has the advantage that it releases its energy in any position and even when in motion. The disadvantage is that the force released is very uneven, viz., with the progressive unwinding of the spring it decreases very rapidly.

A clockdrive that was as even as possible thus required not only a steel ribbon — still difficult to make at the time — but also some way of compensating for the decreasing tension of the spring. The solution was the "snail" (fusee wheel), a conical drum with a spiral winding which took up a cord or chain that was fastened to the spring.[212] With this arrangement the decreasing force of the spring is transferred to the fuse wheel with increasing mechanical advantage and the torque is evened out. Despite numerous scholarly efforts, we do not know when and where the combination of spring and fusee as a clockwork drive was developed. Once again the sketches of Leonardo da Vinci support the conjecture that the solution later in common use was the result of numerous and varied experiments.[213]

As the "Burgundy clock" in the Germanische Nationalmuseum in Nuremberg shows, the technology of spring-drive cum fusee had been perfected by around 1430.[214] As long as this drive for clocks was still a rare novelty, it is recognizable in the texts from the added comment "without weights." For example, in 1459 the French king had five clocks built, among them one "demi orloge doré de fin or sans contre poix."[215] The new and uniformly smaller clockworks were made for different casings.[216] New were the box-shaped table clocks, which were also displayed in the Brussels Seuse manuscript (fig. 24). This drive allowed the construction of clocks that were increasingly smaller in size, and from the middle of the fifteenth century the miniaturization of clocks became a fashionable trend. The astrologer and court clockmaker of the Marchese of Mantua offered in November of 1462 to build a "horologieto" which not merely resembled that of the Duke of Mantua but was even better. Between 1505 and 1507, numerous "horologieti" or "horologini" were built for the same court.[217] We have reports from the court of Lodovico Sforza "il Moro" that around 1490 not only small, portable clocks ("horologini piccoli e portativi") were customary, but also that small chiming clocks were attached to ball-costumes.[218]

This alone is not proof that clocks that could be worn on the person were developed in Italy. It is surely out of the question, however, that around 1510 the famous Nuremberg clockmaker Peter Henlein "invented" the pocket watch, occasionally also called "Nuremberg egg." This has been known for quite some time, but since the legend lives on even in noted reference works and encyclopedias, the issue needs to be addressed once again.[219] What is interesting about this tenacious story is that it does not appear until the end of the nineteenth century. It became firmly entrenched in popular historical consciousness through Carl Spindler's novel *Der Nürnberger Sophokles,* Walter Harlan's tragic drama *Das Nürnberger Ei* (which was made into a film in 1939, with Veit Harlan as director and Heinrich George in the lead role), and finally through the 1942 commemorative stamp bearing the statue of Heinlein issued by the Reich postal service. Heinlein's belated career as an inventor was accompanied by scholarly polemics, the results of which were hardly taken note of, however.[220]

The older sources, in contrast, never explicitly mention Peter Heinlein as the inventor either of portable clocks or of egg-shaped clocks. In 1511–1512, Johannes Cochläus published the *Cosmographia Pomponii Melae* and added his own *Brevis Germaniae Descriptio,* where we read, in the description of Nuremberg:

> Every day they (the craftsmen of Nuremberg) invent finer things. For example, Peter Hele, still a young man, fashions works that even the most learned mathematicians admire: for from only a little bit of iron he makes clocks with many wheels, which, no matter how one might turn them, show and chime the hours for forty hours without any weight, even when carried at the breast or in a handbag.

Johann Neudörfer's 1547 *Nachrichten von den vornehmsten Künstlern und Werkleuten, so innerhalb hundert Jahren in Nürnberg gelebt haben* tells us, "This (Andreas) Heinlein was very nearly the first of those who invented how to put small clocks into little boxes" ("Bisam Köpf").[221] Contemporary appreciative voices thus emphasize that Heinlein made very fine and very small clocks that could be worn on the person and that ran for an astonishing forty hours. In addition, he was supposedly one of the first clockmakers who built clockworks into "Bisamköpfe," small containers of musk fashioned from precious metal and used as fragrances or disinfectants.

Peter Heinlein (born around 1485, died 1542), in 1509 master in the Nuremberg guild of locksmiths, is mentioned in the city's accounts as a supplier of spring-driven clocks ("selbstgehend arologium") and also of a Bisam apple with clockwork ("vergulten pysn Appfel . . . mit einem Oaiologium"). Most of these clocks, which were surely very rare and much sought after, the council gave away as presents. In 1541, Henlein built a tower clock for the castle in Lichtenau, and he also made a name for himself as an expert in astronomical instruments. In no contemporary source was this obviously very talented and renowned clockmaker mentioned by name as an inventor. The "Nuremberg egg" is a philological misunderstanding that goes back to Johann Fischart's translation of Rabelais (1571). We read there of the "Nörnbergischen lebendigen Aeurlein," the last word being a diminutive of the old word "aur" ("clock"). The oval necklace watches, which did not appear until later, led people to read "Eierlein" ("little egg"). In the fifteenth century, Nuremberg, alongside Augsburg, developed into an important center for the manufacture of fine mechanical instruments. What contribution Peter Henlein made to Nuremberg's fame as a center of clockmaking therefore remains unclear.[222]

The rapidly rising production of small timekeepers from the sixteenth century on, the "personalization of time" (D. Landes), opens new chapters in the history of technology, some already written, as well as new, unwritten chapters in the history of time-consciousness.[223]

5

From Prestige Object to Urban Accessory: The Diffusion of Public Clocks in the Fourteenth and Fifteenth Centuries

In his comments on the first public clock in Milan, Galvano Fiamma praised as exceedingly useful for all walks of life not just any time-indication, but the public differentiation of the hours according to their sequential number by means of an automatic mechanism. The reference to the social usefulness of public clocks is notable insofar as it anticipates something contemporaries still had to discover during the second half of the fourteenth century in the cities of Europe.

According to contemporary accounts, the situation had changed completely by the beginning of the fifteenth century. The public clock had now become an attribute of cities; life in the cities was equated with life by the clock.

Social utility could play an effective role in the acceptance and diffusion of this new device only over the medium and long term. That public clocks were useful, and later indispensable, was something that could reveal itself once people were living with them. Here we are concerned, first of all, with factors that, independent of experiences of usefulness, promoted a transformation that was already at the time felt to be an innovation. We are interested in the motives and interests of the actors and in the impulses and stimuli that led in each case to this innovation. In contrast to invention, innovation describes, on the one hand, the datable process of the introduction of something new at a certain place, and, on the other hand, the totality of such events or processes. With regard to geographical aspects and the pace of the process, innovation will also be called, in a broader sense, diffusion.

The introduction of public clocks was not only a technological but also a social innovation. Even though non-automatic striking works were also put into operation at the beginning of the fourteenth century and subsequently remained in use for a long time, the diffusion of public clocks began only after the development of usable hour-striking works. However, this technological development was the prerequisite for the spread of modern hour-reckoning as a new social practice only in a general sense. As a rule we find evidence for this practice in a city only after the installation of a clock. But the new practice also spread independent of the technological prerequisites. Many cities adopted it before the new technology arrived. They did so by making do for a long time with a tower warden clock and manually tripped hour strokes, until such time when the architectural, financial, or personnel conditions existed for the installation of an automatically striking tower clock. In Montpellier, for example, it was not until 1410, relatively late for a city of this size, that a tower clock was acquired, the reason being that the manual striking of the hour, attested since 1398, had proved too unreliable. An hour bell was in operation at the cathedral of Xanten since 1375, at the latest. But only in 1383 do we also hear of a clock. The early striking works were also technically more costly and more susceptible to malfunction than clockworks. If they broke down the public hour signal had to be kept up by other means. We can recall the hour striking with the help of sandglasses in Barcelona, Perpignan, and Zaragoza mentioned in the previous chapter. But such procedures were stopgap measures and very rare. For the discussion that follows I have therefore disregarded the special technical circumstances of the innovative process in individual cities. For example, if the "horglock" is mentioned in the date of a document, as in Eßlingen in 1395, this is considered a case of innovation, regardless of the technical means used and even though the next, comparable time-indication does not appear in this city until several generations later.[1] And while numerous setbacks can be observed in the transition to modern hour-reckoning, I shall deal with the innovation or diffusion of public time-indication as an irreversible process.

All of the notions advocated in the scholarly literature about the factors that were decisive for the spread of public clocks go back, irrespective of many nuances, to the fundamental work of Gustav Bilfinger. These views hold that the striking clocks and modern hour-reckoning became a necessity for the increasingly complex way of life of the city-dwellers, whereby Bilfinger placed special emphasis on the need of urban trade and commerce to coordinate their activities. Striking clocks and modern hour-reckoning were the product of the rising urban bourgeoisie and of the laicization of education it promoted: "Thus the entire evolutionary process of urban life was pushing, with a certain natural inevitability, towards the invention of the striking clock, and it was then born in fact from the needs of urban life."

The social locus of this innovation is not clearly known. However, since the early striking clocks are so far attested only in the urban sphere, Bilfinger's hypotheses are certainly still worth debating. Bilfinger also points out some of the factors favoring the process of innovation in the first phase: "In the countries we are accurately informed about (Italy and the Netherlands), we find as a rule that the impulse for the installation of striking clocks came from princes and wealthy urban communities." We shall briefly recall the other hypotheses I outlined in the introduction. Werner Sombart, formulating it pointedly, interprets the acquisition of public clocks by "modern princes and progressive city administrations" as a sign of the level of mercantile rationality attained in Italy. Marc Bloch, too, relies on Bilfinger's work when he describes public clocks as financed with funds from the urban bourgeoisie. Yves Renouard, in his work on late-medieval merchants and bankers especially in Genoa, Milan, and Florence, subsequently emphasizes modern hour-reckoning as a manifestation of the transformation in mentality of which these groups were the main vehicles. Christian Bec has made the merchants into agents of the transformation in consciousness: "The merchants of the fourteenth century, by building the first public clocks, created the regular, laicized hour, the urban hour, the hour of the opening and closing of shops and workshops."[2]

Jacques Le Goff picks up these hypotheses and attributes to the merchants a need for more precisely measured time. The Church's lack of interest in the striking clocks that Bilfinger noted was reformulated by Sombart into resistance from the Church—something that is today occasionally imaginatively elaborated.[3] In Le Goff this shift in emphasis eventually turns into the now widely used slogan of the new "merchant's time" which supposedly displaced and replaced the old "Church's time." In addition, Le Goff has also pointed out the interests of the merchants in their role as urban employers, calling the communal clocks instruments of their political and economic domination. This naturally gave rise to his conjecture that the main regions where public clocks were to be found may very well coincide with the regions of the crisis-ridden urban textile industry.[4]

Internal urban differentiation and coordination, mercantile interest, employers' interests, princely luxury, and ecclesiastical resistance—with these key words we can characterize the factors that have until now been introduced as significant for the diffusion of public clocks, whereby it should be noted that the hypotheses of the various authors are in each case only partial arguments of more comprehensive statements about the transformation of time-consciousness.

Before we can test these hypotheses against known and newly uncovered sources, it is necessary, first, to define more clearly the object of our inquiry. In terms of the history of technology, only the differentiation between clocks with and without striking work would be of importance. Both kinds of works can be designed larger or smaller in size. The differentiation between public and non-public clocks, however, turns out to be important, because the change in the practice of determining the time of day, and with it one of the aspects of the transformation of consciousness that can be reconstructed, is traceable only in connection with the spread of public clocks and not merely of mechanical clocks as such. This differentiation, the necessity of which was clearly pointed out already by Bilfinger, has repeatedly been overlooked and disregarded in scholarly and popular literature. Notwithstanding unequivocal statements in the sources to the contrary, Pierre Pipelart's "horologium regis" is made into the first public clock in Paris, while the clock in the Strasbourg Minster appears in lists of early tower clocks. Dondi's Astrarium is turned into a church clock.[5] To prevent such fallacies from being passed on, a list of the criteria of differentiation will be followed by a brief review of the sources on the earliest public clocks.

What is a public clock?

The term "public clock" was already used in 1353 by Petrarch to describe an innovation of his own time, but it remained very rare until the end of the fifteenth century.[6] Only those clocks should be called public which indicated visually or aurally the sequence of hours of the full day. The kind of technology used—tower warden clock or tower clock—is not important. "Public" should be taken to refer to any larger number of people, especially city-dwellers, but also members of a princely residence, the neighbors of a monastery, or the members of a university. One or more of the following criteria should be met in the texts, and the content and context of the sources must not exclude any of them: contemporary description of the clock as public, as the communal clock, as a "large clock," and so on;[7] contemporary references to continuous time-indication, for example, to the "twenty-four hours" or "the hours of the day and the night;"[8] a communal construction order or communal financing of construction, maintenance, and repairs; a publicly relevant use of the time signal, for example, in communal statutes or the dating of charters and documents;[9] a public installation site, a high weight of the clockwork or clock-bell, the reach of the time signal.[10]

Even though not all sources can be interpreted with these criteria beyond doubt, their use does yield a picture of the early history of public clocks that is sufficiently consistent and sufficiently secure against the "veto of the sources."

The first public clocks

1307–1308 Orvieto: According to the Riformagioni (council protocols and statutes) of these years, the council decreed the levy of a tax to repair a clock ("ariologium sive campanile") in the town hall, and in the following year it decided to hire a keeper to operate it. We have no information from this time about the technology of the clock and the use of modern hours. The subsequent chronological, and likewise still very early, references to the difficult use of the Italian method of hour-reckoning for the clock on the Torre Maurizio at the construction site of the cathedral (1351) make it likely that its introduction occurred somewhere between these two dates.[11]

1309? Modena: Casting of a bell of 900 kilograms "with which the hours are struck"; in 1343 a clock is set up in the Cathedral.[12]

1317–1318?, 1336 Parma: To signal breaks for wage laborers, a public clock ("quae pulsatur pro horis") is ordered. "Horas pulsare" is an old expression, also used for the ringing of the canonical hours. Still, in

Modena the weight of the clock, and in Parma the context of its use, argue for a public time signal. A clock is clearly indicated in Parma in 1336 when we read: "Another small bell, in the city tower, rang the hours of the day and the night."[13]

1322 Ragusa (Dubrovnik): According to the Riformagioni, the city employed a clock-keeper in this year. Ragusa was at the time under the rule of the Republic of Venice, and during the fourteenth century it got most of its qualified personnel from Italy. It is not entirely clear where the clock warden came from, presumably from Reggio Calabria. It was not until 1389 that the city had a large tower clock constructed by a southern Italian master craftsman.[14]

1325–1344 Valenciennes: In the statutes of the drapers hall, the period of selling is marked off by a clock signal, "arloges, erloges, orloge," or "orloges." When the statutes were reissued in 1344, the clock was also mentioned in a topographical reference. Since there is no evidence in this region for the use of the modern hours, we could also be looking at a bell signal that was tripped at a certain time (in the manner of an alarm). We note, however, that in the communal accounts we find mention, as early as 1352, of a clock in the belfry, and that in 1363 this clock was called "large clock."[15]

1336 Milan: Here the public site, the continuous striking of the hours, and the early use of the modern hours are unequivocally attested. The clock that was built at the initiative of Azzo Visconti thus remains the first securely documented hour-striking tower clock.[16]

1344 Padua: In reference to the tower clock in the palace of the ruling Carrara family, the chronicles emphasize the automatic striking of the twenty-four hours of the day. What is behind the invention ("inventum") mentioned on the commemorative plaque for Jacopo Dondi remains unclear. Very early modern hour-indications complement this picture.[17]

1347 Monza: The clockwork ("orrelorium") that was brought from Milan for the cathedral was built by Magister Johannes, who was in the employ of the Visconti. This Magister was probably Johannes Mutina de Organis mentioned around 1352 as a ducal engineer in Milan. In 1353–1354 he built a clock in Milan which was subsequently brought to the tower of the cathedral in Genoa.[18]

1349? Vicenza: According to a statute for the city's clerics, Matins was to be rung in such a way that the citizens could orient themselves "after the clock."[19]

1352 Trieste: The addenda to the communal statutes provide for the construction of a clock for public use.[20]

1353–1354 Genoa: Giovanni Visconti had annexed Genoa in 1353. At his request Johannes Mutina built the clock for the cathedral tower. This clock was explicitly described as the first communal clock that "strikes every single hour of the day and night."[21]

1353 Florence: The clock in the tower of the Palazzo Vecchio was a communal effort. Gregorio Dati reports around 1400 that the clock's bell could be heard all over the city.[22]

1351–1353 Windsor Castle: The clock in the Great Tower was built in London by a clockmaker from Italy and two assistants and was taken to Windsor by wagon. The hammer of the bell weighed one hundred and sixty pounds.[23]

1353 Avignon: The papal court had several clocks from the beginning of the century and from 1336 on it employed its own clock keeper. In 1353, work was under way on a large clock case which was painted and illuminated on the inside. This was, perhaps, a large clock inside a building, as is sketched in Villard de Honnecourt, or a new type of compact clock tower, possibly like the one that appears in a depiction of the papal palace in a fifteenth-century miniature (fig. 32).[24] The city of Avignon itself did not receive a public clock until 1374–1375, a gift from the pope.

1354? Prague: The existence of a public clock of unknown location is suggested by the reference to a second clock-stroke at evening time in the statutes of a collegiate chapter, and by the mention of the widow of a clockmaker (1359) and of an imperial clockmaker (1361). The adoption of the Italian method of counting the hours in the countries of the Bohemian crown gives reason to suppose that the technology was also imported from Italy.[25]

1356 Bologna: Giovanni da Oléggio from the family of the Visconti arranged for the construction of an hour-striking clock ("arlogium pro horis pulsandis") and the casting of the bell for the tower of the city palace. This clock, too, is explicitly referred to as the first public clock.[26]

1356 Perpignan: King Pedro IV of Aragon's order for the construc-

32. Miniature with a tower clock, from an illuminated manuscript of the Chronicles of Jean Froissart. Paris, Bibliothèque Nationale, Ms. fr. 2645, folio 14 verso.

tion of a clock in the tower of the castle states that "a clock" ("aretlotge") and a bell ("scimbalum") were to be built "for the benefit of all, by the stroke and sound of which clock the hours of the day and the nights can be individually recognized." The king himself saw to it that the project was carried out quickly. According to the surviving accounts, the clockwork weighed about one ton, the clock-bell about three tons. The name of the builder is given as Antonio Bovelli from Avignon, who had previously worked on the Papal buildings as a plumber.[27]

1356? Pisa: Beginning in this year we find several modern hour indications in the chronicle of Rainieri Sardo.[28]

1358? Venice: Giovanni de fu Guidotto de Relogiis received full civic rights ahead of schedule (having lived only twelve instead of the usually required fifteen years in the city). The surname is a likely indication that either he himself or his father had performed meritorious service for some special clock.[29]

1358 Regensburg: Endowment funds were used to pay for a communal building order, the "improvement of the Hours" in the tower at the market square. This notice in the city's treasury accounts, which can no longer be verified, undoubtedly refers to an hour-clock and probably also to a public clock.[30]

1359 Vincennes (residence of the French king near Paris): The dated clock-bell lists the future King Charles V of France as the donor and the person who commissioned it. The clock and the striking of the hours are explicitly mentioned in the bell inscription: "J. J. ma faicte pour orloge suy ordenee les heures."[31]

1360 Siena: The clock for the tower at the Piazza del Campo was evidently a communal undertaking.[32]

This list of the earliest public clocks excludes clocks that are mentioned with figure automata, carillons, and astronomical indications. These clocks represent an older technological tradition and were at this time more widely diffused than the striking tower clocks.[33] When such clocks were installed, not inside princely palaces, but freely accessible inside churches, their indications were meant for a public. Some of these clocks had dials with hour indications long before such dials were installed also on tower clocks.[34] The chimes and automata were set in motion in regular intervals, in some instances probably every hour. Though it was possible to determine the time of day with these clocks, this was at best a secondary purpose. At that time they were not being used as public clocks in accordance with the criteria I have outlined above. Nowhere do we find time indications that can be associated with this type of clock. We cannot detect any connection between the diffusion of these clocks and the spread of modern hour-reckoning. The choice of words in the Strasbourg chronicle makes it clear that the two types of clocks were already at the time clearly differentiated. The automata clock in the minster church, completed in 1354, is described in the chronicle as a "clock with the three kings," whereas the tower clock, which was not installed until 1372, is described according to its function as "time clock . . . which strikes the hours."[35]

We are concerned only with the more recent technological tradition; it alone was of importance for determining the time of day in actual practice. In the subsequent period the two lines of technological development merged; astronomical and calendrical indications and time signals were often linked at the end of the fourteenth century. In 1372, the clock in the cathedral of Frankfurt was given a dial inside the church and a striking bell in the tower.[36] The miniature in the Brussels Seuse manuscript (fig. 24) shows that such an arrangement was technically easy to set up. For the period after 1360, I have therefore assumed that even in cases where the language of the sources is not unequivocal, a coupling with the striking work existed or the time signal was regulated manually according to some other clock. In these cases the earliest mention of a clock has been taken as the date for the first public clock. Where it is possible to differentiate the public clock as an additional or later installation, as, for example, in Strasbourg, its date has been used.

As far as the first phase of diffusion is concerned, a glance at the list of public clocks up to the year 1360 leads instantly to a first finding and an important clue: public clocks and the modern system of hour-reckoning originated in the Italian cities. Petrarch, in the passage I have quoted, speaks explicitly of an innovation that was prevalent in northern Italy. The technology, presumably invented in Italy, was at first exported to other countries only by Italian technicians. Independent of the existence of native clockmakers, Italian tower clock builders were working in Avignon and Perpignan as well as in England. Bohemia, and for a time Austria and the Kingdom of Poland, also adopted the Italian method of hour-reckoning. The places of diffusion are large cities, and a remarkably high number of them are princely residences. The sources generally reveal little about the motivations and impulses behind the spread of this innovation. We can, however, make out a few of the factors that were important in promoting the diffusion.

Diffusion on the initiative of the authorities

The spread of the new technology around the middle of the fourteenth century had been promoted especially by the Visconti, then also by the Carrara, the Este, the king of Aragon, and the pope. Subsequently many other territorial rulers actively promoted its diffusion at their residences and cities.

After installing a clock at Windsor, Edward III of England did the same between 1366 and 1369 at his residences in Queensborough, Sheen, and King's Langley. The first public clock in London was built during the same period in the Palace of Westminster. In the following decades it was maintained by the king.[37]

In addition to his residence at Vincennes, Charles V of France also provided clocks for his palace of St. Pol in Paris and for his residences Beauté-sur-Marne (1377) and Montargis (1380). The first public clock of the city of Paris, the Horloge du Palais, was built on his initiative (1370). The king supported the tower clock of the cathedral of Sens (1375) by sending a bell-caster and the royal clockmaker, Pierre Merlin. In addition he granted a deduction of five hundred livres from the dues of the city. In Noyen he made possible the completion of the city clock in 1379 with a grant of one hundred gold francs. The public clock in Avignon, which the Pope was having built for the city, received support from the king in the form of his royal clockmaker Pierre de Ste. Béate (1374–1375). During the reign of Charles V's successor, the palace of Melun was given a clock, and Pierre Merlin built the large clocks of Angers (1384) and Poitiers (1387). Charles VI lent emphatic support to the construction of a large clock, modeled after the Paris clock, in a bridge tower in Lyons (1381). He also intervened on the side of the citizens of Nîmes, who wanted to place the city clock in the tower of the cathedral against the opposition of the canons (1399). Charles VII continued this tradition of royal initiatives with a gift of two hundred livres for the clock in Blois (1452).[38]

The French princes of blood were not far behind in their activities. The Duke Jean de Berry had the clock of the Cathedral of Bourges built in 1372 for five hundred gold francs, and brought in a clockmaker from the realm for the project. He equipped his residences La Nonette and Mehun-sur-Yêvre with clocks in 1378, and arranged for the building of an enormous clock tower in Poitiers, the capital of his appanage, defraying 70 percent of the cost himself. The city clock in Riom was installed at the request of the Duke (1391), as was that of Niort (1396), where a special tax was approved. The clock of St. Jean-d'Angely (around 1400) was most likely a gift from the duke. In Villefrache-en-Rouergue he supported the clock project with a donation of two hundred francs, in Vierzon the use of an "aide" was approved in 1428 for the city fortifications and the public clock.[39] The picture we get already from the scattered evidence shows the influence that those with political authority had on the diffusion of the new technology. However, the true extent of this active modernization from the top becomes fully apparent only in the very substantial documentation from the realm of Burgundy. Like their peers, the dukes of Burgundy were collectors of valuable and complicated clocks. Several permanently employed clockmakers looked after this collection. They also maintained the clocks in the residences and city palaces, twelve of which can be documented by the beginning of the fifteenth century: Rouvres, Montbard, Villaines, Germolles, Brussels, Bruges, Ghent, Arras, Mâle, La Mantoire, Hesdin, Locquignol, and Le Quesnoy.

Dijon, the capital of the duchy, received its public clock through a spectacular act of plunder. After the victory over the rebellious Flemish in the battle of Roosebeeke (1382), Philip the Bold had the clock of the city of Kortrijk, which was decorated with automata and which the chronicler Froissart praised as one of the most beautiful anywhere, dismantled and taken to Dijon by oxcart. The duke subsidized its rebuilding in the Church of Notre-Dame with one hundred francs, to which the duchess added another fifty. The Jews of Dijon and Italian moneylenders were encouraged to make smaller donations. The larger part of the expenses, however, had to be covered by a special tax levied on the citizens.[40]

In the Netherlandish territories the Burgundian dukes were involved in the installation of public clocks in part as active promoters, in part in their capacity as lords over communal finances. Mons, in the region of Hainaut, did not have a belfry at the time. The duke gave the citizens permission to install the clock in his palace. Afterwards he also paid for half of the expenses of maintaining it. The duchess gave the city of Lens a gift of twenty francs for a clock. In 1394, Aire-sur-Lys received two hundred francs from the duke, also as a gift. This is of interest because the work bell of Aire-sur-Lys that was authorized in 1355 led Jacques Le Goff to call the public clocks the instruments of the merchant class.[41] The duke's donation, which became known after Le Goff's essay had been published, shows that the new "time of the textile workers" could prevail in this city only with strong assistance from the territorial lord. To gifts of this nature were added supportive measures in regard to taxation. In 1396, Sluys was authorized to levy a tax for the building of the town clock; Tamise was allowed to do the same in 1403. Also in 1403, the city of Termonde received permission to sell life annuities to renovate an existing clock.

In the duchy of Burgundy, Beaume, at the request of its residents, was authorized in 1395 to use part of their salt dues for a clock, on the grounds that the clock was of use not only to the city but also to the ducal officials and judges. That was also why the duke urged swift completion of the project. Chalon-sur-Saône, too, arranged its financing in this way with ducal help (1402–1407). In Nevers, the duke in 1399 authorized only the use of one of his buildings; the costs had to be borne by the inhabitants themselves.

Comparable measures can be observed during the entire Burgundian era. John of Burgundy granted the citizens of Auxerre permission in 1457 to move the city clock to one of the fortification towers. The last city to acquire a clock with help from the dukes of Burgundy was Nivelles. The circumstances remind us of the events in Dijon and seem like a kind of revenge for the removal of the clock of Kortrijk. In 1496, Charles the Bold used his festive entrance into Nivelles as duke of Brabant as the occasion to present the city with a large clock with figure automata.[42]

There are many instances in Spain, France, and the Netherlands where rulers large and small followed these examples. The Spanish king authorized a clock and a bell in Barcelona (1392) on the grounds that striking clocks had been invented to allow the citizens of honorable cities to live orderly lives and to call sleepers and idlers to virtuous works.[43] Italian princes continued the active support of urban innovation that the Visconti had begun. Cansignorio, lord of Verona, had a clock and a clock tower built in the city (1371). Pinerolo in Piedmont acquired a public clock in 1379 at the initiative of Count Amadeus VI of Savoy. Prior to that the count had already ordered the procurement of a clock in Chambery; it was installed in a tower of his palace at the expense of the citizens. The clock in Turin was presumably also the result of his initiative. The Duke of Tuscany, Cosimo Medici, donated a clock in Prato in 1571 for the common weal of a section of town, apparently because the town clock could not be heard everywhere.[44]

In many different ways the territorial lords were involved in the acquisition of public clocks in the cities of their realms. The authoritative character of a ruler's support for innovation emerges wherever the authority figure acts in person, or where action is taken explicitly on his orders, "iussu domine." The patronage aspect of this support is evident where gifts or subsidies made the acquisition possible to begin with. The fact that such gifts also came with a political price is not important. What matters is that they influenced the date and possibly also the quality of the acquisition. Where the use of taxes and dues was authorized for financing public clocks, the inhabitants or participants in a city's market paid with their own money; only the levying of taxes and dues was tied to rights reserved to the lord. In the sixteenth century in France, the construction and maintenance of public clocks was dissociated from special authorizations and made a permanent part of the communal budget. At the diet of Blois in 1576, the cities were given the power to dispose directly over appropriations for public buildings from the authorized taxes. The upkeep of clocks was explicitly mentioned as one of the maintenance tasks.[45]

It is striking, moreover, that the financing of public clocks with participation from the territorial lords was customary only in the Netherlands, France, and in a few Italian territories, but not in the cities of the Holy Roman Empire. Participation of the rulers in the investments made by cities in France, Burgundy, and Savoy thus reflected, in one sense, a lower degree of communal autonomy. But the flip side of this was a closer financial dependence and a closer political alliance, which, at least in the smaller cities, created favorable conditions for innovations that were initiated at the top.

The taxes authorized by a territorial lord were used not only to build the public clock. Frequently they were also used to finance, in conjunction with the clock, fortifications, guard and church towers, the repair of damage caused by fire and war—in other words, a whole range of communal investments in infrastructure. At least in the Burgundian cities it was even customary that small cities received support from the ducal treasury for such projects.[46]

To be sure, the acquisition of a public clock was not only a utilitarian undertaking. A clock was also a prestigious project that brought renown to the city as well as its lord. Even though we find only occasional explicit statements to this effect in our sources, the frequent mention of public clocks in the chronicles of the cities and those of the ruling houses makes this sufficiently clear. Even if a part of the authorizations went back to requests and petitions submitted by the cities, the decisions made by the lords and rulers were surely influenced or accelerated by the projected installation of a public clock.

The prestige value of these installations is particularly evident in a specific kind of permission granted by the territorial lords, in which financial considerations played no role. Since the public clock should be audible to everybody, it was, where possible, installed in a communal tower. In cases where the time signal was not struck on the communal bell itself, the hour bell was usually installed next to it. The typical location of the public clock was thus from the beginning a politically highly sensitive site. The communal tower and the communal clock were legal and symbolic expressions of communal autonomy and from the thirteenth century on the center of the city's system of signals (see chap. 7). Communal rebellions against the territorial lord—often initiated by the ringing of the public bell—were frequently punished in France and the Netherlands by the loss of the right to a tower and a bell and often also by the destruction of these symbols. From the second half of the fourteenth century, justifications for restoring the right to towers and bells mention public clocks as an argument, one in which pragmatic considerations protect or conceal events of great symbolic importance.

The residents of Cambrai had revolted against their episcopal lord at the beginning of the thirteenth century. In response the bishop had punished the city by destroying the belfry and the communal bell.[47] In practical terms the citizens could no longer ring the bell for rebellious gatherings; they had been symbolically humiliated. In their struggle against the bishop who denied them a tower and a bell they sought help from the Roman Emperor Wenceslas. Their efforts were successful. In 1395, Wenceslas granted them permission to build a belfry with a tower clock. The bishop's resistance to this undertaking was broken only at the beginning of the sixteenth century, when Maximilian I renewed this permission.

In 1388, the citizens of Béthune in Flanders petitioned the Count of Namur for permission to build a new belfry of stone, assuring him that only the town bells and a striking clock, the sort that was common in the regions' other cities of renown ("bonnes villes"), would be installed there. Following a civic uprising in Rouen, the belfry had been torn down in 1382 and the bell of the community had been taken away.

In 1389 the citizens requested that the permission for a belfry and a bell be restored, on the grounds that they wished to install a clock there. The king granted both requests, and the city chronicle reports that construction for the "grosse orloge de Rouen" began the very same year. Augen, too, had lost its bells after an uprising. When the French king reauthorized the chiming of the clock, the consuls assured him that only two bells existed, one for the clock and one for the school.[48]

Its connection with the city tower and the city bell made the public clock a political object within the lord's sphere of responsibility. This also gave rise to unspoken obligations. As Pope Pius II, Enea Silvio Piccolomini began, in 1459, to rebuild the Tuscan village of Carignano, which belonged to his family, into a modern city. Under the name Pienza there arose a Renaissance city whose plan was based on a uniform aesthetic and urban concept. Around a piazza, Pius II arranged for the building of a cathedral, an episcopal palace, a palace for the Piccolomini family, and finally a communal palace with a tower for the city's bells and a clock. The clock was built in Rome and taken to Pienza. The lord of the city wanted to furnish the town with everything, and in the middle of the fifteenth century that included a public clock, a detail the papal builder expressly mentioned in his autobiography.[49]

Competition for prestige and the clock as ornament of the city: Communal initiatives

The overall contribution that territorial lords great and small made to the diffusion of public clocks in the cities of Europe was undoubtedly impressive. Still, even in regions where such activities are very densely documented they were not the rule. In the vast majority of cases we must assume that the impetus for the acquisition of public clocks came from the communities themselves. The relatively few texts that allow any kind of inference about the nature of communal initiatives confirm what has been variously conjectured: inter-city competition for prestige was a motive of considerable importance behind the drive to get a clock.[50] In addition, we can see the considerable contribution that powerful ecclesiastical institutions made in large cities, and the no less important participation of the churches through cooperative behavior nearly everywhere.

Competition for prestige means that a city acquired a public clock because other cities already had one. It also means that the expenditures involved with the clock were intended to express the rank of the city as compared to the capital or to neighboring cities of a region. Undoubtedly the territorial lords also competed amongst one another for prestige when they helped their cities to acquire what was at times a particularly costly clock; however, the competitive motive is specifically addressed only in the cities.

The earliest testimony to this ambitious race to obtain a prestigious innovation is an addendum—noticeable for its peculiar linguistic pathos—to the dry supplements (comprising hundreds of articles) to the statutes of the city of Ivrea in Piemont, dating to the year 1368. In an almost excited tone, the city boasts that it is distinguished over all the other cities by the light of virtue. In order that the city be fortified also by the light of understanding and honor, one hundred gold gulden, and more if necessary, should be set aside to acquire a striking clock, a "horologium pulsatile." All this should happen quickly and without the least delay.

With less pathos, though speaking more clearly to the point, the small town of Schweidnitz in Silesia tried, as early as 1370 (the year in which Paris obtained the first public clock), to outdo the far larger and wealthier city of Breslau. Representatives of the city council requested from a Breslau clockmaker "a clock equal to the one in Breslau or better." In 1391, Lucca commissioned its clockmaker to fashion the clock for the Palazzo Pubblico equal to or better than the clock in Pisa. In San Gimignano, the council decided in 1410 that the city should receive a clock just as the other, better towns of the region had. Padua did not wish to be outdone by Treviso when it installed a new clock with elaborate indications. And for all the cities in the region of Venezia, the public clock in Venice was the model to be imitated.[51] The clockmaker in Namur in 1393 demanded an especially high price if the clock was to be as good as the clocks in Brussels and Mons. In Aardenburg a clock was commissioned that was to be as good, as beautiful, and as heavy as that in the neighboring city of Sluys (1397).

The Horloge du Palais that was built in 1370 at the initiative of

Charles V in a corner tower of the Louvre by the Pont au Change, while it was not the first public clock of the realm, was for a long time the influential model in France. In 1381, a clock like the one in Paris was to be built in Lyons in a bridge tower at the Saône river, as a contribution to the common weal of the city and the surrounding territory. When the clock in the cathedral of Chartres was rebuilt in 1392 with the addition of a striking work, the chapter agreed that the clock should become "like the one of the Palace of Paris." At the beginning of the sixteenth century, a stone pavilion was erected next to the cathedral. It was to house the clockwork and be equipped with a dial on the outside. The contract with the bell-caster bears out the continuing rivalry with the model in Paris: he was to make the bell "after the manner of the clock in the Palais in Paris or better. The sound shall be equally as beautiful and harmonious, in the proportion and amount of the metals the bell shall be heavier than that in Paris."[52]

No city wanted to be outdone by anyone else. In justifying a salt tax for the city clock in 1410, Montpellier declared that a French city of some renown simply owed it to itself to have an automatic striking clock. In Holland in the sixteenth century, the clock of the Groote Kerk in The Hague was the measure of all things. In Amsterdam a similar one was demanded in 1516 for the tower of city hall; the inhabitants of Delft wanted one for their town hall that was in every respect better (1570).

Such battles for prestige can be observed also within regions. In 1557, Montélimar commissioned the installation of a costly clock with figurative automata that was to be like the city clock in Romans. The fact that the clock in Romans had very nearly ruined that city's finances was ignored. Clocks became the object of competition for prestige even within some cities. In Reggio Emilia, Bartolo de Comadri made a commitment to the monastery of S. Maria delle Grazie in 1463 to build a clock that was as beautiful as the one his brother-in-law Paolo de Raineriis had constructed for the church of S. Domenico.[53]

The territorial lords lent their support to this "urbanization for reasons of prestige" (B. Chevalier). The citizens of Rennes in Brittany had decided, quite late (1467), to build a clock tower. In the process a quarrel had arisen in the city over how much should be spent on the bell. The contending parties appealed to the duke, who decided that they should make the largest bell they possibly could. The scale of the ambitious project led to technical setbacks during construction, but after a few years Rennes had one of the most famous clock towers and probably the heaviest clock bell in France.[54]

In texts that give reasons for and justify public clocks, there is frequent reference to the common welfare and the general utility of public time indication. Accordingly the entire city was always treated as the intended audience of the time signal. The sole exception is the clock of Preßburg in Slovakia, where a clock with three dials is said to have been built in 1410. According to the account we have, the dial was omitted on the side facing the ghetto, because the Jews of the city did not want to contribute to the costs for the clock. In the small town of Coulommiers not far from Paris, the clock was described not only as practical, but also as a highly useful investment in the future.[55] However, the reference to the common welfare is also an old formula that legitimated public investments against potential objections from special interests, and justified the right to pass the costs on to the general public.[56] It was irrelevant whether these investments came in response to initiatives of the territorial lord or at the "urging of the citizens." The old formulas had long been used for all kinds of public building projects (roads, bridges, fortifications, wells, or canals), and when it came to the hiring of communal doctors or teachers. The clock was the "urban investment par excellence" (P. Wolff), and since it combined practical and symbolic utility the cities usually spared no expense.[57]

"Communal initiative" does not mean that the impetus always came from the political community. Initiative or support in the cities came from the chapters of important cathedrals and from large monasteries more frequently even than from territorial lords with their urban residences. This does not reveal a special interest distinct from that of the community; rather, it reflects the continuing political and economic importance of these ecclesiastical institutions in the urban sphere. In England, cathedral chapters and monasteries nearly everywhere took the place of the poorly developed communal institutions, which in the fourteenth century barely had any representative buildings of their own.

The influence of wealthy chapters and monasteries is even more palpable in the larger French cities. In many cases they arranged for the installation of a public clock before the civic community did so. This is securely attested in Reims, Chartres, and Tours; it is likely for

33. Miniature of the town hall square in Lucerne with the tower clock installed in 1408. Diebold Schilling's *Luzerner Chronik* (1513).

Rouen, Autun, Lyons, and Troyes. To this we must add cities in which the territorial lords did not support the cities as such, but the cathedrals, in making these acquisitions: for example in Bourges, Sens, and Angers. In the Holy Roman Empire, too, bishops and cathedral chapters procured public clocks or supported the cities in doing so: for example in Mainz, Minden, and Xanten.[58] At times this, too, could lead to intra-city competition for prestige. According to a chronicle of the city of Konstanz, the canons of the cathedral angrily suspended their project in 1438 because the bishop had beat them to it: "[T]hen the clock and the dial at Saint Stephan's was made by Count Ott von Röteln, who was the bishop. And the canons of the cathedral had also begun a clock . . . which they immediately abandoned angrily, because Count Ott had placed the clock in Saint Stephan's."[59]

In cities large and small, churches expressed their cooperative attitude by making buildings or bells available, and not infrequently also by their willingness to defray a part of the expenses for constructing or maintaining the clock. Conflicts over the public clock in ecclesiastical buildings were very rare and never concerned the clock itself or the time signal. In many cases the cooperation was fixed in contracts and agreements that outlined the rights and duties of a church and the political community. In these contracts and agreements, mention is made of the fact that the public clock is useful also to ecclesiastical life. Still, the political community was always regarded as the party most interested in the clock and as the principal target of the time signal. Ecclesiastical and territorial law during the following centuries adhered to the principle that the primary beneficiaries of public clocks were not the churches but the communities.[60]

A typical example for the form that cooperation between church and community took in the installation of a public clock comes from Montreuil-sur-Mer. In a document from the year 1377, the abbot of the monastery of St. Sauve authorized the citizens to rebuild one of the towers of the church for the clock in such a way that the tower could be entered without disturbing the life in the monastery. Another condition was that the clock bell not be rung for other purposes. In their petition to the abbot, the citizens of this small textile town had made reference to the usefulness of the hour indication for the common welfare, and the abbot had accepted this argument. What makes this document special is the opening letter "A," which has been designed as a small vignette; it contains what is probably the earliest depiction of a public clock (fig. 34).[61] Above the tower housing the stylized clock with a bell and a jack in the upper level hovers a birdlike demon, who threatens on a speech ribbon to topple the tower. In the

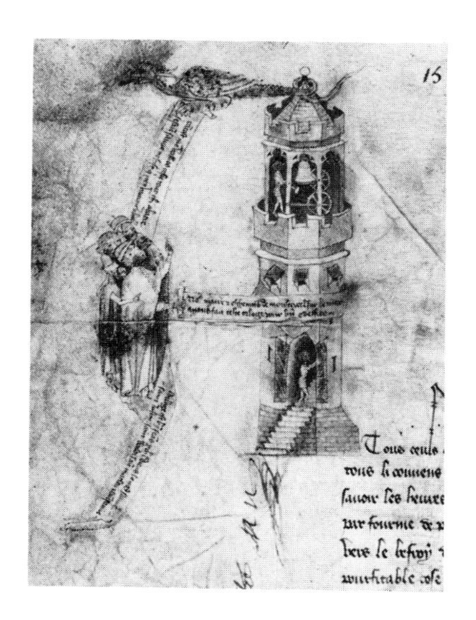

34. Initial "A" with a tower clock, in a document from Montreuil-sur-Mer, 1377. Paris, Archives Nationales, KK 532 no. 2ter.

left stroke of the letter we can make out the "échevins" (assessors) who are entrusting the well-built clock to the protection of the saints.

As a prestige-enhancing project, the public clock quickly became one of the features that distinguished a city as a city and was thus part of the urban decor. From the beginning of the fifteenth century, at the latest, possession of a public clock was part of a city's self-identity—this, too, was a factor that promoted its diffusion. The expensive design of the public clock increased a city's renown and bolstered its self-confidence.

In a case before the parliament of Paris, it was said in regard to the Horloge de Palais that King Charles V had it built for the "ornamentation of our city." A chronicle from Magdeburg described the purpose of a new clock in city hall as follows (1425): "for the honor of the city and the utility and comfort of the citizens." The large new clock in the city hall of Avignon, the installation of which began in 1461, was also intended to increase the "honor and beauty" of the city.[62]

We can trace the process that made the clock part of the urban decor much more clearly and far more frequently in the pictorial sources than we can in the texts. From the beginning of the fifteenth century, we find naturalistic city views with realistic depictions in miniatures, on altar panels, and in oil paintings. At the same time it became customary in the large cities to equip the public clocks with dials. The acoustic time indication was now joined by a visual indication. It is evident that this was a process that was independent of the diffusion of public clocks and began later. In Paris, the keeper of the Horloge du Palais complained in 1419 that the dial that was installed added to his workload. In Siena it was not until 1424 that a dial "que mostra le ore del giorno" was built for the clock that dated back to 1360. In Oudenburg, the clock from the year 1383 was given a "a dial with a hand indicating the hours" in 1402. Lüneburg had a public clock since 1379; in 1445 a "dial on the clocktower" was added.[63]

Miniaturists and painters often documented this new addition to the cityscape with noticeable pride. At the same time it became elevated to the attribute of the ideal city. The iconographic evidence is not mere illustration; rather it supplements or replaces written sources in regard to many otherwise unknown topographical and architectural details. It also bears out the news-value of public clocks and the place they held in the cityscape.

In an altarpiece from the school of Avignon, the dial of the cathedral

35. Miniature with a tower clock. From *Histoire de St. Mitre,* attributed to Nikolaus Froment (end of the fifteenth century). Cathedral of St. Sauveur, Aix-en-Provence. Photo: Ministère de la Culture, Inventaire général des monuments et des richesses artistiques de la France.

in Aix-en-Provence dominates the depiction of a city square (fig. 35). The only other information we have about a public clock in Aix comes only from a very brief mention in a later text.[64] The famous Horloge du Palais appears in a fifteenth-century miniature because the dial in the picture has been enlarged, moved higher, and shifted to a broad side of the tower at the old Louvre. Along with it we see the dial of a second, unidentified clock (St. Eustache? St. Paul?).[65]

The prominence given to the clock in the cityscape is clearer still in two miniatures (figs. 36, 37), which are better grouped among the idealizing city views. A fifteenth-century Paris manuscript of the Chronicle of Jean Froissart contains the scene of the execution of Guillaume des Pommiers and his secretary in Bordeaux in 1375. The miniaturist has taken a real city (Paris? Bordeaux?) as his model for the depiction of a typical city. In the picture of a Cologne painter, the panorama of Cologne becomes the backdrop for the intermingling of real and idealized elements. The oversized dial on the spire of St. Lorenz that dominates the center of the picture may have belonged to a clock for which all textual evidence has been lost. Several of the depictions of the martyrdom of St. Ursula and of the 11,000 virgins, which was popular in the fifteenth century, show similar scenery (fig. 38).

Realistic as well as stylizing city views are found in the woodcuts to

36. Entrance of Isabella of Bavaria into Paris in 1398, from an illuminated manuscript of the Chronicles of Jean Froissart (fifteenth century). London, British Library, Ms. Harley 4379, folio 3.

Hartmann Schedel's "Chronicle of the World." The view of Erfurt is evidently based on a good model (fig. 39). However, the artist uses dials also as architectural props in imagined city views. The woodcut used to illustrate Ferrara (fig. 40) is also found in descriptions of Damascus, Perugia, Siena, Mantua, Naples, Verona, and Krain in Carinthia.

Like towers and walls, the public clock had become an urban attribute. Beginning in the fifteenth century it was also a permanent ele-

37. Execution scene in an illuminated manuscript of the Chronicles of Jean Froissart (fifteenth century). Paris, Bibliothèque Nationale, Ms. fr. 2644, folio 1.

38. Arrival and martyrdom of St. Ursula in Cologne (detail). Cologne Master (around 1450–60). Cologne, Wallraf-Richartz-Museum. Photo: Rheinisches Bildarchiv.

39, 40. Views of Erfurt and an imaginary city in Hartmann Schedel's *Weltchronik* (1497). Munich, Bayerische Staatsbibliothek.

ment of city iconography in the depictions of cityscapes. The October miniature in the *Breviarium Grimani,* a work that was produced in the Netherlands, uses strong colorization to emphasize the clock amidst the grey-blue hue of the city tower and walls (fig. 41). At the same time the clock is singled out as one of the characteristics that distinguished life in the city from life in the surrounding countryside. The pictorial sources confirm the observation of the Black Friar: "In cities and towns men rule themselves by the clock."

Administrative diffusion — Clocks in villages — The public clock as the object of "good administration" ("Gute Polizei")

Beginning in the fifteenth century, the diffusion of public clocks through direct or indirect pressure from territorial administrations became the customary pattern. Initially this third form of active promotion of innovation supplemented the continuing, specific supportive measures by territorial lords and the acquisitions of communes that were driven by motives of prestige. It was aimed more at small towns and villages than at cities, most of which had by now acquired clocks. Only in these small towns and villages did this supportive policy occasionally encounter delaying or recalcitrant reactions from their inhabitants; as a result, these residents were, right up to the nineteenth century, the recipients of special administrative attention and admonitions to maintain the clocks, and the targets of administrative supervision.

The diffusion of clocks through the actions of administrations resembles the previously described measures of territorial lords in its effect, meaning that clocks were installed in places that would not have procured clocks, or not at this particular time, without such support. The policy of the administrations differed from that of the territorial lords in that it was not aimed at specific places. Instead, it encompassed all localities of a region or tried, by means of general regulations, to diffuse public clocks to all localities within a given area.

Once again the first evidence comes from northern Italy, where the territorial lords' policy of procuring clocks seems to have taken on systematic features early on. Initially the policy concerned small localities that were distinguished in some way as military strongpoints or princely residences. In 1422, the Marchese d'Este, lord of Ferrara, sent a clockmaker to Fossadálbero with the order to build several clocks. A

41. Month of October in the *Breviarium Grimani* (beginning of the sixteenth century). Venice, Biblioteca Marciana, Ms. lat. XI 67, folio 10 verso.

petition from a clockkeeper in Castelnovo Bariano reached him in 1436: the man had fallen and injured himself during his nightly check of the clock in the tower, and was asking for overdue pay. In the same year the marchese commissioned the building of a clock for his castle in Voghenza. During the second half of the fifteenth century, the Gonzagas of Mantua saw to the procurement, in some form or another, of clocks in Volta, Goito, Quistello, Bozzolo, Canneto, Marcaria, and Isola Dovarese.[66]

Around 1473, the administration of the Milanese dukes of the family of the Sforza undertook, in the region around Novara, a new kind of systematic clock procurement for small and very small localities that were not in any way connected to strongpoints or residences. Preserved in the ducal archives is a partial list recording all localities in this region that have a clock, those that are to acquire a clock, and those that are supposed to contribute to its financing.[67] If we plot this list onto a map, it becomes clear that the project was intended to provide complete geographical coverage. The average distance between the sites where clocks were located was to be less than five kilometers. The time signal would have reached the entire region around Novara. However, as the financing scheme reveals, the rationality of this plan had its limitations. Individual places were supposed to help pay for clocks that were not geographically closest to them; a few were to pay for clocks they could by no means have heard. Moreover, communal boundaries were irrelevant. The methodical treatment of the area was evidently thwarted by traditional situations of ownership and dependencies. Unfortunately we can no longer determine whether the planned clocks were actually built, and whether such administrative procurement measures were undertaken more often—which would be even more revealing.

As was true for cities in the fourteenth century, relatively wealthy villages in the fifteenth century also acquired clocks on their own initiative. Such acquisitions probably began around the middle of the fifteenth century. From this time clocks appear in small cities and villages. In 1445, Boussoit-sur-Haine in Hainaut financed a clock from its own funds. Sources from Ghent reveal that Nazareth, Lede, Waarschoot, and Evergem employed clockmakers from this city.[68] We are hardly in a position to say anything about the frequency and extent of such acquisitions; because of the particularly unfavorable survival of sources, the information we have is simply too widely dispersed. However, there are indications that Italy in the fifteenth and sixteenth centuries had fallen behind the northwestern countries. Antonio de Beatis, secretary to Cardinal Luigi of Aragon, described Germany and Flanders in his travel account (1517–1518) as regions in which even the smallest collections of peasant houses had "quadrants to indicate the hours without the sun," though these did not strike. L. Guiccardini repeated this observation in his description of the Netherlands (1567): not only cities, but villages, hamlets, and the better houses, too, had clocks. Both men evidently expressed in this way their experience of something that had no counterpart, at least in Italy. In the account of his journey to the baths (1580–1581), Michel de Montaigne, in turn, highlighted Switzerland as a region where even the smallest church possessed a splendid clock. He, too, complained about the lack of public clocks in many parts of Italy.[69] Montaigne's impression of Switzerland is more than confirmed by the miniatures in the Lucerne Chronicle of Diebold Schilling. Schilling depicts not only a number of urban clocks, but on several occasions also village clocks, especially in the Alsace and the Vaud region, which he considered to be particularly wealthy (fig. 42). To be sure, in some instances the documentary value of the miniatures is more than questionable.[70] Still, such testimony shows that around 1500 public clocks were no longer something unusual in the villages of northwestern Europe, but were considered part of the decor of a wealthy village.

In the heartland of the German Empire we can trace the spread of public clocks in the small towns and villages only from the sixteenth century on: at first through scattered references in reports of the destruction wrought by the Peasants' War, later and in much greater frequency in the visitation decrees from Protestant areas. According to these ecclesiastical records the visitors checked not only the punctuality of the chiming and the services, but the duration of the sermon as well. Divine services and school statutes frequently presuppose the existence of striking clocks and sandglasses, and the presence and condition of these clocks was also routinely inspected. Countless admonitions to the sextons and schoolmasters to be diligent about "setting the clock," and regulations concerning their pay—in the cities it came "from the

42. The village of Brunnstadt in the Alsace, Diebold Schilling's *Luzerner Chronik* (1513).

council," in the villages "from the communal treasury"—put it beyond doubt that the vast majority of villages had at least one simple striking clock as early as the sixteenth century.[71]

Inspections of the existing clocks in villages were also done in the Catholic areas. For example, after the Thirty Years' War, a visitation account from the Sundgau (1647) records village by village the condition of the bells and clocks in rigid formulas such as "campanae cum horologio perditae" ("bells and clock ruined"), or "campanae 3 adsunt sed horologium destructum" ("three bells are present but the clock is destroyed").[72] Occasionally villagers had to be forced to pay for the maintenance of the clocks. For example, we read the following in a visitation decree in Arzberg (1618):

> Some years ago in Arzberg they had a clock made which strikes a bell. The residents of Nichtewitz and Kaucklitz are supposed to contribute, the owners of a hide of land 1/2 Reichstaler, gardners 1 local taler, but they are unwilling to

> do so. Previously their excuse was that they couldn't hear the clock. Now they have admitted that they can hear it but still they don't want to pay: they are to pay their share; where they don't the authorities shall make them.[73]

Protestant territorial sovereigns, as holders of the highest ecclesiastical authority, supported these efforts with general territorial and ecclesiastical statutes. The Ecclesiastical Constitution of Saxony that was passed in 1580 (and adopted for Lower Saxony in 1585) required the acquisition of a clock, or at least a sandglass, by territorial law. As the immediate purpose the Constitution mentions church services, but beneficial effects were expected also in other spheres of housekeeping:

> §39. And in villages without a clock, the pastor should admonish the church, and in particular the people who can afford it, to buy one, so that the church-offices can be carried out at the appropriate time in accord with the clock, and the people in other respects, too, should be guided by it in their housekeeping.
>
> But if the parishioners are so poor that they cannot buy a striking clock, the pastor shall give thought to a sundial, which can be obtained at little cost. And until it is installed, the sextons shall learn from the pastor who has a compass, or purchase it themselves, and the sexton shall use it to determine the ringing.[74]

"Portae, pulsus, pueri" ("gates, bell strokes, children") — thus went a familiar saying attributed to Charles V in the early modern period. Solid and guarded walls and gates, proper schools, and orderly bell strokes would thus have been for the emperor characteristics of a well-governed city. However, "pulsus" has also been interpreted to mean that in every city the emperor visited his first concern was the functioning of the public clock. Both variations are frequently attested, even though the origin of the formula can no longer be ascertained.[75]

The reported saying of Charles V was not just a casual wordplay, for around 1400, schools and public clocks were relatively new, and around 1500, relatively typical elements of urban life. The emperor, a famous collector of clocks, may well have occasionally paid attention to the functioning of the public clocks. More revealing is the symbolic equating of functioning clock, orderly administration, and "law and order." Here expectations and experiences with the clock are condensed into an image whose traces we can follow back to the early years of public clocks.

In the sources on the Horloge du Palais in Paris we read that the French king had the clock built to promote the orderly running of the bureaucracy and the orderly life of the citizens: "ut nostra parlamenti curia et habitans ipsius ville melius se regerent et regularent."[76] This can be understood in very practical terms, but like the comments of the English Black Friar quoted at the opening of the book, the reflexive use of "regere" and "regulare" has ethically prescriptive overtones. Beyond the experience of practical usefulness, the clock became an image for orderly political life. This becomes clear, for example, when kings and princes, in anticipation of their visits, admonished cities to pay attention to the running of the public clocks.[77] For territorial lords and the cities, clocks had become a sign of the quality of the government. In Lyons, the need for a usable communal clock was justified with the argument that honorable citizens, merchants, and visitors needed it for a regulated life ("vivre plus reglement"). The clock was a characteristic of well-governed cities ("bien reglez et gouvernez par bonne police") both large and small in France and in other countries.[78]

In the interest of church services and the schools, the Ecclesiastical Constitutions from the sixteenth century on placed public clocks and their running in the care of the territorial administration. The eighteenth-century sciences of public finance (cameralistics) and good administration ("Polizeiwissenschaft") continued this policy. According to these texts, the "division of the time" served the great variety of civic businesses and the ordering of work and leisure (German "Zeitverkürzung," lit., "shortening of time"). The synchronous running of the clocks became the object of "Stadtpolicey," "good urban administration." The cameralistic texts do not mention that the demands of the postal system and the transportation system in the eighteenth century had long since aroused a new interest on the part of the states in public clocks and their running. But they do repeat the old arguments when they highlight the practical and decorative function of clocks. In the eighteenth century, as well, the procurement of public clocks was seen in small towns and villages as something that required administrative help from above: "It becomes plainly apparent that a clock has truly become one of the necessities of our life. Moreover, one would consider it an essential defect of good administration if even in the smallest cities . . . one were to look in vain for a striking clock. Even in somewhat wealthier villages such a clock . . . is an indispensable item, a credit and convenience to them."[79]

The clock boom in the fourteenth century

From the very beginning, the installation of a public clock was considered a sign of a city's openness to innovation, of its wealth and the vigor of its administration. We would not be reading too much into the sources by saying that the public clock was from the outset treated implicitly as an index of urban modernization. This is already born out by the frequent mention of these clocks in the chronicles of cities and rulers. And it applies not only to the early public clocks in the Italian cities, but also to small cities that acquired clocks only relatively late. For example, we can still sense the feeling of pride in the newly acquired object in the city chronicle of Erkelenz in the Rhineland: "In the year 1406 the first clock was built at a cost of 15 Rhenish gulden."[80]

Modern urban historiography has taken over this attitude, in part in casual formulations like "already in the year . . . ," in part very explicitly by describing the process of acquiring a clock as "early" or "late." N. van Werveke was one of the first to describe the clock of the city of Luxembourg. He tells us that it was acquired already in 1390 and thus at a time when only a handful even of the larger cities could boast of having one. A. Rigaudière calls the installation of the clock in Saint-Flor in 1387 an "événement capital" for the urban landscape of the Auvergne. As for the cities of Brittany, for which we have dates only from the beginning of the fifteenth century, Leguay notes explicitly that this region lagged behind the rest of France.[81] Hitherto the problem with such statements was not whether they were right or wrong in individual cases, but that there were no criteria for judgments such as "early," "late," "fast," or "slow." The accumulation of dates, which has become possible also because contemporary as well as modern urban historiography have paid so much attention to public clocks, is one way of remedying this problem. A statistical analysis of the established dates when public clocks were acquired can shed light above all on the temporal structure of the process of diffusion. Moreover, this will also allow us to determine—and possibly explain—regional peculiarities, accelerations, or delays.

I have been able to assemble just under five hundred dated mentions containing first references to public clocks for specific localities for the period 1300 to 1450. These dates were screened using the criteria for public clocks suggested above. The count, however, was of cities, not of public clocks, with each city appearing only once under the earliest date, regardless of the nature and number of its public clocks. Each date stands for one of many very different processes. Only in a minority of the cases is it possible to unequivocally assign the installation of the clock to a specific year. The sources that allow us to do so include above all chronicles that report the completion of the work, contracts with clockmakers, final accounts, and to a lesser extent, dated bell inscriptions. In the sources we can see cities involved in years of planning and preparation, but also cities and towns that took advantage of sheer coincidence. One such example is the town of Tulln in Lower Austria. Konrad, the secretary of this small town, had been bludgeoned to death in the year 1372 by the smith and clockmaker Niclas Swaelbl of Breslau. Following the trial, the clockmaker had to swear an expiatory oath (the "Urfehde"). The judge and council imposed on him as penance a personal pilgrimage to Rome. In addition, he was to build a clock with a striking work ("arloy das an sich selber slach") for the parish church of the town. If both things were accomplished by Michaelmas (Sept. 29) of the following year, Niclas Swaelbl would go free. We don't know what had brought the clockmaker to Tulln in the first place. Evidently he did build the clock, for in the following year he was back at work in Breslau.[82] The city fathers of Tulln used an unusual opportunity to secure the services of an expert, and to procure a striking clock for their city a mere two years after the metropolis of Paris had done so. In nearby Vienna we hear nothing of a public clock at this time.

In none of the cases can we say anything about the interval—important for some theories of innovation—between the spread of information about an innovation and its adoption. Most of the dated reference are only termini ante quem. They included notices of repairs and restorations, maintenance costs in city accounts, wages for clockkeepers and oaths of clockkeepers, mention of public clocks in the dating of documents and statutes, modern hour indications, and, finally, iconographic evidence. Here the date counted and the date of installation might at times be separated by decades. All these cases shift the overall chronological picture. Structural changes in a class of potentially relevant sources have the same effect. Here special importance attaches to the rapid growth of written communal administration, in particular the spread of urban accounting in the second half of the fourteenth century. Circumstances relevant to our question thus lead to considerable chronological distortions. For example, the small cities of Culemborg (1374), Duderstadt (1397), Pegau/Saxony (1399), and the large cities of Leiden (1390), Boulogne-sur-Mer (1415–1416), and Rotterdam (1426) appear in our lists, not with the date of their first public clock, but with the date of the oldest extant city accounts, which mention, as one would expect, also the public clock.[83] Thus the following chronological framework of the diffusion of public clocks should be seen as no more than a cautious assessment.

The chronology and geography of diffusion

The findings for the early period thus confirm once more what was already evident from the list of the first public clocks: the diffusion began between 1300 and 1350, with increasing pace in the larger cities of northern Italy. Between 1350 and 1360 the new technology was exported from Italy to large European residences.

In the subsequent period (to about 1370), we can trace the diffusion of the technology in the German Empire (Augsburg, Nuremberg, Munich, Frankfurt, Breslau, Zurich), in the Netherlands (Brussels, Utrecht, Deventer), and in France (St. Quentin, Paris). The residences of territorial lords became less important. During this decade, at the latest, a transition took place from the outside expert who was brought in to the resident builder of tower clocks, a transition that was decisive for the spread of the new technology. Konrad von Kloten was building clocks in Zurich, Master Swaelbl in Breslau, Schweidnitz, Troppau, and Tulln. But certainly not every large city had local builders of tower clocks. The king of England had three clockmakers brought over from Delft, an expert was brought to Paris from Germany, and St. Quentin had "maistre Pierre des Olloges" brought in, presumably from the region of Lille-Valenciennes.[84] However, we have hardly any clues to the paths of diffusion for this critical transitional period. Above all, it is unclear where the clockmakers working in Prague, in the large cities of the empire, and in the Netherlands acquired their knowledge and skill. Though the sources from the cities of the empire as a whole are striking in terms of volume, in each individual case they are very meager, and for Cologne, for example, they are missing altogether. The "Nuremberg clock," as an intermediary form between the Italian method of counting the hours and the method used in northwestern Europe, does indicate a path of diffusion "through the territory of the empire," but there is no conclusive evidence for this. We are likewise unable to determine whether, and if so when, a center of diffusion that was possibly independent of Italy took shape in the Netherlands. Even though the setback in written administrative records during the great crisis of the plague years makes it unlikely that any substantial amount of material will be found, it is possible that the archives of northwestern Europe might yield more new clues. The period between 1350 and 1370, in particular, deserves more detailed study owing to its importance for the transition to local production.

The most striking result of a quantitative analysis is a remarkable boom between 1371 and 1380. In this decade our lists show nearly eighty cities, about 16 percent of all cases prior to 1450. This means that many cities must have acquired the necessary skills during the previous decade and that the desire for a public clock was translated into concrete plans.

This boom is no less pronounced in the "early" regions of Italy (around 15 percent) and the Netherlands (over 20 percent) than in the "late" regions like France. In the heartland of the Holy Roman Empire (around 15 percent) it appears in the statistics with a small delay, in part because of acquisitions prior to 1371. The height of the boom falls approximately in the year 1376. This third phase in the wave of acquisitions, which is characterized by growth rates in two digits, lasted about three decades. Its peak, the crossing of the 50 percent mark, occurs before the turn of the century. It was not only the less important cities that procured clocks during this period; even cities like Lucca, Montpellier, Aalst, Brunswick, and Lucerne got their clocks only now. At the same time, the proportion of small and very small towns becomes noticeable. At the end of the fourteenth century, the diffusion reached the geographic boundaries of Europe with Santiago de Compostela (1395), Moscow (1404), and the European colonies of the eastern Mediterranean (Caffa/Crimea 1375).[85]

These geographic boundaries were simultaneously cultural barriers. During the High Middle Ages, Europeans were fascinated by accounts about clock automata in the Islamic world. From the beginning of the fourteenth century the situation was reversed, and mechanical clocks and automata were exported from Europe to the Near and Far East. The first piece of evidence is the decision by the senate of Venice to permit Magister Mondinus the transfer of eighty gold ducats for an artful "horologium" from the kingdom of Cyprus. Four years later (1338), the baggage of a company of Venetian merchants included, alongside coinage gold and what appears to have been an automatic fountain, a "relogium" that was intended for the Sultan of Delhi; Carlo Cipolla sees this as the beginning of the transfer of European technology to Asia. Thereafter the tradition of such transports to Oriental courts was never completely broken off. Gillebert de Lannoy left a gold watch behind in Rhodes during a diplomatic mission; the English king had intended to send it to the Turkish sultan, who died while the mission was en route. Following the peace treaty of 1477, Muhammed II, the conqueror of Constantinople, asked the Signoria in Venice to send a good painter, an expert in eyeglasses, and a clockmaker.[86] The yearly payments after the truce between the Sublime Porte and the Holy Roman Empire consisted after 1577 in part of shipments of clocks from Ottoman order lists; the clocks were accompanied by trained clockmakers.[87] In strange contrast to these extorted "presents" or "tributes" for the private collections of Ottoman dignitaries was the reluctance on the part of the Ottomans when it came to the introduction of public clocks. In spite of the fact that the tradition of public time indication by means of water clocks lived on in some cities at this time, the ambassador Ghislin des Busbecq tells us that the Ottomans had no interest in public clocks with European technology. European firearms were being adopted, but not printing presses, because printed sacred scriptures would no longer be sacred; public clocks were likewise not adopted, because they would have undermined the authority of the muezzin.[88] We are told that Murad III (1574–1595) had plans to introduce striking clocks "like in Venice," but gave them up under pressure from the clerics. Only in the Balkan regions conquered by the Turks do we hear in the sixteenth century of public clocks from "the time of the infidels."[89]

The wave of acquisitions following the boom of 1371–1380 lasted until approximately 1410. By this time the larger European cities had installed public clocks. Thereafter the growth rates in all regions drop very noticeably below 10 percent (in each case compared to the previous decade). The curve of the assembled acquisition data thus shows the S-shape that is typical of diffusion processes. However, the flattening out of the curve, which becomes even more pronounced after 1450, does not have to do in this case with the satisfying of existing needs. In actuality, as we have seen, a phase of procurement aimed at broad geographical coverage began after 1450. As a result, the growth rates should be rising, not declining. Their decline is caused by a number of factors: public clocks lost their news value for contemporary as well as modern historiography of cities; communal sources have been poorly studied from the Late Middle Ages on; finally, there is a large number of small and very small localities for which no written sources exist any longer. Under favorable circumstances, regional studies might reveal a very different picture. My survey has also not taken inter-city diffusion into account. Around 1450, very large cities like Paris, Rouen, and Milan had at least four to six public clocks in operation, medium-sized cities like Lüneburg four, and even smaller cities like Moulins in the Bourbonnais had two.

The spread of public clocks can also be examined from a geographic perspective:[90] as one would expect, northern Italy and northwestern Europe, regions unique for their high degree of urbanization, show the greatest density of diffusion over the entire period. Bohemia, Silesia, the large cities of the empire, southern France, and the kingdom of Aragon were reached at about the same time with scattered cases. By the end of the century the European urban landscape seems equally represented. The English possessions in France, Brittany, and the Scandinavian area appear to lag behind. However, is this map not merely a reflection of urban Europe around 1400, with its highly urbanized and less urbanized regions?

To test how representative this map is we would need data sets on the European cities. Relevant would be population numbers, regional hierarchies of cities, and classifications according to economic, administrative, cultural, or ecclesiastical characteristics. However, qualitative classifications of cities that cover large areas and can be compared, which would be useful for our purposes, do not exist for this period. Considerable controversy attaches also to the term "city": what defines it? legal status? walls? the predominance of the tertiary sector? We are therefore not in a position to test the hypothesis that functional differentiation, the diversity of different functions of a city, had a decisive influence on the pace and direction of the diffusion. However, the accumulation of functions is contained in the very concept of the old European city, and we know from experience that the degree of accumulation depends on size, though this correlation should not be understood in too narrow a sense.

Classification of city size, long the concern of comparative urban history and more recently also of historical demography, is thus an indirect way to arrive at general statements about favorable and less favorable conditions for innovation.

Questions of this sort were not asked around 1400. People knew particularly wealthy or particularly powerful cities or the most important sites of fairs and episcopal cities. However, at the time territorial lords neither counted their cities nor listed them according to size.[91] The ideas that circulated about the number of cities and communities tended to be on the fantastic side. For example, in 1404 a monk at St. Denis reported on a project to simplify the levying of the taxes needed to carry on the war against England by collecting twenty ecus d'or every year from every church spire in France, each of which stood for a community or village. The number of spires was pegged at 1,700,000. After subtracting the 700,000 communities exempted from the tax burden because they were particularly hard hit by the war, the money to be raised from the remaining one million solvent communities would have been more than enough to cover the financial needs of the crown.[92] Machiavelli was still circulating such figures in 1512.[93] Today the real number of communities in the kingdom of France at the time is estimated at between twenty and twenty-five thousand. There are no estimates for Europe as a whole.

Determining the population of the cities throws up a host of problems. The demographic sources that can be quantitatively analyzed record only a segment of the urban population, depending on the fiscal, military, legal, or socio-political interests that led to their compilation.

Taking into account the always numerous exemptions from taxes and dues, that segment must be multiplied by some factor we can only know approximately. It is even more difficult to determine the multiplication factor for cadastral sources, since the height of buildings, the density of population, and the relationship of walled to inhabited space are quantities rarely known to us. In the Late Middle Ages, data we seek to compare for a given city are often separated by more than a century. In many cases the effects of demographic changes can therefore only be conjectured. Finally, when it comes to drawing up a hierarchical classification of cities, we must bear in mind that "large city" and "degree of urbanization" are relative concepts that vary from region to region. In northern and central Italy, a population of 10,000 was the minimum size for large cities, whereas in England it was the maximum size, which was rarely reached. In 1973, L. Génicot published a survey of large European cities around 1300, in which he summarized and compared the information in the older literature.[94] The data come from a range of different sources dating from the twelfth to the fifteenth century. His list contains about seventy cities with a population of more than 10,000. For my own comparative purposes I have left out the Spanish cities in the Islamic sphere, with their fantastic population figures, which are in most instances the result of calculations based on physical size. I have added a short list of cities that most likely reached this size. Thus Europe around 1400 had about one hundred large cities with over 10,000 inhabitants. A comparison of both sets of data yields the surprising result that seventy-five of these cities are recorded to have had a public clock prior to 1400. Looking at the approximately forty cities with more than 20,000 inhabitants, we can see that we lack only a few dates from cities where the available sources are particularly bad when it comes to the question we are pursuing (Bordeaux, Toulouse, Marseilles, Messina, Rome). However, it does not follow from this that only large cities were vehicles for the process of diffusion.

Since lists of medium-sized and small-sized European cities covering large areas are not available, a methodologically more difficult method has been used to assess the role of these cities in the diffusion process. H. Amman created a map on "Economy and transportation in the Late Middle Ages around 1500" for the *Handbook of German Economic and Social History*.[95] The map shows the territory of the Roman Empire and some adjoining regions of eastern France, northern Italy, Poland, Hungary, and Scandinavia. Nearly six hundred cities are indicated according to five different categories of size. Undoubtedly these quantitative classifications can be substantiated only in part, and the map also reflects the knowledge of an experienced social historian. Comparing Amman's map of cities with our data set reveals that of the nearly six hundred cities on the map, about one hundred and forty can be assigned a date for a public clock before 1410. Nearly all of the cities with a population exceeding 10,000 are included, and nearly half of the cities with a population between 5,000 and 10,000. We can see once again that our data set on the larger cities is remarkably representative. If we calculate the average probability of the groups of cities to appear in the data series and then use this result for a projection, the findings can be further differentiated.

If we ignore possible shifts in the city landscape during the fifteenth century, and assume, for the purpose of constructing a nearly ideal set of data, that cities with a population between two and five thousand are not represented by only 18 percent, and cities with a population between five and ten thousand by 53 percent, but that all groups of cities are represented like the large cities with about 90 percent, the significance of the larger cities for the diffusion of the public clock between 1371 and 1410 would decline considerably. Even during the boom decade the share of the large cities in the acquisitions prior to 1410—speaking of the area encompassed by the map—was only 3 percent, compared to about 8 percent for the medium-sized and small cities and towns. By the time we reach the decade between 1390 and 1400 the latter had long since become the actual carriers of the diffusion process. If we examine the list of "early acquirers" between 1356 and 1377, small cities like Fano on the Adriatic Sea, Troppau, Brieg, and Schweidnitz in Silesia, and Ivrea in Piemont stand out with dates before 1370; Tulln, Görlitz, Stade, Stadthagen, Culemborg, and Castres stand out with dates during the boom phase. The Regensburg archivist Carl Theodor Gemeiner described this development as early as 1812; using the city of Donaustauf as an example, he pointed out that still in the fourteenth century even less important market towns followed in the footsteps of trading cities in the acquisition of public clocks.[96] Without wishing to overestimate the indicative value of such calculations, we can note the following as the second important result of a quantitative analysis of the dates: after 1370, cities with a population between two and ten thousand were no less receptive to innovation than large cities; they were substantially involved in the great wave of acquisitions.

Given the lack of comparative studies of cities and of comparable diffusion studies for large regions, a discussion of the regional aspects of diffusion can only be cumulative and based on selected examples.

With respect to the region of the Netherlands, we have seen that the very large cities (Utrecht, Deventer, Ghent, Bruges, Ypres, Lille, Douai, St. Omer, Arras) appear on the lists in a very early phase of the diffusion. But here, too, small cities (Culemborg, Montreuil-sur-Marne) are represented in numbers and with early dates.[97] The carriers of the diffusion are here primarily the communes; support from territorial lords came into play only after 1390.

In the medieval French monarchy, the number of cities "deserving of the name" is estimated at between 220 and 250. However, to my knowledge no lists of cities exist. For some time now, scholars have been trying to get a grip on the problem of classifying and ranking the French city landscapes with the help of contemporary sources. In this effort two approaches have been used to reconstruct contemporary perceptions. In the first approach, a list of those cities which at some point were important from a political and administrative perspective can be drawn up using the summons of representatives of the estates to parliament. According to the research of C. H. Taylor, between 1302 and 1335 507 cities were invited to one diet, about 230 to several diets. There are two drawbacks to these lists. First, they are the products of an administration that was not very developed. Second, outdated perceptions, carelessness, and favoritism have left their marks on them. Thus traditional ideas concerning the importance or rank of a city led to the inclusion of tiny urban agglomerations, some of which can no longer be located today.[98]

The second approach involves the use of a seemingly objective criterion for determining the degree of urbanization. From the twelfth century on, the establishment of mendicant monasteries had been following the trend of urban development.[99] Explicit criteria for the mendicant orders in the establishment of a house were the size of the public that could be reached by preaching and the possibility of supporting a convent economically. This wave of foundations was planned, and each foundation was preceded by a careful inspection on site of the prospects of providing for the community's livelihood. The Papal Bull *Quia plerumque,* issued by Pope Clement IV on June 28, 1268, defined a minimum topographical criterion. The distance between the convents of the various preaching orders was to be no less than five hundred meters (three hundred "cannae" [rods]). Regardless of local custom, a rod was to be calculated at eight handbreadths ("palmae"). This was the first time a universally valid index of urbanization combining a minimum size and economic productivity had been formulated. The modern historian can then derive the relative importance of any given city from the number of Dominican, Franciscan, Augustinian, and Carmelite houses. Jacques Le Goff undertook extensive research in this area and presented as his findings an *Inventaire de la France urbaine médiévale;* in the meantime work on the urban history of other countries, using a similar methodology, has appeared. Compared to the map of cities invited to the diets, the map of the houses of the mendicant orders in France offers the advantage that it is neutral with regard to the political situation. It highlights the importance of southern and southwestern France and accentuates its historiographically neglected city regions. In addition, it reveals the growing importance of the smaller cities in the Late Middle Ages. The list of cities with four or more convents correlates well with L. Génicot's list of large cities. However, does the map provide a truly objective picture? It is apparent, at the least, that the "fossilized" diocesan structure of very small divisions, which had survived in the south from late Roman times, was evidently reproduced, and that, once again, localities we can no longer find are treated as cities.[100]

Neither the list of the 216 cities invited to the diet of Tours in 1308 nor that of the 222 cities invited to various assemblies in the year 1316 can be compared with our data, because they include too many cities that are too small. By contrast, of the twenty-eight cities that had four or more convents around 1335, our list has fifteen dates prior to 1410; for the twenty-four cities with three convents it still has eleven. Likewise, more than half of the cities invited to the various Estates General during the fifteenth century can be furnished with a date. That is true for the approximately seventy "bonnes villes" in the year 1468, as well as for the approximately one hundred cities at the diet of Tours in 1484, which had been systematically selected according to administrative districts.[101] If the compare the lists, we find that the cities of the French kingdom are fairly well represented, though with considerable regional variations. Speaking of the territory encompassed by the borders of modern France, we can differentiate three zones: the historical regions of Flanders, Artois, and Picardy, as well as Normandy, Champagne, the Ile-de-France, Orléanais, Berry, and Burgundy were reached early and broadly by the wave of acquisitions. As already mentioned, in these regions wealthy cathedral chapters were, in addition to the communes, the carriers of the innovation. Poitou and Anjou stand out through somewhat later dates and through what tend to be discrete processes of diffusion, meaning that the date for a central residential city (1384 Angers, 1387 Poitiers, 1396 Nioret) are followed only much later by other dates in the region. In these areas we are probably looking at a pattern that is typical of diffusion at the initiative of the authorities, and we can confirm R. Favreau's observation that Poitiers, for example, would have had to wait many more years for such a "nouveauté" without ducal intervention.[102] In Brittany the development of cities occurred later; accordingly all the dates of this region visibly lag behind.[103] Leaving aside the exceptions — Perpignan as the residence of the kings of Aragon, and Avignon with the county of Venaissin as papal territory — the south of France (with Guyenne, Languedoc, and the particularly well researched Provence) presents itself, in spite of relative dense documentation, as a region that was reached late.[104] Even though the lack of usable dates from Toulouse, Aix, Arles, and Marseilles could be distorting the picture, we see once again that by the fourteenth and fifteenth centuries the primary focus of economic and cultural life had long since moved to the north. While the cities in this region appear on all lists, they no longer play any role as centers of innovation.

The building of tower clocks in Spain was in the early period an imported technology independent of the highly developed instrument making at the Spanish court. Apart from Antonio Bovelli, the Italian specialist who came to Perpignan from Avignon, we find Italian and Flemish-German tower clock builders in the large Spanish cities in the fourteenth century.[105] To that extent the early public clock in Perpignan shows itself to be another example of a discrete diffusion which had no repercussions for decades.

The early and broad wave of diffusion in Savoy and in northern and central Italy continued during the fifteenth century. Although city lists are not available, it is clear that everywhere even small cities, since 1360 at the latest, shared in the diffusion movements that were carried by communal and princely initiatives (Ivrea 1368, Fano 1366). The observations of travelers quoted earlier about the relative scarcity of public clocks cannot be confirmed for the time period under examination. This does not mean, however, that rural Italy did not in fact begin to lag behind after the beginning of the sixteenth century.

The yield of dates for the comparatively very large cities of southern Italy is meager. However, it is certainly not possible to speak of delayed modernization with respect to this region. We hear of a public clock in Palermo in 1374. Modern hour reckoning was in use in Naples since 1381, and in 1389 Ragusa (Dubrovnik) hired a clockmaker from Lecce in southern Apulia.[106]

The situation in late medieval England is more difficult to assess. Alongside numerous references to a native clockmaking tradition we find the importation of tower clock builders from Italy, Flanders, and Germany, a practice that lasted into the fifteenth century. However, this does not exclude the possibility that there was also indigenous tower clock building. The development within the urban landscape is not clear, since usable dates are missing for the most important cities after London (Norwich, Bristol, Southampton).

Scandinavia is the only region where contemporaries themselves noted a distinct delay in the introduction of clocks, the division into hours, as well as the use of bells. In his description of the northern peoples, the archbishop of Uppsala, Olaf Magnus (died 1544), attributed this lag to the unlettered simplicity of past centuries. In his view, though, the late transition to this "usus modernorum," as compared with "alia nationes," was not really surprising, for even in innovation-hungry ancient Rome the division of the day into hours had been introduced only at a late date. He added that in his time the gap had been closed and that good and accurate clocks for both public and private use were being built all over the north.[107] Like Pliny, the archbishop counterposed to rural life ("unlettered and simple") a historically later urban development. He thereby indicated delayed urbanization as the cause of delayed innovation, and in this he was probably correct.

Overall the quick tempo of the diffusion of striking clocks is conspicuous. However, this tempo—as the dates for the diffusion of windmills, at least, suggest—was not unusual for the Middle Ages. Only the diffusion of printing, which was even more strongly linked to the mobility of persons, was distinctly faster. The example of another "key machine," the steam engine, is not the only one that shows that technical innovations by no means occurred more quickly in the modern age than in the Middle Ages.

The final question is this: is it correct to assume that when it came to the procurement of clocks, special interests within the city prevailed over opposition to such a move? There is, first, the frequently repeated thesis that traders and merchants in the cities had a particular interest in this innovation and thus actively promoted it by bringing about corresponding political decisions or by financing the clocks. A weaker version of this thesis states that the spread of public clocks suited the rationalistic ethos of this group. This goes along with the parallel thesis that the network of the large trading cities, in particular, determined the pace and direction of diffusion. Based on what we have so far discovered, all these theses must be rejected. To be sure, it is true that the group of "innovators" includes some of the most important mercantile metropolises in Europe at the time (Milan, Genoa, Florence, Bologna, Siena), and that such cities are also prominently represented in the group of "early acquirers" (for example, the large cities of the Holy Roman Empire and the Dutch cities). However, they share that role with cities that were far less important in mercantile terms, cities like Orvieto, Zurich, or St. Quentin. Most of all, nowhere did special mercantile interests come into play in a visible way; nowhere can we detect a special financial engagement on the part of the merchants. In Milan and Bologna it was the city's lord who made the decision; in Bologna the residents even complained about the costs it imposed. The same holds true for the later period of the great wave of acquisitions. Merchants, non-locals, and participants in markets and fairs are occasionally mentioned, but no more and no less frequently than clerics, students, officials, or judges. Nowhere do the merchants make their appearance as a separate interest group; nowhere can we detect a special pioneering role of the mercantile cities. To be sure, these theses could be generalized to the point where they become true but trivial: there was certainly no quarrel between the Italian city princes, the kings of Aragon, England, and France and the merchants in the cities when it came to the establishment of public clocks. The political power in most European cities lay in the hands of mercantile oligarchies; consequently the territorial lords acted largely in the interests of the merchants, and the clocks in their residences were a symbolic expression of this alliance of interests. The investments made by the cites also served above all the interests of merchants and traders. All this is certainly true, but any inference we draw from it becomes banal. Markets and trade were simply an integral part of the concept of the old European city, and the European network of cities was the sphere in which the diffusion of the public clocks occurred. Similar problems arise with the supplementary thesis that the geography of the diffusion of public clocks was largely identical with the crisis regions of the urban textile industry. This implies the supposition that the exacerbation of the conflicts in the cities over worktime forged merchants and traders, in their role as employers, into a group with an interest in the clock as a means of stricter control of worktime; it made them the initiators behind the procurement of clocks (cf. chap. 9, Increasing precision). However, none of this is of any consequence for the geography of diffusion. This thesis rests on an error in perspective. The highly urbanized regions of the European urban landscape were identical with the regions of the urban textile industry in crisis — assuming we are correct in speaking of a broad crisis at the end of the fourteenth century.

The introduction of the public clock can be seen — taking into account, however, the temporal framework I have sketched out — as part of the process of urban modernization. In this context modernization should be understood broadly as an abandonment of traditional ways of life and unrelated to the paradigms of development in industrial societies and democracies. The public clock is suitable as an indicator of modernity. However, one cannot call a city "modern" or "technologically progressive" for the sole reason that it acquired a clock early on, because the dates of the various acquisitions are subject to the influences and vagaries I have described. A city's level of modernization can be assessed only in conjunction with other technological and social innovations. These innovations include hammer mills, mills, installations for supplying water, foundries, as well as church organs. They include also the establishment of schools, the hiring of doctors, teachers, and engineers, or uniform financial administrations. Other innovations could be added to the list. Only the examination of more complex bundles of roughly contemporaneous innovations would produce truly usable indicators of urban modernization in the Late Middle Ages.

Late Medieval Clockmakers

Visual depictions of late medieval clockmaking appear only from the second third of the fifteenth century. They remained a rarity and are found almost without exception in a single iconographic context, the so-called pictures of the planet-children.[1] These pictures show the influences that, according to astrological beliefs, individual planets exert on the character, health, and abilities of humans. The children of Mercury, for example, were considered to be poetically gifted, eloquent, hungry for knowledge, and open to new and strange things, to mathematics and astronomy as well as to magic and fortune-telling. The talent of passing on acquired knowledge was also attributed to them. Among the sciences they were particularly suited for geometry, astronomy, and practical surveying.[2] Their professional talents were seen to lie with the "subtili ingegni et ciaschedun'arte bella," that is to say, in the artistic-artisanal sphere.[3] In contrast to the planet-children texts that drew on Arabic sources, beginning in the fifteenth century we find in the planet-children pictures clockmakers alongside organ makers, painters, wood-carvers, art metalworkers, weapons smiths, teachers, writers, clerks, and cooks. According to contemporary perception, clockmaking was thus among the higher specialized artisanal work, and it owed the esteem in which it was held to its proximity to both science ("scientia") and art ("ingenium").

A painting of Lombard provenance from an astrological manuscript in the library of the dukes of Este in Modena emphasizes the artisanal aspect and shows the clockmaker with a journeyman in an open

43. Clockmaker's shop. Detail from a picture of planet children (fifteenth century). Modena, Biblioteca Estence, Ms. lat. 209, folio 11.

storefront-workshop (fig. 43). The clockmaker in the Housebook from Castle Wolfegg in Upper Swabia is shown as an artisan with astronomical knowledge at work setting his clock (fig. 44).

Of course clockmakers, too, benefited from the growing appreciation one can observe in the Late Middle Ages for the "artes mechanicae" as activities that were useful and necessary for life.[4] In addition, Parisian natural philosophy at the end of the fourteenth century honored clockmakers by comparing the cosmos or creatures with artful clockworks and the creator-God with a clockmaker.[5] As constructors who designed and built their products, clockmakers thus took their place alongside architects, who were also highlighted in these comparisons.

It is easy to make these sorts of comparisons plausible with references to clockmakers who acquired fame as brilliant constructors or

44. Clockmaker with quadrant. Detail from a picture of planet children in *Mittelalterliches Hausbuch* (around 1480). Schloß Wolfegg, Bibliothek der Fürsten von Waldburg-Wolfegg-Waldsee.

sought-after technical experts. As an occupational group, however, clockmakers cannot be characterized this way, since the vast majority of them did not in fact enjoy such high esteem. When it comes to this majority of clockmakers, the extant accounts from all countries usually do not contain more than this typically sparse note: "To Hans, the smith, for the clock . . ."

Evidently there was no such thing as a typical clockmaker or an occupational group—let alone a professional category—that could be characterized by sufficiently shared features. The technical and artisanal competence for building clocks existed in various occupational groups.

The compilation of sources on the history of clocks up to the end of the fifteenth century has brought to light over 1,200 names of clockmakers and people who were involved in the repair and maintenance of clocks. This has more than doubled the prosopographic material previously available for this period in country-specific lists of clockmakers (use of which I gratefully acknowledge).[6] When it is not concerned with social elites, prosopographic research becomes easily suspect as the obsessive accumulation of isolated, meaningless data on people. And little is in fact gained by recording the name of a clockmaker—often only a first name—that appears nowhere else. In this regard the situation with clockmakers is different than it is, for example, with artists or builders. After only a few generations the creations of clockmakers were "scrap iron." Securely dated clockworks from the fourteenth and fifteenth centuries are very rare today and have not been preserved in their original condition. It is thus hardly possible to use personal names for attributions or for reconstructing technological developments. Still, it is worth collecting the names, for in regard to a small group of clockmakers they allow us, for example, to make statements about their movements.

Until now, remarks concerning the beginnings of clockmaking and its development during the first two centuries have tended to be hypotheses advanced in passing. They could not and did not wish to be more, since even the known material had never before been compiled in one place. Naturally we are frequently told that the appearance and spread of clocks also brought about a surge in clockmaking. This was no doubt the case. Linked to this is the conjecture that this surge in clockmaking led to a quick increase in the number of specialized artisans who were able to fashion complicated machinery from metal.[7] This plausible hypothesis of the clockmaker as the prototype of the versatile technician or precision mechanic was derived from the history of early industrialization and extended back into the Late Middle Ages. It is based on the widespread observation that in the early stages of the machine age, clockmakers, in particular, emerged as inventors and constructors of new machines and precision tools.[8]

Hypotheses about the direction in which medieval clockmaking developed can be assigned, in simplified form, to two types. One focuses on the early tower clocks and describes the artisanal development as a path from something that was primitive, crude, and large to something that was complicated, delicate, and small: a path of increasing precision and miniaturization. Accordingly, the development of craftsmanship is characterized as a path from blacksmiths to precision mechanics.[9] The other hypothesis looks at the initial rarity and early reports about complicated astronomical clocks and argues that clockmaking was at first pursued only by a small elite of experts (astronomers, precision mechanics, goldsmiths).[10] The small number of such experts and the initially weak demand for their product gave rise to a kind of nomadic existence among clockmakers, similar to what was the case with builders and artist craftsmen.[11] Both conjectures are based on somewhat vague interpretation of the reports concerning the nature and diffusion of the early clocks. We shall see that neither of the two variants is tenable.

Early clockmakers

If it is correct that the clock escapement was not an invention but a development resulting from various efforts in the same direction, it would be fruitless to search for *the* clockmakers from whom all others learned their skill. Since the escapement probably grew out of monastic and originally water-driven alarm and striking mechanisms, one ought to assume a number of different monasteries as the sites where this development occurred, and no single monastic order should be singled out. This is not true, however, for the additional invention of the automatic striking work. Its development might just as likely have occurred outside of the monastic realm. All early reports about clockmakers are consistent with this. The first medieval European clockmaker known by name is Hermann Josef from the Premonstratensian monastery of Steinfeld in the Eifel region (around 1200). His vita describes him as a man who, despite a physical handicap, was a reliable sacristan and a capable mechanic who deserved praise for his efforts in the installation and repair of clocks in the surrounding monasteries. The occupational designation "horologiarius" appears, to my knowledge, for the first time in 1269–1270 in the beer accounts of the Cistercian monastery of Beaulieu (Hampshire), where it describes what appears to be a permanently employed clockkeeper. In the beer accounts from St. Paul's Cathedral in London, a larger sum was paid in 1286 to an outside "Bertholomeus orologiarius." In 1271, Robert Anglicus speaks as though of a well-known profession when he reports of the "artefices horologiorum" who tried their hand at building a clock for the astronomers.[12]

The related question of whether the new occupational names already indicate the existence of a new device, as C. F. C. Beeson and D. Landes are inclined to believe, or whether we are still dealing with late forms of water clocks, cannot be answered.[13] What the specialized occupational designation does show is that these devices were elaborate constructions that could not be maintained by just anyone. However, had not William of Auvergne in the middle of the thirteenth century said in regard to the "horologia, qui per aquam fiunt et pondera" that they required "renovatione frequenter et aptatione instrumentorum suorum"?[14] Since the development of the mechanical escapement itself has left no semantic traces, there is no reason why those who built the new wheeled clocks should have been described any differently than those who built the horologia.

Once the diffusion of mechanical clocks set in at the beginning of the fourteenth century, the picture changes noticeably. Clerics and monks, who earlier may have held a sort of monopoly, remain an inconspicuous minority after this time. As an occupational designation, "clockmaker" is taken over into the national languages; moreover, it becomes apparent that the multiplicity of instruments with mechanical clockwork drives (monastic alarms, chiming works and musical automata, astronomical simulations, striking clocks) has its counterpart in the diversity of qualifications of those who built them. The situation up to the middle of the fourteenth century presents itself as follows in the individual countries:

The "Jehan l'aulogier" who is mentioned in the Paris tax roll of 1292 (as residing in the "Grant-Rue-Saint-Benoict") raises doubts for linguistic reasons.[15] We might be looking at a misspelled form of "alogier" (to give lodging), since the rolls of the years 1296 and 1297 mention several "osteliers" and "taverniers" in this street.[16]

The accounts of the royal French household document that the goldsmith Pierre Pipelart—whose name indicates that he came from northern France—worked on a "horologium regis" from May 1299 to March 1300 and was paid six livres tournai. In 1308 he was again given a larger sum for this clock or for a new project. Undoubtedly this was an elaborate clock for the royal court or for a church, though, as we have seen, it was not the first public clock in Paris. Gilebert de Lupara—according to the tax rolls of 1296 and 1297 a crossbowmaker who lived at the Louvre—was working on an "opus quorundam horologiarum" in the summer of 1301, a heavy clock of iron and other metal which the king wanted to donate to the new abbey of Poissy. Between 1322 and 1336 there was evidently a permanently employed court clockmaker ("varlet de chambre et ollogeur au Louvre"), Girard de Juvigny, who was involved in a dispute over debt and in a suit against Count Robert of Artois.[17]

The clock in the cathedral of Sens was repaired in 1319 by Robertus

Anglicus from Paris. The last name is not necessarily indicative of English background, since it was very common in Paris at that time.[18] In Noyon in 1333–1334, the clock of the cathedral, which was equipped with bells, was inspected by a "magister orlogiorum" who was passing through town. Among the automata ("engins") in the castle park of Hesdin was a clock ("orloges") that was presumably connected to a musical work via bellows. The required metalwork was done in 1304 by Henri "le serurier." In 1333, a Bartholomeus "orologiarius" reset and adjusted the clock in the cathedral of Cambrai.[19]

The first clockmakers at the papal court in Avignon were also from the region of northern France and Flanders. In 1329 master Nicolas of Bruges built a clock for the Carthusian monastery in Chorges at the request of the pope. After 1335, Jean Bequet from the diocese of Rouen was "magister horologiorum papae." In Italy, few names of clockmakers are recorded for the first half of the fourteenth century. The evidence from the Holy Roman Empire is even more meager. The first name during the period in question is Thideman "seyghermaker," who became a citizen of Stralsund in 1341. Like the man who sponsored him, he was probably a smith.[20]

A comparatively large number of clockmaker names, among them also a (widowed?) woman, have been extracted by G. Fransson from English tax and court records: Adam le-Orloger (Colchester 1311), Thomas Orloger (Essex 1319–1327), Cecilia le Orloger (1328 Lincolnshire), Simon Orloger (1322 Sussex).[21] There seems to be no doubt that these were occupational surnames, and confusion with a type of ship called "orlog/horlok" is unlikely.[22] Apart from a Thomas "orologiarius" mentioned in Lincoln in 1324, during this period clockmakers are referred to at the construction of the great astronomical clock in Norwich and in the abbey of St. Albans. In Norwich, in addition to the clockmaker Robert, the brothers Laurence and Roger Stoke were working as "horologarii" in 1323–1325. Roger was also praised in the chronicle of St. Albans for fashioning a metal cross. Both brothers later (1340–1350) collaborated with Abbot Richard of Wallingford.[23]

The information that is obtainable for the period up to the middle of the fourteenth century makes it clear that clockmaking did not emerge from any particular metal-working trade. Smiths, weapons smiths, and goldsmiths had mastered clock technology at the beginning of the fourteenth century and used it for a variety of clock types. We can speak of an elite of experts only in regard to the builders of astronomical instruments.

The occupational name "clockmaker," indicative of specialization and probably also intended as a form of advertising, appears in all countries around the same time. However, it is highly unlikely that we are already dealing with permanent specialization, meaning that those called clockmakers in fact earned their livelihood predominantly by building clocks.[24]

Learned constructors

The prominent fourteenth-century clockmakers who are mentioned in histories of the mechanical clock were, strictly speaking, not clockmakers at all but learned constructors of astronomical instruments who were also capable of building clocks. As Petrarch already commented in regard to Giovanni Dondi, it is almost disparaging to call them clockmakers. One thing all members in this group—which counted at most two dozen people in the fourteenth and fifteenth centuries—had in common was a university education. The learned constructors worked for the most part as astronomers or astrologers, many of them also as court physicians.[25]

Richard of Wallingford (1291/1292–1336), twenty-eighth abbot of the Benedictine abbey of St. Albans (Hertfordshire), was the son of a smith, "faber ferrarius."[26] A patron financed his studies at Oxford. His unfinished life's work was the construction of a complicated and elaborate astronomical clock in the church of his monastery (fig. 13). He was not an original thinker in the field of theoretical astronomy; rather he concerned himself with designing and building observation instruments and mechanical calculating devices. A fifteenth-century miniature shows him working on a metal plate with his bow compass (fig. 45). His *Tractatus horologii astronomici* (the title is of a later date) contains some brief and difficult notes concerning the clockwork and some rather clearer notes concerning the drives for the astronomical simulations. This work was presumably written during the construction of the clock and in part by the hand of his students. The astronomical clock he had started to build in 1327—"magno labore, maiore sumptu, arte maxima"—was not entirely completed when he died. Richard's

45. Abbot Richard of Wallingford in his workshop. Miniature from around 1440. London, British Library, Ms. Cotton Claud. E IV, folio 201.

contemporary fame is illustrated by a report that he responded to the reproaches from the English King Edward III that he was neglecting the upkeep of the monastery in favor of his clock by saying that a successor could just as easily look after the buildings, while nobody else in England would be able to complete such a complicated work. Compared to the other learned constructors, his fame had receded into obscurity until quite recently, because his manuscript had been unused. Today there are a host of modern reconstructions of his clock.

Jacopo de Dondi (born around 1290) had studied in Padua and was town physician of Chioggia near Venice. He is believed to have been the constructor of the public clock that was built in Padua in 1344 for the city's Prince Umbertino Carrara; the clock may have been equipped with astronomical indications.[27] It was perhaps already through him that the epithet "dall' Orologio," with which clockmakers were distinguished in the fourteenth century, became hereditary in his family.[28] Jacopo Dondi's clock was destroyed in 1390, and in the fifteenth century the clockmaker Novello Dondi dall'Orologio replaced it with a new one. His famous son Giovanni Dondi dall'Orologio (died 1388) became professor of medicine in Padua around 1350; later he also taught philosophy, astronomy, and logic. Sometime after 1365 (when he completed his calculations) Giovanni Dondi built the astrarium, occasionally also called planetarium, for his patron Gian Galeazzo Visconti.[29] In 1372 Dondi moved to Pavia, presumably taking with him the astrarium, which found its permanent home in that city's ducal library. The astrarium was not a clock in the sense of an instrument for observation and indication; rather it was specifically described as an aid for illustrating the works of the ancient astronomers. The *Tractatus Astrarii,* presumably begun during construction of the clock, deals with the making of the gearing of the planets and of the calendric indications; it presupposed the clockwork drive as something that was long since known (see chap. 4: The clock escapement). Even though the *Tractatus* contains instructions for maintaining and setting the instrument, the clock rarely worked correctly after the death of its builder. However, this did not impair Giovanni Dondi's fame. On the contrary: precisely because nobody could be found to keep the contrivance running or repair it, the creator of this device—in the late Middle Ages praised as one of the wonders of the world—was considered an exceptional genius who was venerated by many contemporaries, among them Petrarch and Philippe de Mézières. At the end of the fifteenth century, the Nuremberg astronomer Johannes Müller von Königsberg (Regiomontan), who had seen the astrarium in Pavia, began to build a copy, which he did not complete in his lifetime.[30]

Giovanni Fontana, too, had studied medicine in Padua around 1420 and had become town physician in Udine. Though he had not systematically pursued studies in astronomy, throughout his life he gave attention to machines, war materials, and mechanical musical works, among them water clocks and hydraulic organs, in the tradition of Greek and Arab mechanics. His manuscripts—only parts of which are still extant—contain sketches of clockworks "newly" invented by him and used to drive mnemonic apparatuses, and of sandglasses that were reversed automatically and turned hands on dials.[31]

We still know too little about the astronomers who made the calculations for the astronomical clocks of the fourteenth and fifteenth centuries. The accounts for the astronomical clock in Cambrai (1348–1349) contain a reference which might indicate that the Paris astronomer Jean de Linières had participated in the construction of this clock at least in an advisory role.[32] Another Parisian physician and astronomer, Jean Fusoris (1365–1436), son of a tinsmith from the Ardennes, combined learning and a talent for construction with a certain business sense. In addition to the papal court, he supplied the French, Burgundian, and Aragonese courts with instruments, astrolabes, equatorials, armillary spheres, and complicated clocks. Fusoris ran a workshop in Paris where specialized instrument makers along with other artisans and journeymen built the expensive devices. His social advancement was reflected in the fact that he came to hold canonical posts in Reims, Paris, and Nancy. In extant manuscripts he not only describes the construction of these instruments; drawings and descriptions of a spring-driven clock, some of whose details resemble the so-called Burgundian clock, are also attributed to him. Banned from Paris during the confusion of the Hundred Years' War for treasonous contacts with the English, he directed and supervised the construction of a large astronomical clock in the cathedral of Bourges from March to November of 1424.[33] Some of Fusoris's notes passed into the possession of one of his students, Henri Arnout de Zwolle, who was also an astronomer, astrologer, and physician. He lived in Dijon and was a busy and well-paid expert in the employ of the Burgundians; he also made charts and inspected saltworks. Between 1447 and 1455 he received the enormous sum of one thousand livres for a clock with planetary indications ("vray cours de sept planets") he constructed.[34] He is said to have died in 1466 during a plague epidemic he had predicted.[35]

At the residences of the kings of Aragon we find a whole group of constructors of astronomical instruments (primarily astrolabes) and various types of clocks at work after the middle of the fourteenth century. They were not university-trained scholars but belonged to the Jewish scientific culture that flourished in Spain into the fourteenth century. Since the middle of the thirteenth century, Jews, at the request of the Spanish crown, had been translating—sometimes directly into Catalan—the most important texts of Islamic science. Although Christians were forbidden to seek treatment from Jews, in the fourteenth century over two hundred Jews were employed at the royal court as physicians, astronomers/astrologers, translators, and financial experts. Majorca was the center of an important school of cartography.[36] Nearly all the clockmakers working in Spain in this early period—though not tower clock builders—came from this Jewish elite of experts. Mosse Jacob, silversmith from Perpignan, supplied and maintained clocks and astrolabes at the behest of the king between 1345 and 1347. In 1352 the king demanded the punishment of those guilty of trespassing on the house of Bernat Ferni ("magister dels alarotges"). Protection was also extended to the clockmaker in service to the court, Nathan del Barri, when he returned to his native city of Carcassone (1372). Isaac Nafuci from Majorca was also supplying the court with clocks and astrolabes. In 1360, in appreciation of his services, he was appointed rabbi of the Jewish community in Majorca; he received assurances that the office would pass on to his son and was granted the privilege of performing kosher butchering for a fee.[37] The court privileged and protected these Jews, who were appreciated as specialists, for as long and as best as it could. But already before the great persecutions of 1391 we lose sight of them. In the list of Majorcan Jews forced to undergo baptism in that year we find one Isaac Nifosi (= Nafuci?) identified as an astronomer; he supposedly escaped to Palestine and there returned to his original faith.[38] The list also contains the name of the cartographer Yehuda Cresques, who has been linked to the creation of the *Catalan Atlas* (before 1381). After this time no more Jewish clockmaker names appear in Spain and southern France.

While not every constructor of the late fifteenth century can be shown to have had academic training, we always find the combination of sophisticated theoretical knowledge and the practical skills of artist-craftsmen. This is true of Bartolomeo Manfredi in Mantua (died 1478), and of Guillaume Gilliszoon de Wissekerke (died after 1494), who supplied the court of the Roi René with many kinds of clocks, the Duke of Milan with a "celestium motuum speculum," and Charles VIII of France with a "sfera regalis." It applies also to the priest and astronomer Johannes Stöffler von Justingen (died 1531) and to the Florentine clockmakers of the della Volpaia family.[39] While we cannot yet see the emergence of narrow specialization, we can gradually detect the outlines of the—no longer academic—profession of instrument maker.[40]

Engineers

Not theoretical training but technical-artisanal versatility characterizes the members of the group of engineers who could also build clocks. The occupational designation "ingeniarius/ingeniator" appears in the twelfth century and describes the creative technician, who could also be called "architectus" or "machinator." However, what matters to us is not the occupational name, which was used especially in England and Italy for architects and constructors of war machines (in the German-speaking region they were often called "Werkmeister"); what matters is only the technical versatility, and here specifically the ability to build clocks.

In this group we find internationally sought-after, free-lancing, and highly paid experts, versatile craftsmen in the employ of cities, as well as artists in the narrower sense of the word. Since our view of the Renaissance has a pronounced artistic imprint, attention has been focused almost exclusively on those representatives who were in some way closely connected to "cultural activities."[41] In addition, those who left behind technical manuscripts of course also came in for consideration. But these precursors of the "genius of the Renaissance" are quite atypical representatives of a "technical intelligence" whose history could certainly be written from the available material.

Trained clockmakers were the exception among these engineers; their main areas of work were building machines, canons, war machines, mills of every kind, hydraulic constructions, as well as musical instruments. The question (which can rarely be answered with concrete examples) to what extent reciprocal stimuli in the figure of the engineer influenced the development of various technical-constructional tasks becomes relevant as soon as large clockworks appear.

The already mentioned chronicle of Galvano Fiamma, with the first secure reference to a striking tower clock, reports also the first application of clockwork technology. In 1341, Galvano describes as a new invention ("adinventio") a type of mill in which the use of many wheels, weights, and counterweights—"like in clocks"—made possible a considerable savings in labor.[42]

Also in Milan in 1352, the monks of S. Maria dell'Valle were given permission to divert water from the new canal, the Naviglio Grande. The monks asked for technical assistance from "magister inzignerius Johannes de Mutina dictus de Organis" (Giovanni degli Organi), known to us as a clockmaker and probably from a family of organ makers.[43]

The motif of the labor-saving technical invention appears again in the vita of the architect of the Florentine Duomo, Brunelleschi. As already mentioned, in his youth he had built small clocks and had experimented with spring drives. His biographer emphasized that these experiments proved useful to him in his work as an architectural engineer, specifically in his designing of pulling and hoisting machines. Work on the "ingegni multiplicate" had been "grandissimo aiuto al potere imaginar diuerse macchine ed da portare e da levare e da tirare" ("a great help for the power of imagining various machines for carrying, lifting, and dragging").[44]

At the construction site of the cathedral in Milan, the clockmaker Francisco Pessono in 1402 suggested a stone-sawing machine that would be driven, like a clock, only by weights ("solum per contrapensibus"). Supposedly the machine would be able to do more work than five men charged with the same task. His suggestion was provisionally accepted, and the estimated expenses for a small wooden model (ten Fl) were advanced, as was customary with such projects, by the cathedral workshop. In September of the following year Francisco Pessono was still at work on the saw and received twenty-five Fl for a large iron wheel. Later, after he had left the construction site, we encounter him again in the sources between 1418 and 1420 in ducal services in Como as a master of canons ("magister a bombardis"). Another clockmaker, Giovanni di Zellini, received a contract in May of 1404 for the construction of a machine for transporting and lifting marble blocks which would require only a third of the personnel then needed to perform the same work. The cathedral workshop authorized fifty Fl, though it passed the risk of failure on to the constructor. The project appears to have been a success, however, for in November of 1404 he was appointed "ingegnere della fabbrica." The engineer Filippo da Modena degli Organi (probably of the family of the above-mentioned Giovanni degli Organi) who was working at the cathedral workshop between 1400 and 1405 was also a clockmaker.[45]

Engineers worked as clockmakers also in the employ of cities. The dial for S. Giacomo di Rialto in Venice was built in 1422 by an engineer from Apulia; in Verona in 1425, an anonymous engineer, having proved himself in some hydraulic constructions, was given the post of clockkeeper.[46]

Because Giovanni Dondi's astrarium was malfunctioning once again, the duke of Milan asked his envoy Antonio de Tritio (da Trezzo) in 1456 to prevail upon the royal court to send master Guglielmo de Parise to Pavia for at least four months; the master had helped out once before, "al temp de la bona memoria." Since Michele Savanarola also mentions such an incidence of repair around 1440 ("de Francia nuper Astrologus et fabricator magnus fama Horologii tanti ductus Papiam venit, plurimisque diebus in rotas congregandas elaboravit"), it would seem only reasonable to look for the expert in question in Paris.[47] So far the attempts to identify this astronomer have been unsuccessful; however, scholars have overlooked the fact that while Guglielmo de Parise may very well have hailed from Paris, he did not necessarily live there between 1440 and 1456.[48] And while the duke's letter is unquestionably addressed to a king, it is not explicitly addressed to the French king. Lastly, Antonio de Tritio was in all likelihood never in Paris, but from 1455 on he was the envoy of the Sforzas in Naples.[49] This makes it likely that the person in question was the canon caster and clockmaker Guglielmo de Parise from Naples, who may have come to Naples via northern Italy and Ragusa. Of course this leaves open the question whether Michele Savanarola is also talking about this Guglielmo when he desribes him as "Astrologus de Francia."

The king of Naples summoned gun casters from northern Italy and France and also ordered canons from Catalonia. The duke of Milan summoned clockmakers from Paris and Naples. Here we catch a small glimpse of the movements of an international elite of technicians about whom we still know very little overall. What mobilized the combination of "ars" and "ingenium" represented in the group of engineers was the need for technicians especially at the great construction sites and the rapidly growing need for weapons in the territorial states; if mechanical clocks had a causative role in this it was only incidental. One thing that stands out is the new diversity of qualifications of these technicians. They hardly seem tied down by any social bonds; instead they have a much closer contact with the public authorities in cities and states who were increasingly dependent on their services.[50]

Specialists and occupational migration

In July of 1452, lengthy court proceedings concerning the salary of the keeper of the Horloge du Palais on the corner tower of the Louvre at the bridge over the Seine were concluded before the parliament of Paris. The man who held the post of clockkeeper was demanding the six sous per day which the king had guaranteed when the clock was being built in 1370. The city for its part pointed to its empty coffers and argued that by now Paris, like other cities, had several public clocks whose maintenance costs were much less. The documents of the proceedings allow us to reconstruct the history of this clock with considerable precision. The historical part of the "arrêt du parlement" describes how the then-king Charles V had the clock constructed for the city, which at that time had no tower clocks. Since no clockmaker suitable for the project could be found in Paris, the king sent for Henri de Vic from Germany as an expert ("in scientia et industria horologiarie expertissimus") at considerable expense. In addition to the salary, the German clockmaker was promised free lodging and a workshop in the clock tower along with various other rights and privileges.[51] A whole group of clockmakers can be attested in Paris from the beginning of the fourteenth century. Evidently, however, none of them had mastered the technology of the heavy tower clocks. But Henri de Vic did not bring this technology to France from Germany.[52] Such clocks, and clockmakers who knew how to make them, had existed for some time also in the region of northern France and the Netherlands. Apparently Henri de Vic was merely a particularly well qualified expert for such a prestigious project. Such specialists were rare, but not so rare that it would have been necessary to summon one from Germany. Only a few years later, Henri de Vic's successor, the royal clockmaker Pierre de Ste. Béate, went from Paris to Avignon to build the public clock in the tower of the papal palace.[53]

The number of specialists working in more than one place and the migration movements triggered by the construction of a public clock are significant also in quantitative terms: we still find about sixty cases in the fourteenth century, and about one hundred and forty cases in the fifteenth century. In addition, we can make out a small group of specialists—like the ones mentioned above—who, owing to their profession, worked on a supraregional level.

There were, to begin with, the Italian clockmakers who exported the technology of striking tower clocks in the second third of the fourteenth century. The clock for the palace in Perpignan was built in 1356 by Antonio Bovelli, who had earlier been employed at the papal court in Avignon as a plumber. An Italian ("lumbardus") was working on the clock in Windsor Castle in 1352.[54] After this time (about 1360) we can no longer speak of a geographic center from which such migration movements originated.

As early as 1358, three clockmakers from Delft received a letter of safe conduct from the English king to practice their art in England without any impediments. In 1389, Ragusa obtained the services of magister Helias from Lecce in Apulia to build a large tower clock. Johannes Alemanus was building one in Valencia in 1378, Antonio Core from Bologna was doing the same in Lerida in 1390.[55] Such extensive wanderings are, however, not typical. Most wandering clockmakers moved within a relatively small region with a diameter of a two to four days journey. The royal clockmaker Pierre Melin worked in Paris, Sens, Angers, and Poitiers (between 1375 and 1397). Girardin Petit from Avignon built clocks in l'Isle-sur-la-Sorgue and Sisteron (1397–1410). The presence of Heinrich Halder can be documented in Basel, Strasbourg, and Lucerne between 1373 and 1419, and that of Claus Gutsch from Villingen in Strasbourg and Rottweil (1389–1400).[56] The construction of a large clockwork was the occasion for a great many journeys and transports for about a year. Before construction began, the city's representatives went on information-gathering trips to look at other clocks and to negotiate with clockmakers. In 1372, a delegation traveled from Mons to Calais ("pour vir et aviser l'orloge de le cloke qui sonne les eures," "to see and apprise the clock which rings the hours"), and in 1379 another one was dispatched to Valenciennes ("pour vir l'orloge, et lassus prendre conseil et avi," "to see the clock and to gather counsel and advice").

More important were the journeys undertaken by clockmakers to give advice on a project, to gather know-how, or to inspect the work of colleagues. Jacob zum Kircheneck, "orglockener" in Frankfurt, journeyed to Cologne in 1372 to inspect a "werg der orglocken," at a time when he was building a clock for the Frankfurt city council. After the procurator in Orléans had negotiated with clockmakers and bell-founders in Rouen (about 200 kilometers away) and had informed himself about prices, Louis Carel from Montluçon was commissioned to build the clock in Orléans. Carel and the clockmaker Jehan Menin, who arrived from Nevers for just this purpose, went to Chartres to look at clockworks. Shortly thereafter Carel had his tools brought from Moulins, where he had built a clock, to Orléans (1352–1354).[57] The sources from three cities allow us to reconstruct the travels of these two clockmakers who specialized in large clocks and worked within a region that was not all that large. Such journeys by experts were not yet organized, controlled, and mediated by a professional organization. Information came from inter-municipal correspondence, on the one hand, and letters of confirmation, recommendation, and self-recommendation for or from clockmakers, on the other. The council of Brunswick twice confirmed in 1385–1386 that master Marquardt had built a tower clock ("eyn gud werk") for the church of St. Catherine. In 1493, Werner Hert von Buchen, city smith and city clockmaker in Frankfurt, was recommended by the city's council to Strasbourg for the construction of the cathedral clock ("uhrwerk mit eym zeiger in der höhe"), on the grounds that he had already built good clocks in Colmar and Worms. Later it was also confirmed that he had built several clocks in Usingen. As municipal clockkeeper in Frankfurt, Lazarus Kreger von Bare from Strasbourg worked on the clocks in the cathedral and the Römer and in addition built the municipal clock in Andernach (fig. 46).[58]

The most famous self-recommendation, and at the same time an

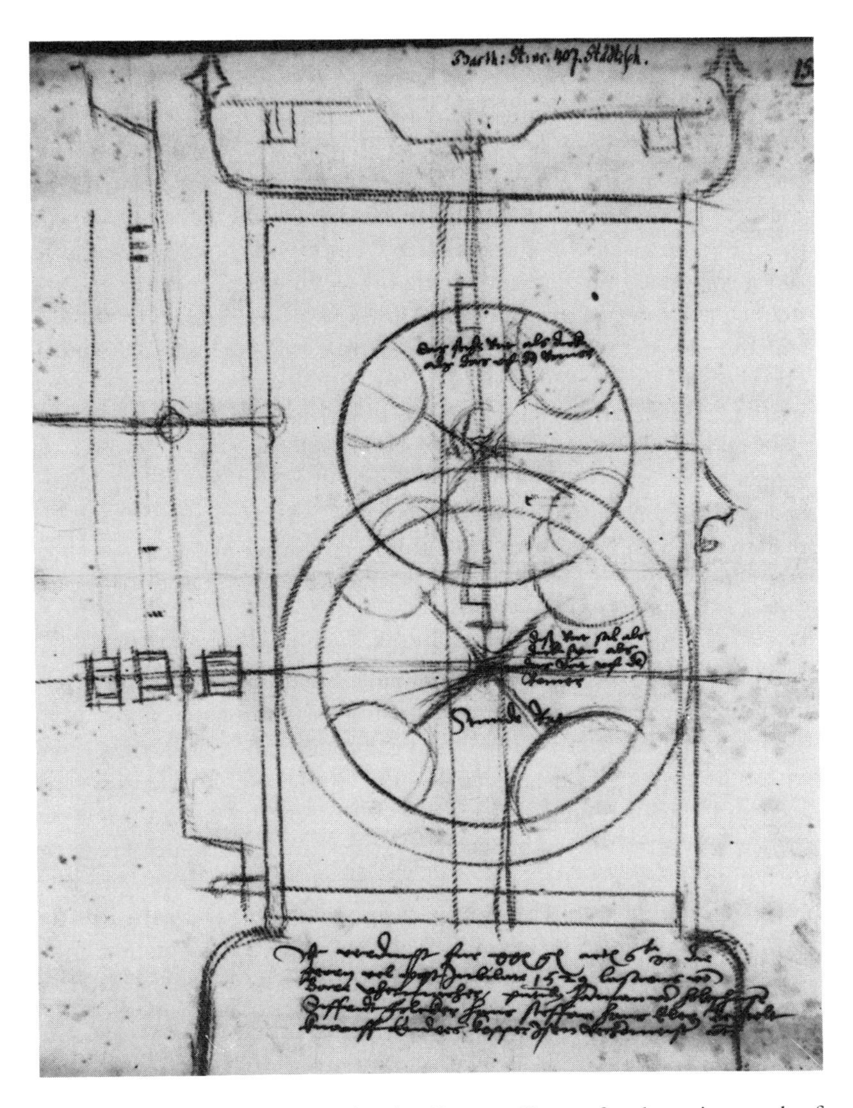

46. Repair sketch of the clockmaker Lazarus Kreger for the going work of the monumental clock in the cathedral at Frankfurt (1529). Stadtarchiv Frankfurt. Text: "This standing-wheel, as thick as that (of the clock) in the Römer"; "This wheel shall be as thick as the wheel on the Römer"; "hour wheel"; "is assigned for XXX gulden, done on the sixth day of Goern or after Jubilate 1529 Lasarus von Bare clockmaker, in the presence of [there follow the names of the members of the commission]"; on the backside: "Renovated the clock in the parish with two wheels."

important source for the history of the clock in four cities, is the letter the evidently aging Don Gaspare degli Ubaldini sent to the republic of Siena in March of 1399. On several occasions he had heard from citizens of Siena that the city was desperately seeking a clockmaker for the clock in the Torre del Mangia at the Campo. He informed the city that he built a clock in the Rialto in Venice with hour-striking jacks and a cock that emerged and crowed three times every hour, and the municipal clock of Orvieto with indications of the courses of the sun and the moon. At present he was constructing a clock for Città di Castello. His clocks were widely known. All he wished for was to serve a city like Siena, to live and die with its citizens. In return he wanted nothing but honor and fame. In May his application was accepted. By August the clock had been renovated, but in the meantime the master had died.[59] Don Gaspare is an untypical case of a free-lance wandering clockmaker who promoted himself. In many other instances the cities who were employing clockmakers as clockkeepers released them only for specific periods of time.

A city's public clock was a device that was prone to breaking down and required a high level of maintenance. Extreme temperature fluctuations, considerable problems with lubrication, the deterioration of the relatively soft iron bearings, and the continual wear on the geared wheels necessitated constant repairs. The cities therefore had to try and keep a clockmaker, once he had come, by extending to him privileges such as exemptions from burgher fees, taxes, and guard duty. Henri de Vic's high salary is evidence of such efforts on the part of the cities. In 1440, Hinrik von dem Hagen was exempted from paying the fees for acquiring citizenship in Hannover; in return he was to maintain the "zeyger." Leonard Wunderlich became a citizen in Cracow in 1456 free of charge, and Hans Graff of Schliers did the same in Salzburg in 1464; in both cases the privilege was extended because they were to become city clockmakers.[60]

As a counterpart to the many occupational migration movements "here and there," the question naturally arises whether, and if so where, we can detect either movements that stabilized in a given direction or local and regional centers of gravity. Generally speaking, nothing of the sort can be observed for the fourteenth and fifteenth centuries. With the exception of the first decades of the fourteenth century, when clockmakers existed in many localities while tower clockmakers were found only in Italy, there were no regional centers of gravity in the Late Middle Ages. Even iron-producing regions or cities specializing in ironworking were not at the same time centers of clockmaking. The primary raw material for clocks, rods and bars of iron, was a transportable, common item of trade. In spite of comparatively spotty sources, we can say that cities in which armament trades were concentrated (such as Brescia and Toulouse), but also cities with centers of fine metalworking (Nuremberg or Würzburg, for example), do not yet, in the fifteenth century, stand out through an unusual density of clockmaker names or frequent mention as the native cities of clockmakers. Only Fabriano, a city known for its paper manufacturing, stands out in Italy for being frequently mentioned as a town from which clockmakers came.[61]

The other noticeable and chronologically stable migration movement is the continual presence in Italy and France of clockmakers from German-speaking regions. Apart from Henri de Vic, other German clockmakers had also gone to France at the end of the fourteenth century, among them Jean de Wissembourg, who in 1377 built the tower clock in the cathedral of Bourges. J.-P. Leguay has found quite a few skilled German craftsmen, particularly from the metal trades, in Brittany in the second half of the fifteenth century. In Nantes, for example, Hansse Dezingue became keeper of the municipal clock.[62] The fifteenth century sees also a growing number of "teotonici"—the Konrads, Herrmans, and Manfreds—among clockmaker names in northern Italy and Rome. However, there is very little we can say about the native cities of these clockmakers or about the special motives behind this migration movement. An oversupply of skilled craftsmen in the metal trades is conceivable, but so is superior qualification or a surging demand in the target countries.[63]

Smiths and locksmiths

We can say very little in terms of general characterization about smiths and locksmiths, numerically the largest group (over 60 percent) among late medieval clockmakers. They are predominantly urban artisans, neither especially well-off nor especially poor, with no particular social weight in their environment; in short: they are qualified specialists with not outstanding characteristics. They are needed, but do not in any way share the prominent status of astronomers and engineers. Some of them are gold- and silversmiths, a few work in the weapons trades. Most of them belong to the guilds of smiths and locksmiths, which since time immemorial had comprised a large number of differentiated metal trades. The written sources of these guilds, however, are for the most part rather meager, and only in the fifteenth century do we find scattered references to clockmakers in the lists of members (in Florence after 1407, in Cracow after 1410, in Basel after 1413, in Andernach, Nuremberg, and Augsburg from the mid-fifteenth century on). We can probably infer from these occasional occupational designations within the guilds of smiths that from this time on clockmaking was no longer a side-occupation but a profession in which one could make a living.[64] The "seyermaker" are addressed for the first time as an occupational group within the guild of smiths on the occasion of a renewal of the statutes in Magdeburg in 1431, and subsequently in Vienna in an ordinance from the year 1451.[65]

It is likely that this period also saw the first attempts by the guilds of smiths and locksmiths to restrict the making of clocks to their members. To my knowledge the first instance is an ordinance from Prato (1451), according to which only members of the guild of locksmiths were permitted to make clocks.[66] However, these regulations concern the emerging trade of makers of small clocks, which, beginning in the sixteenth century, was being organized into separate guilds.[67] It is highly doubtful, though, whether the builders of tower clocks were ever subject to compulsory guild membership. The large number of commissions from outside of the guilds' sphere of influence makes this all the more unlikely. Moreover, we must bear in mind that in the environs of large courts, in particular, many artisans, and especially those highly qualified, were able to escape compulsory guild membership. The numerous clockmakers in the service of the courts were undoubtedly part of these "non-corporative elites" (in P. Sasson's phrase).[68]

Owing to the paucity of the sources, it is also difficult to determine whether and from what time on journeymen went on travels or had to do so as part of their training. The only instance we can trace somewhat is surely not typical because of the unusually extensive journeys it involved. Around 1430, Erhard Eisenros (Girard Yzanrose, Ferrose) from the diocese of Trier is mentioned in the book of the locksmith

47. Initial "I" in the "clock warden's oath" in the oath book of the city of Augsburg (1583). Stadtarchiv Augsburg: "You will swear a learned oath to God and the saints to faithfully regulate, keep, and maintain the clock of the city day and night, with everything that goes along with it, and also to faithfully guard and look after the keys to city hall and the clock, including unlocking and locking, also to be faithful to the city and to warn of harm, all this faithfuly and without guile."

journeymen in Basel. In Avignon in 1445, he took on Menginus Godini from Troyes as a journeyman in order to instruct him in the art of smithing, clockmaking, weaponry, as well as the mysterious "ars artificialiter scribendi." Together with his companion Simon de Troyes he had built a tower clock in Alès (Languedoc) in 1439. In 1448–1451 he was back in this city.[69] Most clockmakers probably began their training as apprentice smiths.

A way of training the next generation of builders or successors that was unique to tower clock building was the training of local smiths by a clockmaker brought in from the outside. A large number of contracts with clockmakers stipulate that locals be involved in the construction of the clock and be trained in its maintenance and repair. In this way, too, technical competence could be passed on and preserved for the city. The smiths who were hired as clock wardens increased the steadily growing group of municipal officials. There are many names we know only from municipal salary lists and oath books.

The late medieval clockmakers were not a homogenous group in terms of training and qualification. The spectrum of the men who carried this innovation reaches from the learned astronomer to the versatile technical expert to the simple smith; it includes articulate authors of building instructions and manuals and illiterates.[70] At the end of the fifteenth century there is still no hint of a trend toward professionalization. The history of clockmaking during the following centuries reveals that it continued to be innovative only outside of the social structures of occupational groups.

7

Clock Time Signal, Communal Bell, and Municipal Signal Systems

Communal tower and communal bell

In the urban realm the new time signal moved into a symbolically and legally important location, the communal tower, where it immediately occupied a prominent place within the system of signals.

The communal clock and clock-bell were almost always installed in the communal tower, which also housed the communal bell. The striking work of the clock was frequently connected with the communal bell, either to make the time signal audible as far as possible or to save the expenses for a separate bell.[1]

Like the keys of the city, the seal, town hall, and the pillory, the communal tower and bell were in the fourteenth century among the legally important and symbolically significant municipal attributes, each of which could stand for the city as a whole. All over Europe the city tower and bell were the bearers of communal identity. In the kingdom of France and the adjoining territories of northwestern Europe, the tower was called "beffroi" and the bell "bancloche." Both marked a city as a relatively autonomous legal entity and thus had to be authorized by the territorial rulers.[2] The granting of a city charter was in these regions tantamount to the granting of the right to a tower and a bell. For instance, in the charter given by King Philip Augustus to the inhabitants of Tournai in 1188, we read that they are permitted "to maintain a bell at a suitable location in order to strike it as they see fit for the affairs of the town."[3]

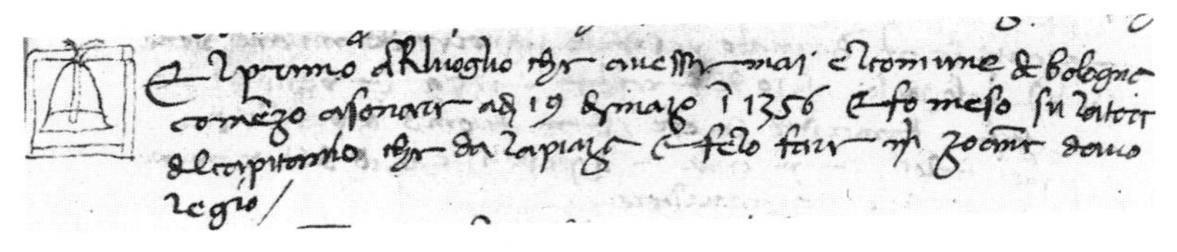

48. Notice in a Bolognese chronicle about the installation of the first public clock. Bologna, Biblioteca dell'Archiginnasio, Ms. 1090, folio 131: "The first clock of the community of Bologna began to strike on the 19th day of March 1356; it was erected in the tower of the Palazzo Capitanato at the Piazza; the Lord Giovanni da Oléggio had it built." Photo: Studio Fotografico Fantini.

"Affairs of the town": these were primarily assemblies of citizens sitting either in council or as a court. The instrument for summoning the citizens was at the same time an expression of the right to hold such assemblies. The formula "assembled in the customary manner at the striking of the bell"—widely used, particularly in Italy—guaranteed the legality and validity of communal decrees and proclamations.[4] The description of municipal assemblies as "consilium campanae," common in Italy, and the decoration of collections of municipal statutes with images of bells and bell-ringers reveal the connection between the signal and the object of the signal (fig. 49). The city bell also warned of fire and danger of war and called the citizens to arms. It was the acoustic delineation of a sphere of authority, and the expression "ban-clock," found in many languages, describes the territory under a given jurisdiction. Sayings that were common into the modern period preserve the memory of such legal spheres and signal spheres. "Unter einer Glocke" ("under a bell") meant being part of a certain sphere of jurisdiction; "unter die Glocke gehen" ("to go under the bell") was during the period of industrialization a vivid way of expressing one's subjection to the authority of a factory owner.[5]

The authorization to ring the city bells was therefore very strictly regulated in medieval cities, and unauthorized use was severely punished.[6] In countless municipal statutes we find time and again the dual legal and symbolic aspect of the communal bell. Unauthorized ringing was not only a "false alarm" and impertinent presumptuousness; symbolically, possession of or access to the municipal bells was a sign of de facto control. In his *Chronicon extravagans,* Galvano Fiammo expressed the link between the possibility of ringing the municipal bells

49. *Liber consiliorum,* a collection of decrees of the council of Turin, 1346. Historical Archive of the City of Turin.

and the de facto rule over the city: "In the palace of the commune is a high tower with four bells; whoever controls this palace and can ring the bells at will can easily rule over the city."[7]

The cult of the clock as the sign of authority thus entailed, conversely, that revolts against the lords of the city were signaled with the ringing of these communal bells: "1368—a large, armed crowd arrived . . . and said they wanted to have . . . the seal of the city and the keys to the alarm bells ("sturmgloggen")" (Chronicle of Augsburg).[8]

Punitive measures by the city lords were accordingly directed against city towers and communal bells. The loss of tower and bell amounted to the loss of legal status and political disenfranchisement, on a symbolic level also to the inability to engage in revolt.[9] The inflicting of punishment against tower and bells could, as "symbolic punishments" in which objects represented the accused, take the place of executions, demolitions, or reparations. Because of their high metal value bells were always highly sought-after spoils, and because they could be recast into cannons they were the target of the authorities' hunger for metals. As late as the two world wars, the expression "to let bells go to war" was in Germany a euphemism for the extensive confiscation of bells for weapons manufacturing. Owing to their symbolic significance, however, bells were also prime targets of aggression in times of war. By hauling off the bells the enemy rendered the citizens honorless and defenseless, incapable of expressing their political and religious identity (particularly during the religious civil wars).[10] As one way of preventing internal wars, the political theorist Jean Bodin recommended taking the bells from the citizens; primarily for this reason, he claims, Muslim princes had blocked the introduction of bells in their realms.[11] L. Fèbvre reports that as late as 1737 an intendant in Bourbonnais had the bells of a city taken down and, in medieval fashion, whipped by the executioner because the inhabitants had sounded the bells against the royal guards.[12]

The restructuring of provinces into departments in the early phase of the French Revolution led to fierce quarrels between the administration and the small towns and villages, because the latter, with active support from the local priests and pastors, fought against the removal of the symbols of the old faith and of local autonomy. The revolutionary wars subsequently saw extensive confiscations of bells for the purpose of procuring metal for coinage and cannons; in addition, these

50. "To hang something on the big bell." Woodcut from Thomas Murner's *Die Mühle von Schwindelsheim und Gredt Müllerin Jahrhzeit,* 1515.

confiscations were legitimated as part of the campaign to rid the land of the Christian faith, for example, through prohibitions on bell ringing (in 1795–1796). At least in some cases, however, locals were able to save their bells by pointing out the importance of the clock for the life of the community.

The communal tower was also in many respects a central site for daily life inside the city. As one of the few massive stone structures, it served the community as watchtower, arsenal, archive, and prison. In the city tower, the "place of the norm," the authoritative weights and measures (for example, the iron ell or the stone bushel) were accessible to all.[13] Later, city towers also became the zero points in the grid of highways. The communal scales were usually near the tower. Every year the communal statutes were read from it. It was also the place where, after an announcement by the ringing of the bells, death sentences, banishments, and the dates of auctions and due dates for interest payments were proclaimed. Idiomatic expressions in German that are still understandable today remind us of the communal tower and the bell as the location and signal of all important news: "an die große Glocke laufen" (lit., "to run to the big bell") (fig. 50) or "an die große Glocke hängen" (lit., "to hang something on the big bell") mean to tell something openly or trumpet it about; "etwas läuten hören" (lit., "to hear something ringing") or "wissen, was die Glocke geschlagen hat" (lit., "to know what the bell has rung") mean the same as "to be in the know."[14]

Architecturally the city tower in northwestern Europe was an urban version of the feudal defensive tower, the donjon, examples of which have survived in Tournai and Béthune, for example. In Italy the communal campaniles were structures that competed with the urban tower of powerful families.[15] We must not be misled by their modern-day appearance. City towers were not enormous gray columns of masonry. A surviving plan of the belfry in Ghent shows the vivid and colorful decoration with banners, coats of arms, pennants, and eaves figures on the steeple which was ornamented with many small towers. The belfry as it looked back then was more reminiscent of a decorated tree than a gloomy fortress tower. But not all cities and towns possessed a separate architectural center of civic life. In smaller places the church tower fulfilled the function of the communal tower; in larger cites the second tower of the cathedral was the communal belfry housing the fire watch, bells, clock, prison, communal weights and measures, and the communal archives. The spires of the minsters of Metz, Strasbourg, and Freiburg are the best-known examples of church towers that were maintained by the civic community as late as the nineteenth century.[16] The public clock shared in the honor that attached to the place where it was located and likewise became, even if to a lesser degree and without the protection of special legal provisions, the bearer of communal identity. The fact that the clock gradually came to replace the communal bell as a characteristic urban element is also reflected in the frequent change of name from "beffroi" or "Stadtturm" ("city tower") to "Uhrturm" ("clock tower"), "Zytturm" ("time tower"), and "Torre dell'orologio." With the passage of time the importance of the communal bell declined more and more compared to that of the clock.

The development of the urban system of signals

The public clock was the final element within the system of signals in medieval towns and cities, a system whose development we can observe since the High Middle Ages. The high point of this development, its fullest differentiation, occurred in the first half of the fourteenth century, and it was the diffusion of public clocks which led in the long run to the simplification, standardization, and—some might say—impoverishment of these complex "acoustic environments" of medieval cities.[17] As important as the communal bell was for the political identity of the communes, the signaling needs of the cities had been continually growing since the High Middle Ages. The exalted status of the communal bell, which was originally probably no more than a practical aid, and the restriction of its use to a limited number of special occasions that concerned the entire community necessitated other signals, above all bell signals, but also fanfares, flags, and pennants. New signals were needed whenever the intended audience was not the entire community but only a specific group, or when the ordinary nature of the event was not in keeping with the dignity of the large bell.

The increase in urban signals repeated in many respects the multiplication of ecclesiastical bell signals. For some time the various church towers of a city had been ringing for many more occasions than merely the hours of prayer or mass. The increasingly complex ecclesiastical ringing arrangements exemplified the tendency to distinguish the ever-more-numerous feasts of the church year not only liturgically but also acoustically. The custom of ringing the bells upon someone's death or—for a fee—at funerals spread ever more widely. Added was the respectful ringing when ecclesiastical or secular dignitaries were in town, ringing to announce ecclesiastical punishments, and the warning ringing at approaching storms. Three bells were soon standard for city churches, a number that was often far surpassed by cathedral churches.

It is likely that law-and-order regulations in the cities were originally linked to ecclesiastical ringing. The prelates and manorial lords of Normandy probably sanctioned what was already an older custom when they decreed at the council of Caen in 1061 that the curfew—the call to cover the fires in the stove and to refrain from leaving the house—should be announced by the evening prayer bell.[18] Over time the opening and closing of the city gates, the changing of the guards, and the beginning and end of the market were regulated in similar fashion. For these and many other purposes the cities gradually acquired a growing number of their own bells. For cities, too, three bells (for example, great bell, council bell, gate bell) were soon a kind of standard accoutrement. Added were, depending on the size of the city, bells for the various council and court assemblies, for the dates on which interest and taxes were due, for individual quarters of the city, for the market (and possibly separate bells for the corn and the fish markets), a "Mus-Glocke" to summon the poor for the distribution of alms, bells for schools and universities, "Kehrglocken" ("sweeping bells") to remind homeowners of the obligatory cleaning of the streets, and on and on.[19] The last great wave of bell installations prior to clock-bells were the "Werkglocken" ("work bells") to regulate the working time of urban workers.

The "emancipation" of the urban ensemble of bells from the ringing of the churches could be more or less pronounced, but nowhere was the break complete. In smaller cities the ecclesiastical bells frequently also took on the function of signaling police regulations. The ringing for early mass (Prime ringing) was often the sign to open the gates; the time for market could be indicated by the mass bell and the end of working time by the ringing for Compline. The connection between the signal for prayer time and the civic time signal is particularly pronounced in the case of the so-called Angelus bell that spread throughout Europe in the Late Middle Ages. This evening call to prayer, since the turn of the fourteenth century rung with a three-fold bell signal (it was also called Ave Maria bell), became in many cities the civic evening or curfew-bell ("ignitegium"). In the course of the fourteenth century an Ave Maria prayer was added in the morning, and in the fifteenth century, another one at noontime.[20] The division of the day into three hour segments, derived from the canonical hours, was replaced by a simple two-part division of the civic day. In 1456, Pope Calixtus III made the noontime Ave Maria prayer, long since customary in many cities, obligatory for the entire Church as a prayer against the Turkish threat. Thereafter it was in Germany frequently called Pacem-ringing. Louis XI of France (1461–1483) supposedly decreed the "sonnerie du midi" for the kingdom, but this is an old legend which one cannot use to argue that it promoted the spread of modern hour-reckoning.[21] We find no mention of such an order in the fifteenth century. The midi-ringing that had long since been customary in French cities had in most places probably been regulated for decades according to the public clock.

A large backdrop of bells was thus part of the large cities around 1300. R. Davidsohn could write in his great history of Florence: "A wanderer approaching Florence, say on the morning of a great feast day, must have heard the tremendous ringing of more than eighty bells already from afar."[22] And this estimate of the number of bells is probably much too low.

The multiplication of bells brought it about that by the end of the thirteenth century their sheer number had become an index of the size and importance of a city. Bonvesin de la Riva's city description *De magnalibus urbis Mediolani,* written in 1288, is considered a milestone in the history of statistics, because it sought to prove the greatness of Milan with a wealth of numbers and measures. Bonvesin lists the number of inhabitants (200,000), the length of the city wall, the daily consumption of grain and meat, the number of solid houses (12,500), of wells (6,000), of mills (over 900) and their wheels (over 3,000), of surgeons (150), doctors (28), and on and on. He pegs the number of churches at about 200, of altars at 480. More than 200 bells rang from 120 church towers (chap. 2, IX). Whoever wanted to see this enormous city and its wonders had only to climb up the city tower and let his eyes roam about.[23]

If the number and size of the bells, if the acoustic ensemble made the size and importance of a city more audible than visible, the variety of the ringing also reflected political and socio-economic differentiations. In his *Libellus de descriptione Papiae* (around 1320), Opicino de Canistris, like Bonvesin, calls attention to the large number of church bells in Pavia. According to his account the largest bell, audible at a distance of more than six thousand paces (ca. nine kilometers), hung in the cathedral. Among the monastic bells the largest belonged to the predicant orders and the Carmelites. The guilds had their own large bell, which was rung to call the people to arms. The Council of the Wise was summoned by a certain bell signal ("certum sonus campanae"), the Council of One Hundred by a different one ("alius dissimilis sonus"). Another signal ("diversus sonus") called all of the citizens together, and still another one ("alius sonus") announced the proclamation of judgments and announcements by the city. In addition to the ringing for funerals, the text also mentions the tolling of the Angelus bell in the evening and in the morning, the wine bell ("campana bibitorum"), a small bell ("scilla") for the closing of the gates at night, and another bell that indicated with seven strokes in the morning that one could leave the city.[24] Indirectly the text used the richly developed acoustic ensemble of the city as an indication of the diversity and relative autonomy of the spheres of urban functions.[25]

Despite a wealth of material, the mere attempt to offer no more than a typifying reconstruction of the functional interrelationship of the urban signals is faced with considerable difficulties. Every city developed its own customs, and in each city the name of the bell would frequently change with the time it was rung, the intended audience, possibly also with the duration of the ringing. This applies not only to bells that were used simultaneously for ecclesiastical and civic signals. Added to this is the breadth of variation in the signals themselves, which is all but impossible to deduce any longer from the texts. With a hammer one can produce a great variety of tones and rhythms on a bell that is at rest. If the bell is moved in its suspension additional possibilities offer themselves: turning the bell upside down and letting it drop, sliding the hammer along in a circular motion, stopping the bell, and much more. Finally, various signals can be announced or repeated by a change in striking technique ("appell"). The diversity of signals produced in this way was understood without explanation only by those who had grown up with them. Texts alone allow only an incomplete reconstruction of these acoustic backdrops. For that reason I shall limit myself to a brief account of three relatively well documented urban bell ensembles.

The "bancloche" in Tournai (1188) was initially used for the council, to summon the inhabitants to assemblies and to arms, and to announce judgments. Another bell ("wigneron") was added after 1270. It was described as "wigneron du matin, wigneron du jour, wigneron du soir, tierce (or) cloque du darain" and was used for assemblies of the council and the court, for guild meetings, for the guards of the city gates, and as a wine bell. Another signal that served the same purposes appeared after 1302 under the designation "prime," "nuene", or "cloche du vespres." It is not clear whether this bell was located in the belfry or in some church tower. Finally, beginning in 1300 we find mention of the "cloke des ouvriers" ("cloche du disner," "resson"), which was used not only to regulate work time but also for selling at the market and as a call to arms.[26]

Treviso had only one large bell ("campana") and one small bell ("campanella") at the end of the thirteenth century. When rung with a hammer ("a martello") instead of a rope, the large bell called the men between the ages of sixteen and twenty to arms. It was used to summon the "consiglio maggiore," and until 1314 it served as the Angelus bell which also announced the end of work in the city. The beginning of work was indicated with the "campanella," which as "campanella degli anziani" called to the assembly of the "curia del podesta." The assembly of the citizens was summoned with both bells simultaneously. When struck with a hammer the small bell also called together the five hundred permanently armed citizens. Eventually, in 1315, the commune had a new, medium-sized bell installed, "campana magna nuova [or] la marangona." At first it was supposed to regulate the work time in the city at sunrise and sunset; however, because it was more audible, it was subsequently also used for the council assemblies.[27]

The communal bells of San Marco in Venice were much older than the civic ringing in Tournai and Treviso, which did not develop distinctly until around 1300. We are so much better informed about the four bells in the cathedral in Venice—in order of size: "Marangona, Nona, Mezza-Terza, Trottiera"—because they are frequently mentioned in the many extant guild and council regulations of the thirteenth and fourteenth centuries, and because their tolling is described in detail in the city descriptions of the sixteenth and seventeenth centuries. The bells of San Marco regulated above all the municipal administration and the activity at the largest work site in Venice, the docks (arsenal). The guilds also used them. The tolling was arranged around two fixed times of the day, midnight and midday, and around two movable points, sunrise and sunset. The early modern sources also allow an approximate fixation of the remaining times of the day.[28] The documented duration of the ringing could have been fixed only since the fourteenth century—probably controlled by sandglasses. In the interest of brevity I have drawn up a schematic reconstruction of the somewhat regular signals, which are based on an equinoctial day with precisely twelve daylight hours. The time indications are modern and do not follow the Italian method of counting that was customary at the time.

Systems of bell signals at San Marco in Venice

Time of day (approx.)	Duration of ringing	Bell signal Message of the signal
Dusk 6		MEZZATERZA, MATUTINA Withdrawal of the guards, opening of the church (twelfth cent.: beginning of work for hatters, earliest beginning of work for smiths on Monday).
Sunrise 7	1/4 hour	MARANGONA, CAMPANA MAGISTRORUM Workers at the arsenal leave for work (thirteenth cent.: beginning of work for joiners and carpenters; threat of punishment).
7:30–8	1/2 hour	MEZZATERZA, CAMPANA LUNGA DE MANE, CAMPANA OFFICIALIUM CAMPANA CONSILII Workers at the arsenal enter the workshops; threat of punishment; at the end of the signal officials must be at their workplace; canons commence the offices; at end of signal the "missa cantata" begins; after this the members of the maggior consiglio and the courts go into the palace, where they remain at least until Terce (thirteenth cent.: if there is a large demand on their services, barbers may bleed patients earlier).
9		MARANGONA, TERCIA SANCTUM MARCUM Courts are in session.
12		NONA Beginning of noontime break in the arsenal; permitted end of work for carpenters with deduction of half a day's wage.
12:30–13	1/2 hour	TROTTIERA, DOPO NONA, CAMPANA LUNGA POST PRANDIUM At end of signal: resumption of work at the arsenal and in the administration; threat of punishment.
14		NONA Snack break.
Sunset	1/4 hour	MARANGONA, CAMPANA MAGISTRORUM DE SERO, AVE MARIA End of work: arsenal, painters, smiths, shoemakers, wool-beaters, hatters; goldsmith may not work between now and the next signal of the Marangona.

19		MEZZATERZA Mounting of the guard.
19:30	1/4 hour	NONA, TERCIUM TINTINABULUM, TERCIA CAMPANA Closing of the guardrooms, prohibition of all cooper work using fire and light until Matins, prohibition of carrying weapons.
20	1/2 hour	MARANGONA
Midnight		MARANGONA
24		Changing of the guards.

The regular workday signals alone lasted between two and three hours. Added was the ringing for the masses, which could vary considerably depending on the significance of the day within the church year. Added also was the ringing that signaled the departure and return of the galleys. Special bell signals announced the election, arrival, or death of the doge, the pope, and various other civic dignitaries. There were signals for particularly solemn council meetings and a half hour ringing at the proclamation of death sentences. If we add to this the tolling of a few dozen city churches, among them that of the Rialto, which was important for the markets, there could have been only brief intervals when there was no ringing at all. Everyday life was temporally structured through and through by bell signals, with hardly a day resembling another. Apart from signals for proclamations, prohibitions, and ordinances, the inhabitants of the city also received a wealth of acoustic information about important public civic events.

Very much like modern disputes over broadcast frequencies, the increasing number of bells in the cities led to quarrels over ringing prerogatives, and the large orders had to reduce the number of bells in their urban monasteries to avoid conflicts with the parish churches. Certain church bells were granted privileges; for example, at some localities the episcopal bells were given the right to ring before all other churches.[29] In the civic sphere de facto prerogatives emerged. For Florence, Dante described the ringing of the Benedictine abbey of S. Maria Assunta ("Badia") at the old city wall as the one from which the entire city took its cues, "dond' ella prende ancora sesta a nona."[30] Modern authors have falsely inferred from this the existence of a public clock.[31] Fourteenth-century commentators on Dante know nothing of this, though they confirm that around 1300 the Badia maintained a kind of "standard communal ringing." According to Benvenuto da Imola, the Hours were rung "more reliably and orderly" in the Badia than in other Florentine churches; Iacopo della Lana tells us that the day laborers were guided by it in going to and leaving work.[32] An ordinance of the armorers in 1321 prohibited working with hammer and file prior to the morning ringing of the abbey.[33]

One might suspect that what was, at least in the large cities, a very effective and differentiated system of signals at the beginning of the fourteenth century had, at the height of its development, also reached its limits. Bell signals that are differentiated according to occasion, time, and intended audience cannot be multiplied indefinitely, and perhaps in some cities the threshold to signaling chaos had already been crossed. Presumably thought was given in some cities to ways in which the wealth of signals could be reduced and the signaling system simplified or individual signals made more readily recognizable. One attempt involved specifying the duration of the signal more precisely by linking it to experiential walking measurements. In Bologna the large communal bell was to be struck at the beginning of the day for as long as it took a person on foot to go one mile. However, such walk-time indications did not work out. A few years later the ringing regulation in the Bolognese statutes was changed: the morning signal was to be given with fifteen light and five strong strokes.[34] In the statutes of Piacenza during the brief rule of Galeazzo I Visconti, the father of the donor of the clock in Milan, we also find at the beginning of the fourteenth century a section on the daytime signals of the communal clock. Here the period of daylight is organized by numbered bell signals: at daybreak and at the first hour of the night the podesta should arrange for a single ringing; around Prime, after None, and for Compline, three series of twenty light strokes would be sounded.[35] While the civic day is still the period of daylight, and the times of day are described in the customary way, the stroke sequence follows a counting scheme independent of specific occasions or intended audiences.

For city dwellers all regular bell signals were also time signals from which all or specific groups could or had to take their cues. The timing of the signals was regulated according to daylight or the position of the sun. Possible, but surely quite rare, was also the regulating of the bells by means of astronomical observations and aids, for example, astrolabes. Perhaps the small number of nighttime signals were set off with the help of alarms, as was customary in the monasteries; however, we have no information on this.

The accuracy of the signals that was achieved by observing natural phenomena was undoubtedly adequate for everyday life in the cities. Since the bell signals themselves lasted a certain amount of time, it was possible, should it be necessary, to specify them with greater accuracy. If there was any doubt, nearly all regulations referred to the end of the signal. For example, according to an ordinance of 1385 in Evreux, the negotiations between retailers of fowl and suppliers could begin only when the tolling of the bell had ceased completely: "A heure de grant messe toute sonnée à la grant église et que le son de la cloche soit de tout apaisié et arresté."[36]

The end of mass in church and the beginning of merchant business in the cities mark also the boundaries where timing had to be coordinated. While the temporal succession of the times for masses and markets, council meetings and court sessions did matter, their precise temporal location did not. As a result, conflicts between various signalers were uniformly not conflicts over precise timing but conflicts over questions of status and rank.

At the beginning of the fourteenth century there was surely no need in the cities for an additional bell signal, especially one which, for technical reasons, could only strike an hour sequence known solely to the educated. The difficulties of readjustment were far greater than the—perhaps theoretically attractive—accommodation to the usage of the astronomers. A need could arise only after it had been realized that the new device offered possibilities of dealing with the great abundance of signals, in part by regulating them better, in part by reducing them. The reduction of the diversity of signals is far more difficult to trace than its increase. However, we can observe that, with the exception of a brief boom of work bells at the end of the fourteenth century, hardly any new bells or bell names appear. From what point on a given bell rang in accordance with a clock is almost impossible to determine now, since this did not, or did not immediately, change the name of the bell or of the signal. The gradual dying out of certain ringing customs during the following centuries has left virtually no trace in the extant sources.

Once again the post-Reformation visitation decrees and ecclesiastical statutes reveal the extent to which bell signals, no matter what kind, had already become time signals, particularly in small towns and villages. It soon turned out that the original plan of the reformers to simply get rid of most of the ecclesiastical ringing as being superstitious and of no consequence to the purposes of salvation could not be carried out. The reformers' early claim that true Christians were indifferent to the time and the hours of the divine service proved to be an illusion with no connection to the real world.[37] After only a few years the uncertainties in how to deal with ceremonial issues led to a veritable flood of regulations. Countless pertinent texts reveal that many of the long-familiar bell signals had become practically indispensable time signals. There was no other choice but to strip them of their magical components as much as possible and to explain their changed significance to the people. The lines of compromise were already indicated in the instructions for the visitors in the Electorate of Saxony (1528):

> Many pastors are also quarreling with the parishioners over unnecessary and childish things, such as the ringing of pacem [i.e., ringing of the Angelus bell] and the like. In such things it is appropriate that the pastors, as the more reasonable parties, should yield to the people for the sake of preserving the peace, and instruct them, where such ringing had been used improperly, so that it now be used properly . . . (weather ringing may be kept as an admonition to prayer) . . . Thus in many places the pacem ringing has established itself such that people know what time it is in the morning, also at what time in the evening they shall return from the fields to their homes.

According to the ecclesiastical statutes for Wolfenbüttel (1564), as well, the signal should be reinterpreted:

> In popedom a certain striking of the bell is performed in the morning, at noon, and in the evening, by means of which the people are to be reminded to pray to the Virgin Mary. However, since the highly praised Virgin Mary does not

> wish to have such an honor, which is due to God alone, and since this is also against the word of God, the people are to be instructed accordingly. The striking of the bell itself, however, . . . can be kept to indicate to the people morning, midday, and evening hour, thus reminding and admonishing the people . . . to pray for communal peace and good government.[38]

In Coburg the morning time Ave Maria and the evening time Salve ringing "with three strokes" should cease, "least we be considered popish." The signal, however, was purified. If the civic authorities so desired, one "stroke" could be rung in the morning as a sign of the beginning of work, and one in the evening to signal closing time.[39] In the Calvinist Upper Palatinate, a report on the implementation of a mandate to remove the remnants of "anti-Christian popedom" in the countryside notes that the ringing of the communal fire bell was stopped in order to avoid giving the impression that Angelus was still being rung.[40]

The problem of reinterpreting traditional bell signals into simple time signals remains visible also in later efforts to reduce the number of feast days. As late as the early nineteenth century, in a Nuremberg proclamation on the abolition of feast days on April 1, 1805, the bell signal for the Hours and Vespers was stripped of its original liturgical character and, through simplification, stylized into a time signal: "At the same time . . . the customary ringing of the Hours as well as Vespers . . . should be retained as a lengthy and, for those who live at a distance from the bell towers, convenient signal of the time of day, and should only be shortened."[41]

Clock time signal and "clock hours"

The new time signal meant over the long term that all regular signals rung on special bells and targeted at specific audiences were replaced by an abstract signal that could completely take over the customary signaling functions of these bells.[42] One only had to listen, count along, and know what should and should not be done at what hour. In the beginning, however, the first steps toward rationalization in all likelihood only added to the confusion of signals. Initial difficulties can be observed everywhere. It is frequently mentioned that the signal was to be audible throughout the city. One way in which this could be accomplished was by installing a sufficiently large or loud bell at a central location. Another way was to strike the hour at several locations. The standardization of the urban space with regard to the clock time signal can often be traced directly from the topography of municipal clocks. Lüneburg's first public clock was installed in the fourteenth century in the centrally located tower of city hall. In the fifteenth century the four parish churches that were located somewhat further toward the periphery were each given a clock. The trading house, situated at the edge of the city on the quay along the river, received its own clock in the sixteenth century.[43]

Substituting an abstract signal for the many varied signals could be done by simply regulating the old bell signals by the new clocks, a process undetectable to us. However, in most instances the municipal clocks were, in various ways, given a privileged place within the system of signals. The times of council sessions, of market, or of work could be tied to the clock time signal instead of a bell signal. As we shall see, this was frequently done, and in many cases presumably with demonstrative intent. Another possibility was to single out the municipal clock as the decisive point of reference in cases of conflict. One example of this is the signaling compromise that was worked out in 1531 in Mons between the abbey of Ste. Waudru, the parish church St. Germain, and the city. The abbey gave permission for a tower with bells at St. Germain. In return it forced through monetary demands and excruciatingly detailed regulations of the new municipal ringing. Ste. Waudru was allowed to police the number of bells at St. Germain, which were limited to eight. Among other things, St. Germain was permitted to ring morning mass at seven, and twice a year Nones around twelve o'clock. At all times the primacy of the ringing of Ste. Waudrau had to be maintained, and the bells of St. Germain must not disturb the sermons in the monastery. An addendum to the compromise (1535) determined that the masses in St. Germain could be rung only with three straightforward strokes, and that in all cases of conflict the ringing of the clock of the city of Mons should be the decisive point of reference.[44]

If there were several clocks in a city, there was only one way in which the municipal clock could be given privileged status as the city's time standard: either its signal or that of the other clocks had to be set off at the "wrong time." To this day the municipal clock in Lucerne possesses a prerogative "from ancient times": it is one minute fast.[45] Marperger's *Horolographia* (1723) provides an explanation that is as simple as it is compelling: city clocks must ring in succession "lest they confuse those listening by their many strokes if they all ring at once."[46] The simultaneity of the new time signals, only a theoretical possibility, also ran in practical terms into the limits of the old signaling technology, that is to say, the bell strokes, which took time.

There were other reasons, too, why the new, more abstract rhythm of urban life brought about by the clock proved to be somewhat remote from everyday life. The twenty-four hours of the full day could not simply displace the old division between day and night. To be sure, the chronicles of the cities proudly announced the installation of a clock that "struck the hours of the day and the night," and in some cities the dials of the communal clocks were also illuminated at night by torches or lanterns. In other places, however, the continuous time signal was evidently seen not only as a useless disturbance but also a desecration of the night and was simply turned off.

A chronicle of Eβlingen reports the measures that were taken in the year 1519 to resist the siege of the city by Duke Ullrich von Württemberg. After the duke's men had boasted that they would take the city "by morning soup," the citizens of Eßlingen, informed of this, wound up the "clocks at six in the evening and rang and chimed the entire night."[47] The records show that in Romans in the Dauphiné, the communal clock started to chime also at night only in the year 1625.[48] At the end of the seventeenth century, Duke Ernst August made available to the small town of Arnum, "which had been waiting for many years with particular eagerness," a renovated tower clock from the Marktkirche in Hannover. He sent it along with the recommendation to keep it in operation at least during the day, "from eight o'clock to eight o'clock."[49] At many a place the modern division of the hours thus remained limited to daylight; for the time being the abstract full day had a difficult time prevailing against the old division of day and night.

8

The Ordering of Time: The Introduction of Modern Hour-Reckoning

A legendary decree by Charles V of France

Modern horological literature illustrates the transition from canonical Hours to modern hour-reckoning with a story that seems to fit very nicely at the end of the Middle Ages and the beginning of the modern age. According to this story, the new striking of the hours goes back at least in one instance to intervention by territorial authorities. King Charles V of France, we are told, installed striking clocks in his residences at Vincennes and St. Pol and built the public clock at the Louvre. Subsequently he issued an ordinance requiring all churches in Paris to regulate their tolling by the clocks or, according to some accounts, by the clock at the royal palace, the Horloge du Palais. It is easy to see why this story could become so attractive. In it the elusive and only vaguely observable process of transition to modern hour-reckoning is condensed into a single event with a single actor. Once the event is out in the world it gives wings to the imagination of historians. Carlo Cipolla believes the ordinance was issued because the king was afraid that the sound of the clocks could not be heard by everyone in town. P. Usher and J. Gimpel see in it a progressive contribution to laicization and modernization in the church towers, "a decisive step toward breaking the dominance of the liturgical practices of the Church." David Landes recognizes it as the affirmation of the primacy of royal power. Jacques Le Goff and Krzysztof Pomian emphasize that the new, rational time "thus became the time of the state. The royal reader of Aristotle had domesticated rationalized time."[1] The early modern secular state put an end to traditional practices of the Church and, symbolically and practically, took the lead in the movement toward the rationalization of time.

This story is undoubtedly well suited to illustrating the transition to modern forms of time-keeping and thus indirectly the change in time-consciousness. But did this richly symbolic event actually take place?

To begin with, we note that none of our informants ever saw the ordinance among the royal decrees or those of the city, most of which were known. In her biography of Charles V, written in 1404, Christine de Pisan devotes many pages to the precisely regulated, and therefore virtuous, daily schedule of the king. She does not neglect to emphasize that the king used candles as time-keepers to divide his day into work and prayer, since clocks were not yet very common at the time.[2] Would she have failed to mention this ordinance in the catalogue of his accomplishments? But we find no trace of it either in her work or the other biographers of the king. The last author who, to my knowledge, even mentioned a source for this story was Albert Franklin (1888). He was, at least, so well read in the contemporary Parisian sources that he noticed the apparent failure to observe this ordinance. He noted that the customary indications of the time of day remained in use for centuries to come; his explanation was that this "wise custom" simply passed into oblivion in the troubles following the death of Charles V.[3]

The source Franklin mentions is a translation of Guillelmus Duranti's commentary on the divine offices, *Rationale divinorum officiorum* (end of the thirteenth century). At the request of the king, Jean Golein had translated Aristotelian texts as well as classic liturgical works into French. The commentary on the offices, the translation of which was completed in 1372, deals with the various kinds and names of church bells, which include clock bells, and also gives model patterns of stroke sequences. Jean Golein amplified the text with commentaries that brought it up to date, and among them we find the—accurate—report about the public clocks of Charles V in Paris. However, a look at the manuscripts of the fourteenth century reveals that a variety of misunderstandings also made their way into Golein's additions.[4] The text contains early accounts of occurrences that can be attested also by other, independent sources: the installation of three striking clocks in Vincennes, St. Pol, and at the royal palace; Golein also reports the hiring of expensive foreign experts. That we are dealing with the introduction of something genuinely new is in part explicitly stated ("premier à Paris"), in part reflected in the still unsure terminology for public clocks ("cloches atrempées"). Problematic is the interpretation of the stroke of the bell: "par poins à manière d'arloges." Some authors have read from this that at least the Horloge du Palais also struck quarter-hours. Though this cannot be ruled out entirely, it is not independently attested anywhere for this prominent clock. In the old divisions of the hour, "punctum" is the fourth part of the "hora," though at a time when "hora" did not yet mean equal hour. After the rise of the modern hours, the quarter remained an important subdivision, and clockworks in large cities at a later time often also struck quarter-hours, though as far as I can determine they were no longer called "puncta."[5] The suspicion arises that the author and some of the copyists were familiar with the older literature but not with modern hour-reckoning. This suspicion is confirmed by the account of the hour-stroke supposedly decreed by the king. One stroke was tolled at Prime, we are told, two (or three) at Terce, and so on. Jean Golein and the copyists of the manuscripts did not hear this sequence in Paris at the end of the fourteenth century, but had read it in the text of Duranti as one possible stroke-sequence for the canonical hours. The variations in the manuscripts show that such sequences were not obligatory and certainly no longer customary. Contemporary Parisian time indications leave no doubt that in the city the hours were from the beginning numbered sequentially from midnight to midnight ("neuf, once, douze heures de l'orloge du palais"). Something else that speaks against a royal ordinance about the hour-tolling of the church bells is also the fact that it would be without parallel in the European urban landscape. At least in France the Horloge du Palais was such a formative and frequently copied model that one could surely expect that such an ordinance would have been adopted elsewhere. Moreover, whether the king even had legal authority over the church bells of Paris and their tolling or could have enforced it is also an open question.[6]

The tolling of the bells by the clock and the use of modern hour-reckoning was nowhere decreed. Initially the urban hour-stroke took its place as an additional way of determining the time of day. It then prevailed in various spheres of urban life, sometimes quickly, sometimes haltingly. It became obligatory only in connection with the codification of regulations concerning concrete problems. In fact, it became an established practice only where it proved useful.

Chroniclers and notaries

Contemporary chroniclers noted the installation of the first public clocks not only as technological accomplishments. Reports that "the public clock began to strike" are often quickly followed by the practical use of the new signal. Chroniclers attached modern hour indications to the births and deaths of high-ranking people, the arrival and departure of important embassies, to storms, events of war, and executions. From the end of the thirteenth century we can observe with greater frequency the mention of the time of day, something which had still been rare in the High Middle Ages. With the appearance of the public clocks in the fourteenth century the number of such indications rises noticeably. Traditional forms of indicating the time of day (for example, "at sunrise," "at time of Vespers") were supplemented with, though not replaced by, modern forms (for example, "4 hours after midnight"). Specifying the time of day more precisely lent the reports a greater air of authenticity and created a dramatic effect through the listing of dense chains of events.

The close connection between the appearance of public clocks at particular places and the use of modern hour indications in local chronicles has been observed particularly in Italy. This connection is striking; however, we must also bear in mind that many chronicles very quickly, sometimes after only a few pages, return to the use of conventional time indications or leave such indications out entirely over long stretches. This shows that while many chroniclers demonstratively acknowledged the new device, there was neither interest in nor need for continuous and consistent indications of the time of day.

Leaving aside astronomical and astrological calculations, modern hour indications are practically impossible without the use of public clocks. A modern hour indication is thus a sufficient piece of evidence for the existence of a clock in a particular locality. However, the reverse is not the case: the absence of modern hour indications is not evidence for the absence of mechanical clocks in general and of public clocks in particular. Gustav Bilfinger used the indications of time in the Chronicles of Jean Froissart, which cover nearly the entire fourteenth cen-

51. Jean Froissart in his study. Miniature from the so-called *Breslauer Froissart,* vol. IV, folio 1 (around 1470). Berlin, Staatsbibliothek Preußischer Kulturbesitz.

tury, to trace the appearance of the new hour indications. The first published part of the work deals with events up to the year 1377 and contains only traditional indications of the time of day. The second part describes events in Flanders and Paris between the years 1377 and 1386. Here modern hour indications appear first with the events of 1379–1380 and thereafter with greater frequency.[7] Froissart thus changed his style of dating and timing events quite precisely during the period of a wave of acquisitions I described earlier, in this way confirming our statistical observations. In the account of the events in Spain and Portugal (1382–1386), about which Froissart informed himself at the court of the Count of Foix, modern hour indications appear only for the court of Foix itself. Bilfinger concluded from this that no striking clocks, or only very few, existed in Spain at this time. However, the sources that have been published since Bilfinger's time show that this conclusion was wrong, that the Count of Foix in fact had clocks and instruments sent to him from the kingdom of Aragon. This example, too, reveals that while contemporary chronicles reacted quickly to the new innovation at certain points, these reactions are noticeably scattered and not stable, that is to say, they disappear again.

Neither in the fourteenth nor in the fifteenth century did they lead to permanent changes in the customary ways of timing events in the chronicles.

In the year 1395 a notary recorded the following: "on the nineteenth day of the month which is called September in Latin, at the hour when the clock ("orglogg") struck five hours after midday." With these modern, though apparently still unaccustomed, indications of the calendar day and the time of day, Ulrich von Petershausen, scribe and notary in Eßlingen, met his own exacting requirements for dating a notarial document drawn up by his hand. This date contains the oldest extant reference to a public clock in Eßlingen. "Five in the afternoon" was evidently not yet a usual hour indication. The explicit reference to the "orglogg" was meant to rule out any confusion with "hora quinta," which could also be understood as an unequal hour.[8] Preventing confusion was also the purpose of the dating of a notarial document from Cologne: March 2, 1424, "about the ninth hour in the morning according to the stroke of the clock of the Cologne cathedral."[9] In the Late Middle Ages, the canonical None, originally an afternoon hour, could have designated at most the midday point. The indication "ninth hour of the morning" would thus have sufficed to characterize the hour indication as clock time. At the beginning of the fifteenth century it was also not necessary for the cleric and notary Johann von Frelenberg to point to the existence of a clock in the cathedral. He dated his document this way, first, to avoid any confusion with another clock-time signal, for example, that of the Cologne town hall, and, second, to make clear (probably unconsciously) that the cathedral (or the cathedral precinct) with its bells was not only the legal but also the temporal framework of reference for his notarial documents. The qualifier "about" ("vel quasi") that is attached to the hour indication shows, in any case, that the notary was not solely concerned with specifying the time with greater precision.

The dating of notarial instruments, which was strictly regulated in formal terms, and all the other dating of charters on the notarial model are a rich field of investigation for the question concerning the direction, speed, and depth of the diffusion of modern hour-reckoning over longer periods of time. Notarial instruments ("instrumenta publica") are verifications of legal acts drawn up by authorized notaries ("notarius publicus"/"tabellio"). The adherence to certain legal and formal requirements established the evidentiary force of the instrument "in and of itself and against all challenges." No matter whether witnesses were still alive, seals broken, or challenges mounted by third parties—until it was proved otherwise, the notarized legal act remained valid for ecclesiastical and secular courts. Notarial deeds thus became, in the Late Middle Ages, fully valid and widely used evidence that increasingly displaced sealed documents.[10]

As an office and an instrument of authentication, the public notariate was an institution with origins in Roman law; it was revived in the eleventh and twelfth century and developed further. By way of the learned ecclesiastical courts and the law students at the Italian universities it spread throughout Europe after the thirteenth century and became an important medium for the reception of Roman law in central and northern Europe.[11] From the thirteenth century on notaries and notarial instruments are found in nearly all European cities. The notariate had considerable importance above all in the region of the "written law," that is to say, outside of Italy in Spain, southern France, Bohemia, Silesia, western Switzerland, Tyrol, and the southern German cities. In these areas, in particular, many thousands of notarial documents and protocols have survived.[12]

The formalities required for the recognition of the validity of notarial instruments included—in addition to the invocation formula (appeal to God), the full name of the witness, the signature of the notary, his unmistakable sign (signet), and the precise location—the detailed date consisting of several parts. The date could be placed at the beginning of the document (in the protocol) or at the end (in the eschatocol). To make it forgery-proof, it should be given in letters instead of numbers. The date naturally included the indication of the year. Roman legal tradition demanded the naming of the regnal year of the emperor, the consular year, and the indiction (cycle of Roman tax years). Already in the twelfth century the Roman imperial years were replaced by the indication of the year since the birth of Christ (incarnation year), supplemented, however, by the pontifical year of the current pope. In the Late Middle Ages the indication of the year was often joined by the regnal year of the emperor, which replaced the pontifical year in the fifteenth century. The name of the month was to be followed by the day, which was frequently indicated in dual fashion after the Christian festal calendar and the consecutive counting of the days.

In the fifteenth century, the latter largely replaced the counting of the days according to the Roman calendar (Calends, Nones, Ides). Some notaries also gave the day of the week.

Beginning in the thirteenth century we find, at first only in Italian notarial dates, a new element foreign to the tradition of Roman law: the indication of the time of day.[13] It is given according to the unequal hours ("hora sexta, h. nona"), the canonical offices and services ("hora primarum, hora in qua missa . . . celebratur"), or the corresponding ringing ("pulsante ad vesperas"). Next to ecclesiastical times, natural and civic division of the day were also used ("hora crepusculi, hora prandii, umb mittag zyte"). In use at the curia in Rome and Avignon were, in addition, indications of the time of day according to the rhythm of the bureaucracy.[14] Such indications specify the time of a particular legal act ("actum"); they do not concern the drafting of the document itself ("datum"). The formulas used therein ("inter terciam et nonam," "hora tercia vel quasi," "circa hora terciam," "umb mittag zyte oder noch doby") show that it is not a point in time that is being specified but a brief period of time in proximity to a signal. There was neither a need for more precise indications nor feasible technical means for them. Already by the middle of the thirteenth century, indications of the time of day in notarial dates were regarded at least as customary. The "Mirror of Government" by Orfinus Laudensis expects the authorities to possess the requisite chronological competence for drawing up "acta" and "pacta."[15]

That these new, additional elements in dating originated in Italy is undisputed; the only open question is their provenance.[16] Some scholars have traced them back to the customs of papal notaries.[17] This cannot be entirely ruled out. However, my own observations, based on the earliest evidence, suggests that they were rather an innovation of municipal notaries; what motivated them can be no more than conjecture.

Even the works on the "artes notariae" do not clearly reveal whether and since when the indication of the hour of the day was obligatory. Salatiel (around 1250) does not yet mention it. Baldus de Ubalids, in the second half of the fourteenth century, declares that it is not essential for the validity of a notarial instrument.[18] It was, however, recommended for special cases, such as authentications at nighttime (permitted only under particular circumstances), since the day period from midnight to midnight was customary among lawyers. To be sure, the *Reichsnotariatsordnung* (Imperial Statutes on Notaries) of 1512 mentions among the formal requirements "the year of our salvation, the Roman tax year, . . . the name of the highest prince; then the month, the day, the hour."[19] A few books of notarial formulas adhered to these requirements. Others, however, declared that they were not essential, though they recommended their use in order to avoid possible conflicts concerning the sequence or temporal precedence of relevant legal procedures.[20] The cautious and careful notary should use them. The sequence of the time indications in the files of the above-mentioned notary from Cologne also shows that he sought to protect himself against conflicts over temporal precedence by listing the order of the petitions of appeal by day and hour. However, it is rarely reported that precedence as to the time of day did in fact lead to legal conflicts.[21]

The records of the episcopal consistory in Prague are among the few large, published collections of notarial documents from the Late Middle Ages. The edition contains over 5,000 dated pieces for the period between 1373 and 1406. The results of a survey of the indications of the time of day are as follows: between 1373 and 1381, the time of day—usually "hora quasi tercia"—is given with considerable regularity. The first unequivocally modern hour indication in the Italian-Bohemian form appears in 1381 explicitly as an indication of clock time: "hora quasi XX orlogii." In the subsequent years, however, all indications of the time of day become more rare again. Scattered clock-time indications are found in 1383, 1386, and 1396. In the years 1406–1407 there is a noticeable cluster of about a dozen modern time indications in approximately 1,800 documents.[22] It would appear that indications of the time of day were important to notaries and scribes only periodically; their use may also have depended on personal predilections and habits. We can also see how slow and scattered the use of modern hour-reckoning still was decades later. Finally, it is noticeable that clock-time indications appear at times when the public time signal in Prague must have been the talk of the town. The first report of a clock at the Altstädter City Hall comes precisely from the year 1381, and in the years prior to 1410 Nikolaus von Kaaden renovated and rebuilt the clock at city hall.[23]

A select series of Genoese notarial protocols from the period between 1400 and 1440 likewise shows, without any visible connection to the public clocks in Genoa, how halting and scattered the reception of the new hour-reckoning was. After 1416 it was occasionally used, though the same notaries employed old and new ways of counting the hours side by side. But at least we learn from this source that clocks were also on the ships used by the defenders of Constantinople in their flight before the advancing Ottomans.[24] In Silesian notarial documents, modern time indications did not appear until late—around 1400—and remained quite scattered during the fifteenth century; in many other collections of documents and deeds they don't appear at all.[25] The possibilities of dating important occurrences more precisely were used only occasionally; the possibility of establishing a more abstract system of reference for the time of day was not used at all. During at least the first two centuries of public clocks, these possibilities went far beyond what the chroniclers and notaries needed and were interested in.

"Assisi, in the year 1525, January 15, in the morning at dawn, at the fourteenth hour in church according to the signal of the bells that are rung at the beginning of day":[26] this time indication from the beginning of the sixteenth century clearly bears out that abstract hour indications were only one possibility which had taken its place *alongside* the observation of natural passages of time and the orientation according to the civic and ecclesiastical tolling; it had not replaced them.

"Merchant's time"?

As we have seen, when it came to the procurement of public clocks, merchants and traders did not stand out as a distinct group from the larger political community. Moreover, the preeminent mercantile centers of the fourteenth century did not play a prominent role for the overall statistical picture of the course of diffusion. These findings still leave intact Jacques Le Goff's theses that the development of commercial networks and the spread of money created the need for more accurately measured time, and that commercial documents in the narrower sense of the word—accounts, travel diaries, manuals of commercial practice, and letters of exchange—imparted a growing importance to "measured time." Now, the importance of temporal factors when it comes to trade and commerce, interest and credit, storage and transport, speculation and insurance, is so obvious that we could confine ourselves to the observation that "measured time" is an expression that can be easily misunderstood, since today such time is always understood to be technically measured time, that is clock-time. Still, the question remains open whether merchants or commercial practice made special use of the new timekeeping devices and/or their product, the new hour-reckoning, for their own purposes. We should remember that Jacques Le Goff did not intend his reflections to be a final statement, but a stimulus for more thorough studies.[27]

Commercial records, and in particular the so-called manuals of commercial practice, have by no means been neglected by scholars during the last decades. Vittore Brance, editor of important texts in the genre of the "ricordi" left behind by the "mercanti scrittori," believes he has found confirmation for Le Goff's theses: in one introduction he notes that these "children of Mercury" grew accustomed to timekeeping through their hourly recording of their activities, thus helping this new, measured time to achieve a breakthrough.[28] However, if one reads the texts themselves, one notes that while the texts written by merchants, traders, and entrepreneurs contain modern time indications in many passages and at a relatively early date, in no case does the use of "measured time" have anything at all to do with the commercial practices of these people. Another early specimen of this genre are the records of the Nuremberg patrician Ulman Stromer. Beginning with the year 1374, the account of his ancestors and his family (which was composed between 1360 and 1407) notes the time of birth of his children and grandchildren in the Nuremberg form of hour-reckoning; incidentally, this is the earliest attestation of this method.[29] The politically influential Florentine cloth merchant Giovanni di Pagolo Morelli kept "ricordi" between 1393 and 1411. Beginning in 1365, he recorded births in his family with the time of day in the modern form. Occasionally he also indicated the ringing of city churches (e.g., 1377 Nov. 27: "e fu il mercoledi notte, vegniente il giovedi, a ore otto e mezzo, presso alla squilla di Santa Croce"), or he gives the time of day both in the modern form as well as in accordance with the canonical hours (e.g., 1404 July 20: "sonate di poco le sedici ore, ciò fu al tocco di nona, nacque . . . un fanciullo maschio").[30] His contemporary and fellow Florentine Lapo di Niccolini de' Sirigatti did much the same. In his "ricordanze" (1397–1427), a sort of counterpiece to his commercial books proper, he recorded, apart from his most important commercial contracts, also births, marriages, deaths, and testaments in his family. Beginning with the year 1386, these records give the time of day in the modern form, particularly for births, occasionally for deaths, and in one instance for a wedding.[31] The actual commercial books themselves, as far as one can judge from the published texts, do not contain any time indications according to "measured time."

In their private records, the merchants were following a maxim which Leon Battista Alberti soon after elevated to the norm. In his tract *I Libri Della Famiglia* (1434), he discussed the economical use of time, the technique of assigning a given time to many tasks, the nightly accounting of how time was used, and the duties of every father to conscientiously register family events: "as soon as the child is born one should note in the family records and secret books the hour, the day, the month, and the year as well as the place of birth. These records should be kept with our dearest treasures. There are many reasons for doing this, but, all else aside, it shows the conscientiousness of a father. If it shows a kind of conscientiousness for a man to keep careful note of the day and the agent from whom he bought an ass, is it less right to make note on the day on which you became a father, when a brother to your children was born? Occasions may arise when you will want this information."[32]

When it comes to the reasons for such records, Alberti is obviously deliberately vague ("many reasons," "many cases"). What he means, however, is clear: the private chronicle with hour indications is morally valuable; moreover it is also legally necessary, for example, in conflicts over inheritance.

In addition to these texts, modern time indications are found at the end of the fourteenth century also in merchants' records that are more personal and more autobiographical in nature. The voluminous correspondence of Francesco di Marco Danti, an international merchant from Prato near Florence, contains letters to his wife which he occasionally signed with "In haste" ("in fretta"); he presents himself as a man hurrying from one appointment to the next, for example: "tonight, in the twenty-third hour, I was called to the collegi." He lets his associates know that he hasn't slept more than four hours a night in years, and on occasion he comes up with truly modern ways of describing the feeling of constant time pressure: "I don't have any time, it is the twenty-first hour and I have had nothing to eat or drink."[33]

52. Astrologer with astrolabe and striking clock, from the *Chants royaux sur la conception* (1519–28). Paris, Bibliothèque Nationale.

With the beginning of the fifteenth century, new elements appear in the time indications of some merchants, elements Alberti may have also had in mind but did not explicitly mention. Bonaccorso Pitti, another of the great Florentine merchants, had been writing his "ricordi" since the year 1412. Modern hour indications are found in historical and chronistic observations for the period after 1378, and in familial events after 1408, whereby the indications of the time of day are accompanied by relevant astrological specifications.[34] At the beginning of the sixteenth century we read something very similar in the diaries of the Augsburg citizen Lucas Rem: "Berchthold, my son, was born Monday, January 11, 1529, in the afternoon a quarter past two . . . One day earlier the new moon had been in Aries at 9 o'clock 2 minutes . . . He died on October 14, 1530, in the evening around 9 o'clock. The day before the moon's last quarter stood at 2 o'clock 44 minutes in the afternoon, 12 in Leo, the sun in Scorpio."[35] Rem's records of births and the fate that befell family members contain, alongside indications of civic clock-time, very precise astrological time indications. However, the so-called age of the moon and the position of the planets in relation to the zodiac were at that time not data gathered by measurement or observation; instead they were calculated time indications whose precision was feigned. To be sure, the growing interest in astrology did stimulate an interest in clocks with astronomical indications, but such interests were neither typical of nor limited to merchants or city dwellers.

Insofar as it is understood as time measured by clocks, "merchant's time" turns out to be a misunderstanding in the more recent historical literature. However, this by no means solves the problem of "merchant's time" as a formulaic expression for a special consciousness of the temporal aspects of the movement of goods and capital. Independent of the history of "measured time" we still need to ask whether, and if so how, the temporal aspects of commercial activity and commercial rationality came to consciousness in the Late Middle Ages, and whether it is not likely, therefore, that in the group of those most affected by this, "time" became a topic of special interest. A beginning has been made by F. C. Lane with his study of the "timetables" of the Venetian galleys and the speculative transactions connected with them, and by U. Tucci with his examination of commercial-speculative model calculations involving harvest dates, available stocks, transportation prospects, and local prices and interest rates.[36] Despite many references to relevant sources, these aspects of "merchant's time" have still been too little studied.[37] Merchants as participants in urban markets, as entrepreneurs, and as employers will be discussed later.

"Resistance from the church"?

The counterposing of the catchphrases "Church's time"–"merchant's time" imparts a graphic and seductive drama to the historiography of the Late Middle Ages. Once one has identified progressive, modernizing factors, their contribution to the rationalization process can be highlighted much better if one can also identify the forces of resistance and adherence to the status quo. We have earlier rejected the notion that the churches and monasteries resisted the installation of public striking clocks and thus opposed the adoption of a technological innovation. To be sure, as spiritual institutions they rarely took the initiative, but they were frequently cooperative when it came to such projects. What about the claim, then, that churches and monasteries were not interested in the social innovation?[38] Did they hesitate or did they oppose progress?

It is true that churches and monasteries, on the basis of the needs of their internal operations, had no compelling reasons to adopt striking clocks and modern hour-reckoning. It is false, however, to say that they were "late" in deciding to go along with this innovation. How did this misunderstanding come about? Bilfinger had taken from J. Martène's commentary on the Rule of St. Benedict the notion that the Cistercians, for example, had retained the medieval Hours until the year 1429.[39] However, the statutes of the General Chapter of the order in the year 1429 do not contain any such regulations. Instead, because of complaints about late ringing and rising for the morning offices in many monasteries, it was ordered that in the future morning offices should be rung at the second hour of the day after midnight on weekdays, and at the first hour after midnight on feast days.[40] The decrees merely specify the always controversial beginning of nightly offices with the help of modern hours, which of course presuppose the use of mechanical clocks. This decision was only one of many comparable efforts at achieving greater precision and was by no means an early case, let alone the first of its kind.

Because of their distance from the classical rules of the Middle Ages, texts of liturgical practice from the fourteenth and fifteenth centuries have been rarely published. But even a brief list of random examples conveys a clear picture concerning the historical development and its geographical framework. In 1379 the statutes of the monastery cum theological college of St. Martial in Avignon, home to monks and students, fixed the beginning of the offices according to modern hour-reckoning. The fourteenth-century revision of the regulations for ringing the bells at the cathedral in Milan organized the ringing for mass, the tolling of the Angelus bell, and the ringing for Vespers according to the modern hours. At least for the Vespers bell this had already been done in Bamberg prior to 1380. Not only was the liturgy for the cathedral in Colmar (1404) regulated by the cathedral clock, details were even arranged by the half hour and the quarter hour. The innovation also gained quick admission into monasteries and outside of the urban environment. The statutes of the Abbey of St. Mary in York, revised prior to 1404, arranged not only waking time but the entire monastic daily routine explicitly by the hours of the clock.[41]

Thus churches and monasteries, no differently than secular institutions, did not hesitate in introducing and making practical use of the new technology as soon as it was available. It might even be possible that at least in some places churches had a kind of pioneering role and furnished their urban environment with time signals as long as the latter did not have a clock.[42]

Given that the mechanical escapement had presumably developed in conjunction with mechanical bell works, and that large astronomical clocks and carillons had been built primarily for churches in the first half of the fourteenth century, there is no reason why ecclesiastical institutions should now have opposed striking clocks and the new hour-reckoning or obstructed their introduction.

The change of the temporal order in the cities

The change in time consciousness that was brought on by the introduction of clocks in the European cities from the end of the fourteenth century was a complex process. We can trace it through various details that are inconspicuous at first and whose interconnectedness becomes visible only in retrospect. As far as we can tell, contemporaries were rarely aware of the largely anonymous process of change in time-consciousness. One of the rare examples is the observation by the English Black Friar at the beginning of the fifteenth century that in towns and cities men ruled themselves by the clock.[43]

How city-dwellers came to rule themselves by the clock can hardly be traced directly via changes in everyday conduct. Only innovations in everyday speech reveal the ubiquitous presence of the clock in the urban sphere since the end of the fourteenth and the beginning of the fifteenth century, at the latest. A short language primer for Englishmen traveling to France includes, for example, a model dialogue for polite questions, such as "What has the clock struck? What time is it?" and so on.[44] However, the reason why it was possible to speak of the rule of the clock at the beginning of the fifteenth century cannot be traced by fortuitous linguistic evidence, but only through changes in the urban ordering of time itself. Here the term "ordering of time" (German "Zeitordnung") does not describe a kind of chronology, as it did in the seventeenth and eighteenth centuries;[45] rather it refers to the totality of time-ordering rules and specifically to the temporal order of everyday life in the city.

In the following attempt at reconstructing typical specimens of these kinds of temporal orders, the only distinction I have made is between weekdays and feast days. Not considered is the temporal patterning of the week, which means, for example, that all regulations that were applicable on only two weekdays were treated like everyday regulations. I have also ignored the festal calendar with its considerable local variations, the rhythm of the markets and fairs, as well as all other seasonal elements.

Our sources, particularly for the fourteenth century, are municipal statutes of every kind, which have survived in large numbers. The frequent recodifications and redactions make it possible to follow innovations in the organization of time over longer periods. From the fifteenth century on, the extant statutes can be supplemented by the no less numerous school statutes, ecclesiastical statutes, territorial administrative statutes, and territorial laws. In regard to questions of time organization, these later groups of sources differ from communal statutes only in that their range of applicability was either larger or smaller.

I will use the relatively rich sources from Cologne to illustrate the chronology of the advance of new kinds of rules and the areas into

53. Sundial on the house of the coppersmith. Diebold Schilling's *Luzerner Chronik* (1513).

which they penetrated during the period of about a single human lifespan.[46]

We first encounter modern hour-reckoning in Cologne in 1374 as the stipulation of a time limit. A work statute for day laborers and workmen fixed the beginning and the end of the work day by the first bell signal for Prime and the last signal for Compline "at the four monastic orders." However, the break during the day was to be "one hour ("ure") long and no longer."[47] Sometime after 1371, but before 1397, the end of working time for belt makers was set at ten o'clock in the evening. After 1385 the meeting of the lords of the register of city assessors was to begin at one o'clock; a little later the time for the session of the court of assessors was fixed at eleven o'clock. Until 1375 both times still took their cues from the masses in the Marspforten chapel, the church of St. Lorenz, and the convent of Mariengraden. In fact, during the summer the beginning of the session of the court of assessors was set by one of the sundials so common in the Middle Ages but rarely documented in daily use: "before the sun comes to the hole in the stone."[48] In 1385 a decree of the council also regulated the sale of Rhine salmon on the fish market. In the summer the selling was not to begin before six in the morning, in the winter not before eight; wholesalers could be served only on Fridays after ten. After 1391, foreigners who came into the city with safe conduct should no longer be on the streets after nine o'clock (in winter after eight). Beginning in

1398 the same times applied to armed people. General curfew regulations after 1398 stipulated that "no priest, no student, no layman, no woman, and no man" should be on the streets after eleven o'clock.

In the wake of the successful uprising of the merchant companies, working times for the various artisanal crafts were newly regulated in 1397 in a large number of "Amtsbriefe" (official letters). The new time regulations — usually introduced without any explanation — were uniformly restrictive in nature, that is to say, they determined, for reasons of competition and quality safeguards, the maximum amount of working time permitted. Added to this were restrictions for activities that were noisy and posed a fire hazard. Weavers of fustian, armorers, and pursemakers could not begin before five in the morning and could work at the very most until nine in the evening. Coppersmiths and needlemakers had to stop at eight o'clock. Joiners (chestmakers) were allowed to work from four until eight, felt hat makers from four until ten. The shortest permitted work time belonged to the smiths. Though they could start at eight o'clock in the morning, they had to cease working as early as five. The work time of belt makers was also newly regulated. The beginning and end of work was to be determined by the "broad and fair daylight." Only during winter could work continue by candlelight until nine. In the fifteenth century, the beginning of work was then set at six o'clock. On Saturdays and on evenings preceding high holidays, belt makers had to stop at four, needlemakers and smiths in the winter at four, in the summer at six and seven, respectively. Tailors who worked past eight o'clock on evenings preceding feast days had to pay penance of one pound of wax.

To be sure, not all of the work regulations in Cologne were switched to clocktime during these years. The "Amtsbriefe" of the yarn spinners and pewterers delineated their work time by the cathedral mass in the morning and the evening bell at night. However, one phrasing in the "Amtsbriefe" of the needlemakers ("vur doimmetten of vur vuoif uren") shows that this mass was celebrated at around five in the morning. During the same period the soaking of dyer's woad for blue-dying was prohibited before ten in the morning on certain days. Buying and selling on the woad market was permitted only after "nine o'clock has been struck." In 1380 a clock-bell had been obtained for the tower of the Carmelite church. According to similar regulations from the first half of the fourteenth century, the woad market commenced after

Prime mass in this church. Then, in the year 1401, the granting of permits for serving wine was scheduled for twelve o'clock. Beginning in 1406, all eveningtime carousals and gaming had to end by eleven at night. Shortly thereafter the opening of the coal market was set for seven o'clock, and the period when wine sellers could offer their product was also given fixed parameters.

Major areas in which changes took place were, apart from regulations concerning administrative policy, civic council and court sessions, markets and retail shops, and working times. And these continued to be the main areas during the following decades, which saw a growing number of diurnal time regulations.

A preliminary technical remark is in order before we continue our discussion: with regard to the organization of time, clocks can be used in social temporal orders as aids for greater precision or improved coordination. As time-measuring devices, moreover, they can also serve a variety of control purposes. Precision is attained if certain events, whose previous temporal determination was vague or nonexistent, are fixed on a temporal axis, in our case the twenty-four hour day. However, precision can also be attained by fixing the maximum or minimum period of time for certain proceedings. Though this setting of time limits presupposes abstract times that are always of equal length, it does not necessarily presuppose that these proceedings are fixed on the axis of the time of day. And while the continuous operation of clocks makes it possible to set hour limits also by two clock-time indications, the typical device for setting time-limits was the sandglass.

Assembly time and assembly discipline

Beginning in the fourteenth century, regulations for council and court sessions were passed in European cities. They yield numerous clues that the expansion of jurisdiction and the multiplication of tasks did more than strengthen municipal autonomy and civic self-confidence. The new rights and new financial possibilities also gave rise to a host of obligations and responsibilities. Tasks that municipal governments strove to acquire or that were forced upon them — in the areas of town planning and the construction of fortifications, provisioning and waste disposal, armament and public health, and schools and universities — necessitated the rapid development of municipal administrations even in relatively small cities. In part this development could be managed with the help of professional people, that is to say, by hiring lawyers and clerks. However, to a considerable and decisive extent it placed demands on the citizens themselves. In particular the important committees of the council, the guilds, and the courts, because of their political constitution, depended on the cooperation of laypeople from the town's citizenry. Merchants and artisans no doubt aspired to these honorary posts during the early phases of their establishment. In many cases the new powers had just recently been wrested from the old ruling elites, sometimes in bloody struggles. In the fourteenth century these processes had still not come to an end in other cities. During the same time, however, we can see in many cities that the committees, largely staffed by laypeople, had also become a burden. The primary reason for this was that they demanded too much in terms of time and discipline from people who had to earn their living outside of their activities on the committees.[49]

Bringing together contending parties and judges in a court naturally presupposed a court date. That such dates also guaranteed the public nature of the proceedings, and thus the legitimacy of the judgment, is well known and does not need to be discussed again. In many instances old legal custom demanded that court proceedings and judgment in any one court could not exceed the space of one daylight period. In late medieval cities, however, court days were no longer the exception. Messengers bearing summons arrived with increasing frequency at the doors of the lay judges and assessors to remind them of their sworn duty of attendance. In the council committees, the number of "ongoing" decisions grew steadily: messengers and envoys had to be heard, answers deliberated, letters drafted, communal officials hired, levies and their distribution agreed upon, complaints heard. The administrative apparatus itself, and in particular the financial bookkeeping, had to be supervised. "Taking counsel" consumed more and more time.

In order to manage the growing flood of deliberative and decision-making processes, on the one hand, and to ensure the openness and accessibility of communal committees and offices, on the other, the times for their meetings were regularized. Regardless of the workload, committees were to meet regularly on certain weekdays. The times of the day were fixed such that not the entire period of daylight but precisely delineated segments became meeting times. The statutes of the Châtellanie in Bruges from the last years of the twelfth century show that the origins of such specifications lay in the judicial system: pleas were to be presented until noon, objections until Vespers.[50]

In Paris, an ordinance of parliament from the year 1320 regulated the meeting of clerics, laypeople, and notaries by the ringing of the bells of the royal chapel for the first mass. Until those bells rang at midday, members could leave the session only to answer the call of nature.[51]

In Ivrea, the podesta was to be in session on court days from mass until Terce and from None until Prime (1329).[52] Around 1330, the judges in Piacenza were admonished to adhere to the customary hours, and the gates of the palace of the podestà were to remain open from morning (or the "hour of the bells") until Terce, and from None until Vespers.[53]

The court of the brotherhood of woad merchants in Cologne sat on court days from the mass at St. George's until the mass at St. Jacob's. The mayoral court in Zurich was delimited in 1324 by early mass, the council bell, and a high mass, but only if the workload required it: "(beginning of the court) when mass has been sung at the Wasserkirche, and then after the council bell has been rung for all colleagues, and the court shall be in session in name until high mass has been sung (in the Minster), if there is much for the court to decide." In a revised version of this statute from the year 1332, the beginning of the session was now signaled only by the council bell; the end continued to coincide with high mass or with the civic time of midday meal: "until high mass has been sung or until the customary time for eating in Zurich."[54]

Indications of clock-time appear occasionally among hundreds of comparable fixations of meeting times at the end of the fourteenth century and with growing frequency in the fifteenth century. The revised version of the court statutes promulgated for the Châtelet by Hugues Aubriot, provost of Paris from 1367 to 1386, stipulated, following complaints about the disorderly administration and the delay of cases, that the "auditeurs" had to appear "à neuf heures de l'orloge du palais ou environ." The cases should be settled quickly and efficiently. In the winter the auditeurs had to remain in their seats until twelve o'clock; in the summer they should take their seats at eight o'clock and remain there until eleven. Moreover, the "audience du greffe" would henceforth be called "at the eleventh clock-hour and no later."[55] The meeting times of a judicial council in Metz were, in 1393, also timed to begin, depending on the season, at seven, eight, or nine "hour dou reloge."[56] Evidently in both cases the clock, acquired only a few years earlier, was to be demonstratively singled out as the new temporal system of reference. And at least in the first case the progress in attaining greater precision seems to have been considerable: one or two generations earlier one could still read in a "reglement" for the Châtelet that the advocates had to appear "after sunrise or within the space of time thereafter not to exceed a short mass" (1327).[57]

From then on we can trace the specifying of traditional natural and experiential time indications with the help of clock-time over the course of several centuries. The reformation of the "court of appeal (Cammer-Gericht) at Cologne on the Spree" stipulated in the year 1540 that the parties summoned "for early daytime" should henceforth appear at six in the summer and at seven in the winter. A summons for the "correct day-time" should thus be understood to mean "at twelve o'clock." Similar statutes from the years 1621 and 1658 renewed these clock-time specifications, but in keeping with a general development it moved them to a later time in the day: to eight o'clock in the morning and two o'clock in the afternoon.[58]

What appears in retrospect as a long-term development toward greater temporal precision amounted within the purview of contemporary experience above all to many steps of abstraction. The abstraction was away from the natural system of reference, that is, daylight and/or time measures that were strongly dependent on experience, for example, the duration of a short mass.

Over the long run such abstractions away from natural or social frames of reference opened up perspectives for new forms of temporal organization. They made it possible to dissolve the link between temporal sequences that had been historically closely connected. Through clock-time indications, the times of assembly and council meetings, for example, were potentially detachable from the dense bundle of other municipal times and their signals. To be sure, when a council committee assembled at a specific clock-time, this did not mean in and of itself that the customary attendance at mass by the members of the council had already dropped by the wayside. However, the form of the temporal fixation included the possibility of setting the time for the council and the time for mass independent of each other, or, if necessary, shifting them with respect to each other. For example, when the consuls in Tournai decided in 1397 henceforth to sit in session from seven in the morning during the summer so that "the good people can be taken care of more quickly," this change in the meeting times in response to the demands of the office did not also necessitate changes in the times for religious services.[59] Clock-times thus made it possible to disconnect in actuality attendance at mass and the meeting of the council. And for our purposes the important point is not when and how often during the Late Middle Ages people did in fact make use of this possibility.

In the statute books of the city of Nuremberg we can find circumscribed meeting times that were disconnected from all external temporal guidelines (including clock-time) as well as from all intrinsic task-related guidelines; these meeting times were therefore abstract times. In 1388–1389, the meeting times for a Commission of Five that was to relieve the council from the burden of less important tasks during the period of the "Städtekrieg" ("war of the cities") was set at two periods of two hours each: "The Five are to sit every day during the year two hours before lunch and two hours after lunch, with the exception of two hours before lunch on Saturdays; and if the council is in session before lunch, they must not sit before lunch, but they should sit two hours after lunch." The councilors had to observe the sworn minimum meeting time of four "or" ("hours") per day, regardless of whether or not they had a corresponding workload. They were also obliged to supervise their own compliance with their duties by means of a sand-glass. "When all five have come together, he [the presiding councilor ("Frager")] shall turn an hour-glass upside down and the five shall sit until two hours have passed, regardless of whether they have things to do or not."[60] The external arrangement of meeting times was only one of the problems of urban administration. Another was the appropriate use of meeting times. And here the discipline of the lay members proved to be a perennial problem. Absences were the reasons behind a new regulation of the courts in Augsburg: "Meanwhile an honorable council of this city of Augsburg has now found for some time that because of frequent absences of the judges ("oder- und andern richter und urtheiler") at the city court, noticeable shortcomings and impediments in the transaction of legal business occur every day and have become a habit . . . (1519)."[61] Apart from absences and tardiness, numerous complaints also reveal that noisiness, interruptions, and verbal and physical assaults were widespread problems. While the older statutes illustrate this only with sparse disciplinary regulations, we can find in sixteenth-century statutes, which are marked by the pedagogical rhetoric of the Reformation, quite detailed justifications for the tightening of disciplinary measures. I shall cite three examples. The Council Statutes of Überlingen (1551) have this to say about the tension between holding office and making a living:

> Since things have taken such a course . . . and since it is also quite evident that the business of my lords of the honorable council has not decreased but increased, it has been necessary to meet every day. The lord councilors have often complained and indicated that they resent this for various reasons, most especially since no one can pursue any profitable enterprise in his own business or household affairs or deal with them as necessity demands.[62]

The preface to the Council Statutes of Schlettstadt in the Alsace (1561) describes the situations that resulted in part from such problems:

> Given that the lord masters, "constöffler, ratsfründt," and guild masters here in Schlettstadt have now for some time been attending the council somewhat laxly and negligently, some disobediently staying away without permission from a ruling lord, some taking leave for minor reasons and taking on other affairs during the usual council day and time, as a result of which it has occurred on several occasions that barely half the council has come together, or if the council was assembled, everybody would get up again as they pleased and leave: all this to the impediment of the affairs incumbent upon the city (and so on).[63]

Even though such complaints could be heard everywhere, the Constance preacher Konrad Zwick wrote a letter to his council in which he urged reform, since in his view the situation in his hometown was especially deplorable (1541):

> There is little demand on our entire government. Among other governments, which have to deal with much more important matters of city and land, council is held less often

> and not as long, and the councilors have much time left over. Compared to others we have only a trifle to take care of. And yet we have council every day, long councils, statutes every day, many councilors have no time all day, and have no peace or rest, and still the affairs and legal affairs incumbent upon the poor city are again and again neglected and postponed . . . Moreover, on account of minor and for the most part insignificant things ("mersthails klain fügen dingen") we use up all the time, hour and day, and it is a proper punishment when God makes our effort and labor fruitless and unproductive, so that we must consume the precious and irrecoverable time to our own disadvantage and that of the city.[64]

People tried to control at least some of these problems in two ways: with attendance fees and allowances for council members, and with tightened council discipline. Discipline included being present and being on time, and the leverage here was to ensure both by threatening to reduce attendance fees.

Practices meant to ensure punctuality and differentiated punishment for tardiness were already known from monasteries during the High Middle Ages: the three-fold repetition of the call to prayer and punishment according to the number of prayers or psalms missed. Both methods appear also in the urban statutes. In addition, however, from the end of the fourteenth century new techniques of organizing time were tried out, some of which remained in use until the end of the eighteenth century.

In the fifteen versions of the statutes governing the meeting of the Constance council between 1376 and 1434, the threats of penalties for latecomers were linked in part to the bell signal, which was repeated several times, and in part to the progress of the meeting. In one case a monetary fine was imposed if a council member had still not appeared by the third bell signal; in another case the fine increased with every point on the agenda ("frag") that the tardy councilor missed.[65] Both methods can also be found, among other places, in the older statutes for Würzburg, Cologne, and Nuremberg. According to an addendum to the Constance council statutes of 1434, an attempt was made at the beginning of the fifteenth century to disconnect the penalties from the agenda points ("fragen") and to levy the punishment according to the hours missed. After the "first hour," one-third of the maximum penalty for complete absence was due, after the "second hour," two-thirds.[66] A similar attempt was made in Cologne. There the statutes for the high court from 1435 set the beginning of the session in the summer at eight o'clock. Assessors who came after eight o'clock but still before the stroke of nine were to lose half of their attendance fee. No fee would be paid to those who absented themselves from the court before nine or prior to the rendering of judgment.[67]

It would appear that regulations of this sort, which tied not only the beginning of meetings and sessions but also the course of meetings and time penalties to the strokes of the clock, proved to be impractical and fell out of use. In Strasbourg, as well as in other cities, the interval between the first and the second call signal had been long enough to allow everyone to get from his house in the city to the council. In the fifteenth century this old interval, experiential and dependent on the size of the city, was fixed at half an hour. After that time had passed the meeting was to begin, and fines would be imposed on latecomers. To help the tower wardens to observe the time limit precisely, they were given a "sandglass" ("zitglas"), "so that they might know when half an hour has passed."[68]

In Brunswick, the churches of St. Catherine and St. Martin had both been equipped with striking tower clocks. The reform of the constitution of the council in 1408 took this into consideration by stipulating that fines for tardiness would be imposed after the second "clock" had struck.[69]

Around the same time, a solution for organizing time was found that kept the specifying of the beginning of a meeting according to the striking of a clock or council bell but regulated the problem of supervising punctuality in a separate manner. In Cologne, for example, the sessions of the mayor's councilors ("Amtsleute") were to begin after 1450 at eight during the summer, at nine in the winter. Once the clock had struck, the council servants set up "a glass with sand, which shall run for a quarter of an hour." Anyone who appeared after the sandglass had run out would lose part of this attendance fee. This kind of regulation was later also adopted for other municipal committees in Cologne.[70] In Nördlingen it is attested after 1429, later also in Wolfach, Thorn, Riga, Tübingen, Asperg, and Kaufbeuren.[71] The still vague connection between the timed beginning and the grace period that followed became particularly clear in a corresponding regulation in the council statutes of Isny (1482): "When the holy water has been given in the hospital for the late mass, one shall set the quarter part of the hour, and whoever arrives after it has run out . . ."[72] At the beginning of the sixteenth century, similar regulations were passed, renewed, or strongly reiterated for all kinds of committees in numerous cities. A redaction of the statutes of the council in Constance (1537) reveals that the abstract fixing of a time limit also had repercussions for the length of the individual agenda points. Here a fine was due after the "fourth part of the hour has run out and one point in the council is over." After two missed points, discussion of which should last a quarter of an hour each, the fine was doubled. A few years later, in 1543, a differently calibrated sandglass was used ("eighth part of the hour"), and the attempt was made once again to impose penalties according to hours missed. To be sure, without weighty reasons the total duration of the session, during which the lord councilors had to be present behind closed doors, could also not be extended beyond a certain clock-time ("when it has struck ten"). At the end of the sixteenth century, the grace period was fifteen minutes again and the length of the council bell signal was also set at fifteen minutes.[73] Similar steps were taken in Überlingen (1551), though here the preceding mass was retained.[74] The organization in Schlettstadt was particularly thorough (1561). The signal of the council bell was to last half an hour. After that servants set up two sandglasses. Once the first had run out and after a special sign from small bells, attendance was taken from lists. Members who appeared after roll-call had begun but before the second sandglass had run out were not fined, though they lost their attendance fees.[75] In Bruchsal in 1588, candles of predetermined duration were burnt alongside the sandglasses.[76]

Such excessively perfectionist regulations did not prevail over the long run. Most cities stuck to timing the meetings by the clock and using a quarter-hour glass to limit the temporal discretion for the actual beginning of the session. This time limit, known today as the "academic quarter," apparently arose in the urban committees. It probably got its name from the fact that it was explicitly adhered to longer in university committees and lecture times. In the urban statutes the problem of timing and punctuality declined in importance beginning in the seventeenth century. Standards of discipline that had still been problematic in the Late Middle Ages had by this time gradually become habit. However, we now encounter these time-control techniques all the more frequently in the village statutes from the sixteenth to the eighteenth centuries.[77] In most of these cases, as well, the sandglass was called for. In addition regulations with less abstract time limits also appear in the village statutes. In Schönbrunn in the episcopacy of Bamberg, the neighbors showed themselves "somewhat dilatory" after the ringing for the "assembly" ("Gemein"). The statutes from the year 1741 therefore stipulated that the tardiness fine was due once the mayor had gone to a tree at the mill, broken off a twig, and brought it back to the assembly place.[78] In these statutes use was made not of an abstract time limit, but of a measure of time that was empirical and valid only for a specific place. The modern purpose of control was here attained by reviving techniques of time limitation that had long since been forgotten in the cities.

Market time

Among the rare references in the fourteenth century to the practical usefulness of public clocks in the cities is a brief passage in a literary allegory of the founding of a monastery (written before 1380).[79] The anonymous author of *The Abbey of the Holy Ghost* mentions, in connection with the use of public clocks, not only the beginning of work time but also the beginning of the offering of wares at market. In one of the earliest references to the public use of clocks, in the city of Valenciennes (1325–1344), a device called "li orloges," located directly at the cloth hall, limited the time for the selling of cloth at midday. In Ulm, according to the account of Felix Fabri, a striking clock was maintained in the house of the tax farmer for the fustian-weavers, even before the city itself had such a clock.[80] What was the importance of time-organizing regulations for the activity at market, and to what extent did public clocks come to play a significant role in them?

In the Middle Ages, the market, as the organized bringing together of sellers and buyers, was never an abstract concept but always an event that was tied to concrete places and concrete times and governed by specific regulations. The labor market, even if not under this name, was the place in the city where weekly wage laborers and day laborers were hired prior to sunrise; the corn market was the sole place where grain could be bought and sold at certain times, and so on.

Concentrating market-like activities at certain locations and times is an ancient practice. The reasons for such a concentration were manifold. Guaranteeing the peace of the market gave those who took part in it safe conduct and, among other things, protection against having their wares seized; this was, of course, only possible for limited periods of time. Limiting market time to daylight served to ensure the safety of the participants; it guaranteed the public nature or general accessibility of the market, and made it possible to carry out quality and price controls and levy taxes and dues. Thousands of market regulations from the twelfth century on show that all these objectives were pursued by, among other methods, limiting urban markets to specific times of the day. Market time began with sunrise or the ringing of Prime and usually ended around midday. From the thirteenth to the eighteenth centuries, in individual cases also as late as the nineteenth century, market time was also acoustically indicated by special market or corn bells. In many cases there were also optical signals such as market banners or broom-like signs ("Marktwische") that were displayed during the authorized market time.[81]

The specifying of market times by clock time that can be frequently observed at the end of the fourteenth century did not directly influence market activities. A certain need for precision seems to have existed, however, where the inspection and selling of textiles was to be temporally linked. We can add many similar examples to the early example from Valenciennes. The comment in *The Abbey of the Holy Ghost* finds numerous confirmations in the sources.[82]

However, in addition to the setting of the beginning and end of market times, we also find in market regulations of European cities from the thirteenth century on a noticeable cluster of temporal subdivisions of market activities, which were responses to special regulatory problems. The purpose of the urban market was to bring producers (peasants and artisans) and consumers (mostly urban residents) into the most direct contact. As much as possible, purchases were to be made openly and firsthand. Producers should get a fair price at the food markets, and the needs of the urban consumers should be met, the latter being of primary importance. At markets with artisanal products, competition within guilds was to be regulated and the city's artisans protected as much as possible against outside competitors. But by the High Middle Ages this model of the market no longer had much in common with the realities of the market. The need to provision the populous cities from the surrounding territory, a need that was constantly increasing, promoted intermediate trading everywhere, which relieved small producers from the necessity of journeying into the city. In addition, the growing importance of trade with finished products, above all with textiles, in many cases altered the function of urban markets: provisioning markets turned into trading markets.

The urban market had thus turned into a complex entity, where the interests of a growing diversity of participants had to be accommodated and balanced. Moreover, from the perspective of those who held jurisdiction over a market, these interests were entitled to different degrees of consideration. The authorities sought to manage the resulting regulatory problems with a multitude of time-organizing measures.

Intermediate trade existed everywhere under many different names. "Vorkäufer" ("prebuyers") were those who tried to get their hands on goods preemptively, that is to say outside of the city or prior to the beginning of market. "Unterkäufer" ("sub-buyers") bought only to sell again, or were brokers working for someone else. Small-scale vendors (hucksters, street peddlers), finally, sold very small amounts and leftover wares outside of the market or market times. All three groups of intermediate traders were regarded as foreign to the true purpose of the market and were suspected of endangering supplies and driving up prices.[83] Numerous market regulations were intended to draw the narrowest possible boundaries to intermediate trading or stop it altogether: prohibition of pre-buying outside of the city, exclusion of intermediate dealers from the market, prohibition of all side-agreements and invalidation of all purchases resulting from them, and so on. Nowhere was the positive, price-equalizing function of intermediate trade explicitly recognized. Municipal administrations reacted only indirectly to this unavoidable phenomenon, for example by establishing granaries ("Kornhäuser") or hiring municipal brokers.[84] The success of the entire panoply of restrictive measures is something we can hardly assess. It does seem highly doubtful, however, that buying and selling agreements could be prevented.

Among the flood of steps taken against intermediate trade on the urban markets, one group of measures stands out for their extremely wide diffusion and longevity. Urban authorities were evidently convinced that the problem of intermediate trade, and in general that of the variously privileged parties at the market, could be dealt with by an internal temporal ordering of market activities. The internal timing of the market was generally arranged in such a way that from the beginning of the market up to a certain point (indicated by bell signal, lowering of the market banner, clock stroke), only town residents and buyers for the great families could purchase things for their own use. After that point everyone could make purchases or negotiate them. Such pre-buying regulations, which are extremely numerous, differed for the most part only in the length of buying time they granted to the privileged buyers. Most regulations were directed against the pre-buyers of poultry, eggs, cheese, game and venison, and fruit. But other groups of wares were also affected, as were pre-buying and small-scale retailing in general. The market as an event with a deliberate temporal pattern becomes visible in a miniature in a *Codex picturatus* (around 1500), that contains the guild and trade regulations of Cracow. Perhaps unintentionally, though plausibly, the artist has brought a public clock into the picture showing a hawker's booth (fig. 54).

The temporal division of the market, already common as an organizational tool in the thirteenth century, could be further differentiated with the help of the clock-time signal. A grain market regulation in Dresden (1570) stipulated that in summertime only local consumers could buy before eight o'clock, thereafter until ten o'clock also the master bakers, and only after that the "Platzbäcker" (pastry bakers).[85] The regulation was justified on the grounds that the bakers were responsible for inflated prices. Grain wholesalers, hawkers, and small retailers were usually permitted to appear at market only after twelve o'clock. Access by outside sellers and buyers could also be limited through time restrictions. In Frankfurt, for example, Jews were not allowed to show up at the ox market before nine o'clock, the purpose being to prevent them from buying up the best animals (1595).[86]

These time restrictions were based on the notion that in a situation where goods were scarce, the provisioning of the city's own citizens had priority. In times of emergency the regulations were therefore written such that intermediate buyers were excluded altogether from making purchases at markets.[87] Added to this was the hope, no doubt a realistic one only in part, that prices could be regulated. Through de

54. Hawker's shop in Balthasar Behem's *Codex picturatus* (around 1500). Cracow, Biblioteka Jagielonska.

facto compulsory market participation during a limited time period—nobody was allowed to open his booth, bank, or store prior to the beginning of the unrestricted market period—the authorities tried to prevent possible high-price offers and speculation in goods during market time. Moreover, time limitations could also be used to assure quality. Fishmongers, for example, were everywhere forced by especially tight selling times to turn over their goods quickly.[88] Occasionally the attempt was made to speed up the selling of fish even further by prohibiting fishermen from sitting down during their allotted market time.

It was very rare that cities, in the interest of actively promoting trade, explicitly renounced every form of regulation that organized market time.[89] As its wide use in all European countries shows, a temporal ordering of urban markets was the only method that was seen as manageable and promising with regard to the goals pursued by the authorities. The allotment of unequal market opportunities for different groups of buyers, and the privileged provisioning of the town residents or a split pricing in favor of city-dwellers, could perhaps have been achieved better through controls on individuals, volume, and prices. But in view of the fact that it was impossible to carry out such controls effectively, temporal segmentation, while not an optimal regulating tool, was seen as adequate and fair in the eyes of contemporaries.

To what extent the goals pursued with the help of such measures were in fact attained is something we can hardly determine. We would have to know whether the peasants in times of crisis did not, after all, sell outside of the cities, or whether the prices paid by town residents and merchants were in fact different. Whatever the case may be, until the nineteenth century authorities could not see any better regulatory possibilities. Only then were such regulations dropped and confidence placed in the mechanism of free price formation.

The question of when and how the late medieval regulations fell into desuetude or were abolished in the various cities can be answered only in a few cases. In Cologne, pre-buying regulations were repeatedly restated, newly published, or confirmed in numerous council statutes of the sixteenth, seventeenth, and eighteenth centuries. As late as 1816, pre-buyers were still barred from appearing at market before ten o'clock, just as they had been in the fifteenth century.[90] The Prussian statutes of administrative policy ("Polizeiordnungen") of the eighteenth century adopted similar rules and turned them into state laws.[91] The "regulation of the weekly market" for the larger cities in the Duchy of Cleves (1773), for example, enshrined the threefold division into "consumers" who had access until ten o'clock, bakers and brewers until eleven, and hawkers and merchants after eleven.[92]

In Vienna, selling prohibitions existed after 1504. Beginning with the Austrian Police Statutes of 1542, at yearly fairs and weekly markets a "banner" had to be hoisted for two hours, during which period only burghers were allowed to purchase quantities for their own household needs. The Vienna market statutes of the following centuries renewed these regulations on several occasions, in part explicitly during times of "price rises." The eighteenth century saw conflicts on the Vienna market with the small retailers and hawkers, because peasants had received permission to sell their own as well as purchased, cleaned geese and ducks until ten o'clock, later until one o'clock. In the subsequent period the time restrictions for producers were further loosened; intermediate traders, however, continued to be tied to specific times. In 1787, a decree of Emperor Joseph II abolished all buying privileges of the "public." However, this attempt at liberalizing the rules initially met with no success. The market statutes of 1791 differentiated once again between the public, people in the trades, who could buy only after ten, and small retailers, who were not allowed to appear in the market squares before eleven o'clock.[93]

In conclusion I shall cite only two references on the unwritten history of how economic liberalism came to prevail within the cities: in Winterthur the market bell rang for the intermediate merchants from 1481 until 1839. Herzberg in Saxony clung to the old pre-buying rules, according to which citizens were allowed to shop between nine and ten in the morning, until 1869.[94] Only then did the beginning of market signal also the beginning of a free market.

The "time-ordering" ("Zeitordnung") *of schools and teaching*

Clock-time regulations that appear in the statutes of urban schools beginning about the middle of the fifteenth century delimited at first the duration of daily classes and fixed the work time of teachers. In addition, they tied school time to other elements of the urban temporal order. Once the length of instruction had been circumscribed in terms of hours, its various parts could also be divided into equal segments. The precursors of the hourly class schedule commonly used today appeared at the end of the fifteenth century. Its development can subsequently be traced especially through the flood of post-Reformation school statutes for higher and lower civic schools ("Bürgerschulen").

While the history of humanistic and Reformation pedagogy and the content of its instruction and its school reforms have been frequently described, little attention has been paid to the hourly schedules that were a new phenomenon at the time.[95] The fact that they were so ubiquitous and self-evident makes it easy to overlook that the patterning of time with modern means brought about, beginning in the fifteenth century, changes in the schools that were not only formal in nature.

Josef Dolch has described the transition from the medieval curricula of the artes liberales (trivium, quadrivium) over a period of three hundred years as the transition from a type of "ordo legendi" to a type of "ordo docendi." The old curriculum structure, which shaped the operation of universities and schools, was organized around the graded arrangement of a canon of Latin authors who were required reading. The sequence of the texts was indicated by the rank they held in the accepted hierarchy of knowledge, and the student, by reading and studying, was to acquire progressively higher degrees of knowledge. Cracks appeared in the old structure of the curriculum contents through the increase in the stock of knowledge and especially through the reception of the Greek classics. Even if the concept of limited periods of study was still largely absent, the multiplication of books that had to be read did produce time pressures that one could not fail to notice. Encyclopedias, indexes, and a variety of mnemonic devices were responses to the problem of how to access the growing body of knowledge more quickly and more effectively.[96]

Time pressure became a perennial topic among humanist authors from the end of the fourteenth century on. It was precisely in the context of education, learning, and studying that we find in Petrus Paulus Vergerius, Battista Guarino, and Leon Battista Alberti a growing number of admonitions and words of advice on how to make economical use of one's time—understood both as the time of one day as well as a lifetime, how to avoid wasting time, and how to give a constant account of how time has been used.[97] Concretely the issue was not only interest in and curiosity about texts, but also the development of rhetorical and literary skills and the cultivation of music and athletics. As we can see from the suggestions of these authors, time pressure could only be met through order, method, and planning.[98] Education and self-cultivation had turned the scarcity of time into a prominent topic. Though allusions to the fleeting nature of the time of life are not absent, the more concrete problems related to the organization of time. The suggested solutions frequently amount to arrangements resembling hourly schedules: the hours of work and reading should be carefully allotted, gaps of time should be avoided or made use of, authors or themes should be assigned to specific hours, and an account given to oneself whether they have been adhered to.

In this milieu sandglasses appeared as a means of setting time limits and supervising oneself. Petrus Paulus Vergerius, in his tract *De ingenuis moribus* (written around 1404), recommended that devices "with which one measures the hours and the time" be installed in public and visible places in the libraries, "so that we can see time itself flowing and fading away, as it were." The wording reveals that these devices could only have been sandglasses.[99]

That one should observe the flow of time was surely also meant as a general moral admonition. However, the time-measuring device was above all an organizational aid. Vergerius does not suggest dividing the study of an author's work by chapter or section; instead, a theme or author should be allotted a specific period of time. Equal periods of time are neutral in regard to the diversity of the contents of books, their varying importance, or different degrees of difficulty. Limited "periods of study or instruction" allow, with respect to contents, a more abstract and freer organization of studies under the conditions of time scarcity. The other didactic techniques that appeared and which are not discussed here acquired a temporal framework and possibilities of control through the setting of limits.

It would appear that the technique of the abstract limiting of instructional periods was in use in the Jewish schools of the Middle Ages — whose curriculum development was very different — at least at the same time as in Christian schools and possibly even earlier. The oldest depiction of this known to me is an instruction scene with a sandglass in a Pentateuch created in Coburg in 1395 (fig. 55).[100] Scholarship on the history of Jewish schools has emphasized not only the intimate relationship between the house of prayer and the house of learning, more intimate than within the Christian communities, but also the outstanding importance which the teaching of children had in the Jewish communities. It has also stressed the abundance and strictness of regulations that were concerned with time-discipline in the broader sense. "Determine a time for lessons; give your students a specific time for coming and going": these words described a time which, according to an expression from the Talmud, could not be interrupted even "if the building of the temple in Jerusalem were at stake."[101] The succession of the subjects of instruction, which varied with age, the techniques of presentation and repetition, and the number of students per teacher were regulated with the same exactitude. The sandglass suspended above the student in the miniature is therefore not an incidental or decorative element, but the depiction of modern ways of carrying out very ancient regulations. This finding is confirmed by textual

55. School scene with sandglass, in a Pentateuch written in Coburg around 1395. London, British Library, Ms. Add. 19776, folio 72 verso.

sources from the beginning of the fifteenth century. In the collections of responses, that is, written opinions of recognized religious authorities, timekeepers are prescribed aids in school from this time on. Jacob Levy (rabbi in Mainz, died 1427), his student Jacob Weil (rabbi in Nuremberg, fifteenth century), and Israel Bruna (Regensburg, ca. 1450), made the procurement of sandglasses and wheeled clocks for determining school time obligatory.[102] This tradition can still be observed at the end of the seventeenth century, and later regulations show that the setting of abstract limits was used not only for controlling the sequence of instruction, but also for calculating the teachers' pay. In the school regulations of the community of Nikolsburg in Moravia (1676), we read: "No teacher may teach without a sandglass." Since the daily lessons, the number of students, and the maximum hours a teacher should instruct were fixed, a salary that also took into account levels of difficulty could be calculated by allowing that "the boys who are learning the alphabet count for half an hour, those studying the Pentateuch for three-quarters of an hour, those studying the Talmud for one hour. Those who, in addition to the Talmud, are learning the commentary Tosafot count for five quarters of an hour."[103]

The Christian schooling that was vigorously promoted after the Reformation frequently took up the pedagogical concepts of the humanists.[104] The goals, however, were more ambitious: education should ideally reach everybody, for without a certain degree of literacy and education the freedom of a Christian could not be attained. Theologically this endeavor was also driven by the desire to set the "house of God" on earth—which encompassed church, state, family, and schools—"in order" as best one could in view of the end of the world, of the earthly time that was felt to be winding down. The alliance between the Reformers and the territorial princes was intended to safeguard the Reformation ideologically and politically, and the recruitment and training of suitable pastors, teachers, and officeholders was seen as part of that safeguard. In addition to the enormous task of providing a financial basis for the schools in the cities and villages, the Reformers were also concerned with expanding the content of what was taught. Religion, music, mathematics, some history, and the ancient languages became subjects in the higher civic schools. Here "subject" meant that—in the long run—the focus was less on reading certain books and more on covering certain subjects, drawing on books to study them. Organizational techniques were also adopted and developed further—and this applies also to the school reform of the Counter-Reformation.

These organizational techniques included the division of students into age and achievement groups, as well as the structuring of classes into an hourly rhythm.[105] The latter is already clearly developed in the Ulm school regulations from the end of the fifteenth century, where it was described—as an "old custom" no less—as follows:

> As soon as the bell has struck vi, the teacher goes into the school and sings the Veni sancte with the students. After that he reads the (student)register and then begins exercises in grammar . . . (in the afternoon) when the bell has struck xii, the master himself comes into school and reads a poet until it strikes one . . . and all students must pay attention, just as they do in the morning between six and seven in grammar.[106]

The Nuremberg statutes of 1505 also divided the subject matter on an hourly basis.[107] In his tract to the lord councilors of all German cities (published in 1524), Luther demanded that boys and girls should attend school for at least one to two hours every day.[108] In accord with Ulrich Zwingli's *Leerbiechlein wie man Knaben Christlich unterweysen und erziehen soll* ("Handbook on the Christian instruction and upbringing of boys"), Zurich in 1523 set up school with an hour each of Hebrew, Greek, and Latin every day.[109]

The title page illustrations of the first edition of both tracts (figs. 56, 57) show that instruction was not only framed by the stroke of the clock; in addition—and into the eighteenth century with growing frequency—classes followed the running of a sandglass. Comparable illustrations show that the sandglass time measure was also supposed to be used for instruction at home. The pictures, too, prove that instruction was arranged by abstract periods and no longer by books or chapters.

These time periods—that is, hours—became the work standard for teachers, even if their pay was not as directly related to the number of hours as was the case in Jewish schools.[110] Time limits made the various subjects into planning units that were outwardly equal. Varying levels of difficulty could still be accommodated, for example, through a skillful arrangement of demanding and less demanding subjects. The different places subjects occupied on the scale of importance were also not abolished. They merely had to be translated into time differences: the more important a subject, the more hours per week were allotted to it. Finally, the temporal framework of one hour could, for pedagogical purposes, be further subdivided. According to the school statutes of Nordhausen (1583), for example, in each case the first half hour

56. School scene with sandglass. Title page to Martin Luther's *An die Ratsherrn aller Städte deutschen Landes, daß sie Christliche Schulen errichten und halten sollen* (Wittenberg, 1524). Heidelberg, Universitätsbibliothek.

57. Teacher with sandglass. Title page to Ulrich Zwingli's *Lehrbüchlein wie man Knaben christlich unterweisen und erziehen soll* (Zurich, 1523). Munich, Bayersiche Staatsbibliothek.

should be used for reading and presenting the lesson, the third quarter for repetition, and the last quarter for a new assignment.[111]

Of course this discussion does not cover all the effects that abstract limits had on the content and style of classes. What is certain, however, is that they proved a useful means of pursuing at one and the same time a range of quite different goals.

In the sixteenth century, which has been called the "century of school statutes" because of the large number of such texts (often very detailed because of the lack of trained teachers), the formal aspects of old school schedules were developed further.[112] This formalization amounted to a dissociation from specific books, specific teachers, and local peculiarities. In this way these abstract plans could also be made valid on a territorial level. As is shown, for example, by the regulations governing the furnishings of classrooms, the combination of striking clock and sandglass continued unchanged.[113] In this regard, too, questions of time remained an important theme in early modern school statutes. To my knowledge the modern German word "Stundenplan" ("hour-schedule") does not appear before the end of the eighteenth century, though the word "Zeitordnung" ("time-order") that is used in this chapter heading does.[114]

The transition to modern hour-reckoning at the late medieval universities had much less of an impact on teaching. Since lower and higher civic schools were for the most part new establishments, or at least thoroughly reorganized ones, little stood in the way of the adoption of relatively modern organizational forms. Comparable fundamental changes did not take place at the universities either in terms of the content of what was taught or in the way teaching was carried out. To be sure, university statutes dealt with questions of time organization in no less detail than the school statutes. But time arrangements at the universities contained a large number of old elements that were strongly obligatory, and they were significantly more complicated. The curricula of the various faculties continued to be "ordines legendi" that were tied to specific books. A lecture generally had an intrinsic timing, that is to say, one dependent on the material in question. Books were divided into "points," which provided the measure for individual lectures. Progress could be checked by the wardens; teachers who exceeded the time were penalized by cuts in their fees.[115]

It is difficult to give the time duration of daily lectures. The three or four daily lectures differed according to the rank of the texts, the instructors authorized to deliver them, and their duration.[116] The morning lecture ("lectio de mane, l. in Primis, l. in aurora, l. in ortu solis") was given by doctors and masters and was therefore also called "hora doctoralis." In the winter it began when it was still dark, and without candlelight it had to be recited by heart. In Paris in 1386 it was described as not so "large an hour," and was therefore reserved for the overburdened doctors.[117] This was the time given to the "lectiones ordinariae" on the more important books. The two other morning lectures, usually named after Terce and the None, and one afternoon lecture ("lectio in vesperas," "hora vesperorum") were allotted to the "lectiones extraordinariae" and the "lectiones cursoriae." With three lectures each for the masters and the baccalaureates, the medical faculty had a different schedule than the law faculty. Disputations, repetitions, and university sermons interrupted this rhythm at irregular intervals. Added to this were the church feasts and ceremonial acts listed in the university calendars and celebrated differently at the various faculties.[118]

In the fourteenth and fifteenth centuries, the universities still had very few central buildings of their own, and lecture rooms were usually spread over a section of the city.[119] The signal systems and time indications were correspondingly complex and are difficult for us to understand today. Some universities had their own bells, for which special dues were charged; others oriented themselves after the time signals of major churches. In addition, the bells and services at smaller churches and chapels are also mentioned in time indications.[120] It is therefore hardly possible to reconstruct satisfactory schedules on the basis of the statutes.

Despite these, for the most part, unfavorable external conditions, the universities, too, quickly switched their activities over to clock-time (Prague 1348, Vienna 1385, Ferrara 1391, Padua 1395). In this way universities in the cities were unified in temporal terms even before they got their own buildings. Some procured their own clocks, as, for example, Merton College in Oxford (1387).[121] Some oriented themselves by the existing public time signals, as, for example, in Pavia, where from 1408 on the municipal tower wardens had to give all signals "causa studii papiensis." The case of Pavia also shows that the switch to clock-times was an irreversible process that created dependencies: in 1465, the rectors of the faculties of law and the arts complained that the main clock for the town, located in the ducal palace, struck the hours in such a disorderly fashion that "it was unclear at what hours the doctors were to read and the students to attend." The duke promised to rectify the situation by hiring a suitable clock warden "in order that doctors and students might know what they should do in regard to classes as well as other public duties."[122]

Adherence to the traditional, unequal lecture times was the reason why the new organizational possibilities could not be more extensively used. This becomes clear, for example, from the statutes of the university of Avignon (1441). Here the beginning of the lecture times is indicated only vaguely in accordance with the Hours, while their duration is given with greater precision than was possible with a public clock:

the "hora doctoralis," which began in the winter before, and in the summer after, Prime ringing, was to last one and a half hours, the "horae tertiorum et vesperorum," an hour and a quarter each, and the "hora none," one hour. In each case these times did not include the signal time ("sine tractu").[123] We know from pictorial sources, though not directly from texts, that sandglasses were also used in the lecture halls to observe these time regulations. The earliest example is probably a miniature from a statute collection (compiled in 1500) concerning a newly established university (fig. 58). In the house rules of the Freiburg Collegium Sapientiae, a kind of students' hostel, we read this concerning non-public disputations: "Lest knowledge of the most highly renowned arts is spoiled in dullness and indolence, we order the following: since the disputation is the file of reason, by which the rust of all ignorance is scraped away, each week, either on Sunday or Thursday in the evening after supper, a disputation in the liberal arts shall take place for about one full hour."[124]

Preaching time

In hundreds of ecclesiastical constitutions between the sixteenth and eighteenth centuries, at first only in Protestant constitutions, later also in Catholic ones, precise and detailed regulations were given concerning not only the times of services on feast days and weekdays, but also of baptisms, weddings, funerals, and the hours of school and catechism. Insofar as the ecclesiastical constitutions were simultaneously administrative regulations of the territorial rulers ("landesherrliche Polizeiordnungen"), they also contained a wealth of time regulations of public life, for example on curfew, the opening times of inns, and the like.

By its nature and extent this regulatory offensive in the areas of pedagogy, liturgy, and social policy was far removed from the original intentions of the reformers. The latter had initially believed that by abolishing all external, "popish," and "superstitious" ceremonies and practices, they were promoting an invisible church. For the purpose of salvation they considered all "external order" as irrelevant. However, the unexpectedly rapid advance of the Reformation brought a series of problems and created an unsuspected need for regulation. There were far too few Reformed preachers, and many reform-minded

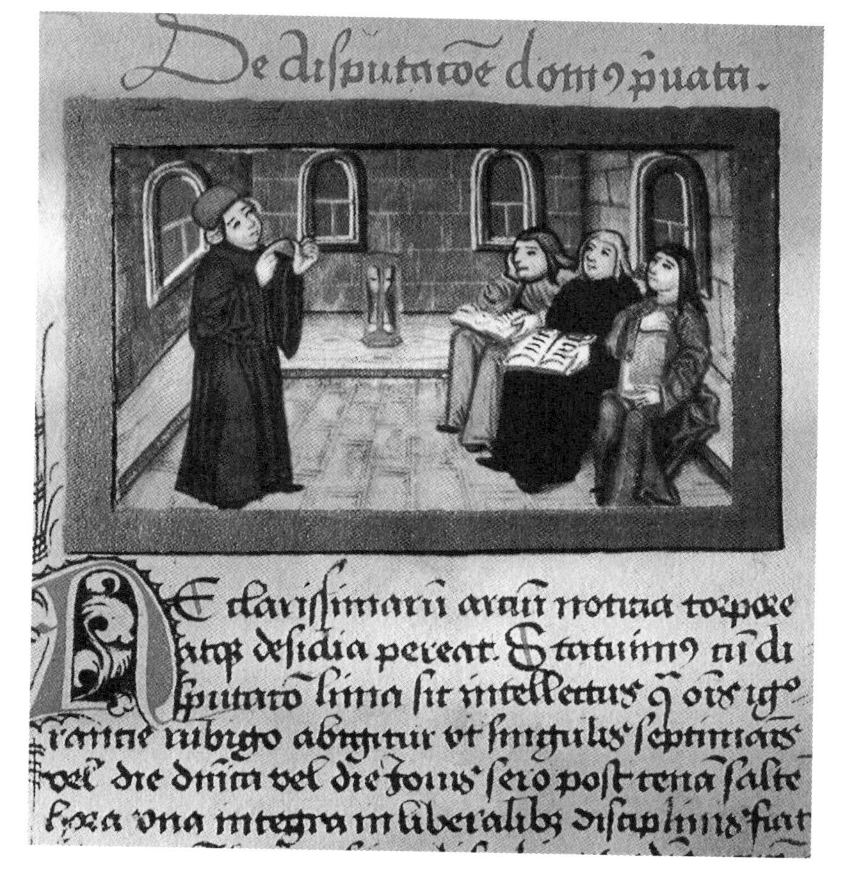

58. Disputation scene from the manuscript of statutes of the Collegium Sapientiae in Freiburg (around 1500). Universitätsbibliothek, Freiburg, Coll. Sap. 2a, 30 recto.

parishes and priests did not know how to celebrate mass and what to preach. The politicization of the confessional conflict created an additional need for new regulations, for after the territorial princes had acquired supreme ecclesiastical authority, they demanded ecclesiastical constitutions for the implementation of the Reformation in their territories. The various radical reform movements gave rise to additional problems of differentiation and control. With emphatic reluctance and hesitation—"because there is a push everywhere for German masses and services, and there is much grumbling and annoyance about the diverse forms of the new masses, as a result of which everybody does his own thing"—Luther addressed himself to the designs for the liturgy and the order of the masses.[125] In the process he repeatedly emphasized "Christian freedom" in regard to all external order of the exercise of religion. The flood of ecclesiastical constitutions (school statutes were frequently included in them) that were published in the interest of spreading and securing the Reformation can also be seen as a comprehensive attempt to diffuse the Christian message for the first time everywhere and to everyone through preaching and teaching. The past appeared to the Reformers as a great chaos of ignorance, blindness, and disorder. At the same time, however, the emphasis on "Christian freedom" involved the danger that reformationist ideas would take on a life of their own with certain teachers and in individual communities. The ideal of order as such developed a certain dynamic of its own. On the whole rules were made for far more things and in far greater detail than was possible to observe in real life. However, new regulatory areas or new regulatory techniques did not yet appear at this time. New was only the density and breadth of the regulations.

The subject of ecclesiastical constitutions was, first of all, the extensive complex of the financial underpinning for maintaining the church buildings, the schools, the charity funds, the pastors, sacristans, and teachers. The second major area was the liturgy, the festal calendar, and the time of masses during the day—all this combined with elaborate justifications, explanations, and admonitions. In this way the constitutions at times grew so large that shortened versions had to be drawn up for the councilors, "so that less time is spent with them and other necessary business is not impeded."[126] While local peculiarities were certainly taken into consideration, the new order did lead to a considerable measure of uniformity in regard to temporal regulations.

Although it was repeatedly emphasized—following the original conceptions—that Christians were not bound to specific days and hours in their devotions and prayers, in practical terms all religious ceremonies were given a "particular time and hour." The number of divine services was drastically reduced, though those remaining were consistently regulated by the stroke of the clock. This measure was directed against the Catholic "secret masses," masses that were cele-

59. Sermon in a reformed temple, called "Paradise," in Lyon (1564). Geneva, Bibliothèque Publique et Universitaire. Photo: François Martin, Geneva.

brated only because they were endowed and usually without participants. The intention of the Reformers was to secure the greatest possible attention and participation for a smaller number of masses by fixing them precisely in time. Outward order—not a matter of conscience and also not necessary for salvation, though definitely advantageous to the common good—was also part of the "education through the church" in Geneva under the Reformer John Calvin. He demanded "specific hours" for all religious ceremonies and had weekday masses limited to one hour. The picture of the spartan temple (fig. 59), decorated "without any foolish display and striving for human fame," shows it in its "original simplicity" but furnished with the modern instrument of measuring time limits, the chancel sandglass.[127] The Catholic ecclesiastical constitutions adopted the techniques of time organization and time limitation with only a short delay.[128]

The concern for the preservation and maintenance of the clocks played an important role in many ecclesiastical constitutions and visitation reports, in part in connection with the question of salaries for sacristans and schoolteachers, in part in the regulations concerning the times for services. Among the wealth of regulations concerning the location of services at specific times of day, numerous limitations on preaching time stand out, since there is hardly a constitution between the beginning of the Reformation and the end of the eighteenth century which does not have them, whereas they nearly disappear completely thereafter. They also stand out because the duration of preaching was limited in addition to and independent of the clock-time regulations concerning services. Sunday sermons were usually limited to one hour, workday sermons to three-quarters of an hour, and other sermons to half an hour. For this purpose sandglasses were frequently required or recommended.

Luther often complained about excessively long sermons, "out of concern for the children," whose curiosity, attention, and intellect should not be overtaxed. He demanded simple and understandable sermons and recommended one hour as a reasonable length. The following is recorded in the *Table Talks* as a "dictum Lutheri": "lengthy preaching is not an art, but preaching, teaching, rightly and well, hoc opus, hic labor est."[129]

These primarily pedagogical reasons for placing time limits on preaching were joined in the ecclesiastical constitutions of the subsequent period by a number of other considerations. In part these can still be traced back to ideas of the Reformers, in part they were new; as a whole, though, they reveal that excessively long sermons were a perennial problem. A random selection of examples from regional ecclesiastical constitutions can illustrate the variety of reasons why limitations were imposed on preaching:

> Limitation of the Vespers sermon to half an hour, also so as not to deprive the fathers of the house and the servants of the time for household chores. (Königsberg 1568)[130]
> Limitation on preaching so that listeners are not inundated, pregnant women are not discomfited, the poorly dressed do not freeze, and the poor can prepare their meals. (Thorn 1560–1570)[131]

"In order that the councilors and officials can come to coun-

> cil at their appointed hours and the common artisan is not kept from his work for too long, also that all listeners are the more enthusiastic about coming to church and looking forward to the blessing." (Regensburg 1572–1588)[132]

> Limitation on preaching with the admonition "not to bring in inappropriate stories or examples." (Electorate of Mainz 1656)[133]

> A question in the visitation instructions: does the priest "preach too long, especially during the week, does he lay out unnecessary points of dispute or controversies of the adversaries all too broadly, does he speak at length in unknown languages, Greek, Hebrew, and the like, which is an obstacle for the simple people?" (Electorate of Saxony 1671)[134]

> Limitation on preaching lest the listeners lose the necessary attentiveness and proper devotional attitude; in particular, the sermon should not be dragged out through unnecessary repetitions and so-called tautologies. (Brandenburg-Prussia 1714).[135]

> Additional fines for those preachers who "criticize such regulations from the chancel and complain about them." (Brandenburg-Prussia 1717)[136]

> Limitation on preaching with the admonition to refrain from all vicious attacks, fairy tales, and controversies that delight in the abuse of others, and to present instead a short and apt oration. (Prince-bishop of Mainz to the heads of the orders, in particular the Capuchins, Nov. 1789)[137]

What we note from this small selection is that attention shifted from concern for the mental and physical comfort of the listeners to the necessary limitation (which was also in the interest of the listeners) on the polemical and didactic excesses of the preachers. It is striking, moreover, that all limitations as well as inquiries and complaints by parishioners concerned excessively long sermons.[138] The reverse case, the setting of minimum preaching times, is reported only for the obligatory conversion sermons for Jews that were ordered by the popes. Here is a description from Rome in the Baroque period:

> Jewish sermons (degenerate into comedies), to which mischief the Jews are encouraged all the more because the monk

> in his sermon often gives very poor reasons and dwells only on the outside rind and shell. Often he also merely reviles and harangues the Jews with pointed and shameful words, which only incites them to laughter and other mischief; and when the hourglass has run out the monk gets back into his carriage and drives off."[139]

Limiting of sermons to an abstract time measure first of all makes all sermons equal without regard for the occasion of their delivery, their content, or importance. It draws boundaries to the eloquence, ambition, and vanity of the preachers and forces them to draft their presentations carefully, with other aids also being furnished to help accomplish this goal. The purpose was to raise the level of the sermons and adjust them to the capacity of the listeners to absorb what they heard. At the same time one could reduce the risk that zealots would leave the solid ground of proper doctrine or that controversial topics would get a discussion that was too drawn out.[140] This could also have been accomplished—though with considerably more effort—by controlling the content of sermons. Abstract limits, in any case, simplified decision-making, because this way one did not have to argue about contents, the evaluation of which was a matter of opinion. Still, arguments were possible, and setting time limits was thus not an optimal mechanism of control in the pursuit of the stated objectives: in the case of bad or heretical sermons, time limits at best contained the damage but were not able to prevent it. However, time limits did make possible practicable controls, they facilitated decision-making, and as a controlling device they could not easily be applied in an arbitrary manner. Since adherence to the prescribed time limits could also be checked easily by the listeners, their existence strengthened the position of the latter vis-à-vis the preacher. Thus during visitations the length of the sermon was among the points "on which the parishioners should be questioned in the absence of the pastor."[141] Whether the hourglass at the chancel had run out was a question the listeners could surely answer more easily than questions about theological deviations.

In addition, abstract time limitations, just like clock-time regulations, made possible the pursuit of a number of secondary purposes, above all regard for other kinds of time demands or the physical frailty of the listeners. As minimum times they would also have been suitable as a kind of standard working time for preachers, but there are no reports about such a use. On the contrary: the preachers felt cramped by the chancel sandglasses, and the theologians and preaching teachers supported them on this. One of their arguments was that the word of God that became audible through their mediation should not be subject to any external limitations.[142] Following the saying of Peter, "I labored for an hour or an hour and a half but caught nothing," they "often believed that the quantity of words preached . . . was directly related to their success."[143] They complained about the public, which was willing to listen for hours to operas and comedies but objected to sermons lasting one hour. The latter was a sign of "an irresponsible revulsion for the word of God."[144] It was all in vain. In Prussia the limitation on preaching was reinforced by royal decree, and punishment was even threatened for those who criticized these measures.

The limiting of the sermon, its proper temporal duration, thus became in the seventeenth and eighteenth centuries a topic of theological and academic treatises, which throw light on many aspects of the pros and cons.[145] The line of argumentation seeking to legitimate limitations was faced with difficulties arising from history. People were not entirely unaware that they were picking up on the classical tradition of limiting court speeches with the help of a similar instrument. But they did not know of the time measures used (and couldn't have since there hadn't been any), and thus the decisive difference between the classical and modern technique of time limitation was invariably overlooked.[146] New about the modern form was the objective and precise time limit (a defined part of the full day), and an instrument that could be much more accurately calibrated. It was only this that made it possible to generalize such regulatory techniques and also fix them independent of specific instruments. The search for pertinent supportive references in the church fathers led to another difficulty that the authors did not recognize. To be sure, the fathers did speak of "hora" in connection with preaching, but they did so in the old meaning of "time period," "limit," and at best (unequal) "hour of the day."[147] However, the equal hour, which came into use only in the Late Middle Ages, had already become so self-evident in the sixteenth and seventeenth centuries that contemporaries no longer remembered the novelty of this method of time organization. Even though the church fathers occasionally spoke about shortness of time ("angustia" or "brevitas temporis"), the length of the sermon became a problem only at the end of the Middle Ages, since the tradition of homilies that were theologically and rhetorically elaborated arose outside of the Catholic mass. One strand of tradition was the Latin sermons for a clerical audience, which were among the prescribed exercises at nearly all universities. Another strand was the vernacular penitential, conversion, and Lenten sermons delivered by the monks of the predicant orders. Numerous accounts from the fourteenth and fifteenth centuries report that they, in particular, could hold the people spellbound for hours at a time.[148] The concern of the Reformers that excessively long sermons could keep the people from work was consequently directed against such mass sermons.

It is possible that the abstract time limit and the sandglass as the timing device were taken over from the universities or schools of the fifteenth century. The "artes praedicandi" of the fourteenth century did not know either one. Robert von Basevorn's *Forma praedicandi* (1322) defines preaching as the persuading of many in a limited time; it notes the difficulty of setting a precise temporal measure for the sermon but goes on to give at least guiding pointers. The time it took for a solemn mass with chanting or for a simple mass without chanting is referred to as the maximum and minimum duration commonly used in many places. In the middle of the fourteenth century, the predicant monk Thomas de Waley then recommended as a remedy against verbosity that the preacher ask a confidant to give him a sign. One such sign could be, for example, that he tug at the preacher's robe.[149] However, a time measure for preaching that was free of subjective judgment was evidently unknown. The use of a sandglass made its first appearance iconographically, attested once again in the manuscript of the statutes of the Collegium Sapientiae in Freiburg. The technique appears to have been in common use only at the beginning of the sixteenth century, when Johan Ulrich Surgant recommended that one follow the model of eminent preachers and take an hourglass into the chancel as a remedy against verbosity.[150] The reformationist ecclesiastical constitutions made this technique of time limitation obligatory for church sermons and ensured its wide dissemination. The Catholic camp continued for some time to follow the old recommendations. The Provincial Synod of Cambrai (1586) still recommended the careful, nondisruptive reminder from a colleague.[151] But soon thereafter the

60. Hans Sachs, *Inhalt zweierlei Predigt.* Woodcut by Georg Pencz, Nuremberg, 1529. Vienna, Albertina.

modern techniques also spread to the chancels of Catholic congregations.[152]

A reformationist broadsheet with two texts by Hans Sachs played on this shared practice: the *Inhalt zweierley predig* ("Content of two kinds of sermons," 1529) shows the speakers "of God the Lord" and "of the pope" as vastly different concerning their setting and their respective audiences (fig. 60). A public that is highly attentive and reading along listens to the Protestant preacher; by contrast, the Catholic preacher faces richly attired, apathetic churchgoers who are fingering rosaries. However, when it comes to the time that each has to woo the souls of the believers, equal opportunity prevails, as both use the same technique of timekeeping.

From the Late Middle Ages on, the new technical means turned the time limits on speeches that were known from classical tradition into modern organizational techniques. From this time on the ideal of eloquence also included the measured use of time, and contemporary iconography of eloquence combined the parrot as the symbol of the artful mastery of speech with the clock as the symbol of the proper measure.[153]

Mechanical clocks, but especially sandglasses, made their way into monastic cells and scholars' studies. I have already noted one of the oldest pieces of evidence, a fresco (dated 1352) with a sandglass in the

61. Saint Augustine in his study. Fresco by Sandro Botticelli (around 1480). Florence, Ognissanti.

cell of a writing Dominican friar. Francisco Petrarch subsequently became the guiding model for the conception of an intellectual way of life, for the external and internal conditions of learned activities, as well as for the actual way in which studies were furnished and visually represented. In his tract *De vita solitaria* (completed in 1366), in particular, he painted the picture of a praying, meditating, and writing intellectual who, precisely through his learned activities, could arrest the swift passage of time (the "flight of the days") and the brief period of life by producing memorable things.[154] Henceforth, symbols of time and of transitoriness became permanent props in the chambers of intellectuals. From that time on, imagined and real rooms—as we know also from the many inventories—were furnished with sandglasses, sundials, mechanical clocks, and astronomical devices of the most diverse kind.[155] Classic iconographic loci are the depictions of the learned church fathers Jerome and Augustine (fig. 61), but in very general terms these props became typical for the depictions of scholars from the sixteenth century. Beginning in the late fifteenth century, pictures of studies that include a sandglass are almost too numerous to count.[156]

These timekeepers were meant as a reminder of time and transitoriness in general, and especially of the limit to one's lifetime. In this respect they are also part of the early history of the so-called vanitas pictures. But as "pedagogical instruments" and as "tools of the spirit" they also stood for ascetic self-discipline and for the sensible arrangement of scholarly work. We must bear in mind, however, that these were modern aids and demonstration devices also from the contemporary perspective. This becomes clear indirectly from Petrarch's ideas about intellectual work, when he writes that the person in seclusion spends his time with the praise of God, liberal studies, and the recollection of new and old inventions.[157]

I fail to see—as some scholars have—a connection between the depictions of studies with their props and "objections to the time standards set up by usury and merchant capital" and "a feudal-restorationist opposition to the bourgeois concept of time."[158] Rather, the techniques of study and asceticism made use of modern technical means and thus belong in the sphere of modern time standards and time organization techniques. We need only recall the well-known examples of modern forms of asceticism, with divisions according to hours, half and quarter hours, and with the rendering of daily accounts and reviews of progress by weekly and hourly schedules in the diaries and "Spiritual Exercises" of Ignatius of Loyola (died 1556).[159] Insofar as we can speak of new conceptions of time and new standards of time—whether or not we call them "bourgeois"—bringing in the sandglass in opposition to the wheeled clock as an expression of old conceptions of time or time standards is simply anachronistic. Both clock types were in fact devices which, simultaneously and side by side, made possible innovations in the area of temporal organization.

Greater precision, the setting of time limits, coordination

The transformation of the urban pattern of time, which began concurrently with the diffusion of public clocks in the late fourteenth century, was not the result of political decisions. We cannot identify powers or interests that worked either to promote and accelerate the transformation or to obstruct it. Instead, the change in the temporal order was a largely anonymous process that began with many small steps in various and mutually independent spheres of urban life. Temporal order can thus be understood as the process of the modernization and increasing density of regulations concerning the organization of time. As a phrase describing a state of affairs, "urban temporal order" is only a retrospective heuristic concept. Within the imagined interconnectedness of all public municipal time regulations, one can use it to trace the unifying effect of public clocks as the center of a modern temporal system of reference and the development of similar techniques of temporal organization. In the narrower sense one can speak of a temporal order as a state of affairs in those areas in which both of these aspects were applied from the very beginning and in a systematic manner, for example, in hourly schedules for schools.

Calling the transformation of time structure an "anonymous" process means that contemporaries were not aware of the panoply of innovations in temporal organization. Questions of time in a very broad sense became topical during the process of transformation, but the conversion of a great variety of individual organizational problems into time problems that made possible sufficient or appropriate solutions usually occurred unconsciously. Anonymity does not mean that contemporaries failed to note any changes at all, for example, differences between city and countryside. Moreover, the emphasis on the communal clock in pictures and methods of dating can be understood as a symbolic assimilation of this transformation.

If we trace the diffusion of the new clocks and the use of the new hours simultaneously, it becomes clear that the innovations assumed two different, though not unrelated, forms: regulations by means of continuous time indication, and regulations which, independent of the indication of the time, used only abstract time limits. Since the spread of minute-indicating clocks in the nineteenth century had the result that the setting of times and of time limits can be done at any desired point with a single timekeeper, it is easy to forget that up until the end of the eighteenth century two kinds of regulation used two kinds of instruments: the clock and the sandglass.

One side of the innovation lay in the greater precision imparted to customary time indications with the help of clock-times. Indications of clock-time were always steps of increasing abstraction away from the temporal reference system of daylight, on the one hand, and the urban ringing of the old type, on the other, which was dependent on daylight and the division according to the Canonical Hours. Here we must bear in mind that the term precision is relative and that we must not impose on it modern notions of accuracy. Owing to the primitive technology of the tower clocks, there were, up to the end of the fifteenth century, limits to changes in the qualitative experience of time. Until that time we are talking about nothing more than the continuous indication of equal hours in numerical succession. Thus when we hear about a "certain time" or "certain hour," this could mean a number of things: the reliability of the time signal independent of daylight and weather; the fixation of a time on the abstract and always equally long temporal axis of the full day; the setting of a time for situations that had previously not been regulated at all, or in a way that was temporally vague or entirely unrelated to time.

The other side of the innovation was abstract time limits with the help of inexpensive, quiet, and relatively accurate sandglasses. The most common calibrations were one hour or a quarter-hour, though other fractions of the hour or running times of several hours were also used. On the time axis of the hours of the day, sandglass times were "sliding times" with a variety of practical advantages. The beginning of the timed period could be adjusted to any event or activity. Meetings, speaking times, work breaks, production processes, but also cooking times could be begun independent of the stroke of the clock and yet they could be precisely timed and also interrupted.

The following were the decisive advantages for the spread of abstract time limits: transferability or generalizability, since these time measures, unlike those in sundials, water clocks, or candle clocks, were independent of concrete devices; controls that were easy and could be carried out by anyone, which made possible just sanctions free of controversy; neutrality in regard to the organizational goal and hence the possibility of pursuing very different purposes simultaneously.

The new range of organizational possibilities that set the tone of urban life to the end of the eighteenth century was developed, tested, and introduced by about 1450, within a space of two generations.

To the changes in those spheres of life that stand out for the early application and/or the special intensity of innovations in temporal organization we can easily add observations from other areas. Numerous regulations of gate closing and curfew times translated customary rules based on daylight into clock-time regulations, though without considering the conceivable separation from daylight. To be sure, in the eighteenth century a whole series of gate-closing regulations were renewed with clock-time stipulations that were frequently very carefully differentiated, the reason being that the fee ("Sperr-Geld") for opening the gate outside of these times had been discovered as a source of revenue for the city treasury.[160] But it was only the development of housing settlements and industrialization in the nineteenth century that quickly made all gate-closing regulations obsolete. The city as a sphere under one time signal and daylight as the common temporal framework for public city life had ceased to exist by the beginning of the nineteenth century.

The distribution of scarce goods was regulated in many different ways through the allocation of time, whereby the type of regulation was no different from the market time regulations we have discussed. Reference shall be made only to the allocation of time for the use of running water to millers, raftsmen, and tanners, where classical and Islamic customs of distributing water were revived with new techniques.[161]

The urban public excluded some of its members from access to the urban space at specific times. This concerned restrictions on Jews during certain feast days or certain mass times.[162] These measures also included time limits for permitted begging, whereby it was possible, for example, to privilege local, registered beggars over foreign beggars with the help of clock-time regulations.[163]

Vivid examples for the translation of difficult problems of material supervision into simple time-control problems are furnished also by the municipal and territorial festal regulations for baptisms and weddings. Here the authorities, driven by reasons of social policy, morality, and fiscal policy, sought to set limits to the material expenditures and ethical misconduct of the burghers or subjects. Limits were set to the total cost, the number of guests, the value of the gifts, the number of dishes served, and the conduct when drinking or dancing. These regulations were usually quite detailed, but it was hardly possible to police adherence to them effectively. In addition to the translation of the long customary wine bells into modern curfew hours, beginning in the fifteenth century there appeared other time regulations that determined precisely the timing of ceremonies and the time or maximum duration of meals and dances.[164] Instead of counting guests or looking at their plates and into their cups, authorities simply fixed the temporal location and duration of the various parts to the festivities. This measure, which was of only limited effectiveness in terms of the purpose of these controls, proved to be an almost optimal device because of the possibility of policing it. However, it appears to have been part of the wedding customs, at least in the sixteenth century, to bribe the sacristan into changing the time on the tower clock.[165]

Setting time limits on torture

The limitation on the degrees of torture that was widely discussed in learned judicial tracts from the sixteenth to the eighteenth centuries also turns out to be the translation of material control problems into procedures of time control. Beginning in the thirteenth century, the accusatorial trial, the purpose of which was to prove the accusation of the plaintiff by means of witnesses, oaths, or ordeals, was replaced by the inquisitorial trial, the purpose of which was to investigate the facts and determine the material truth. That is why the confession of the accused acquired increasing importance if the evidence was only circumstantial. From the thirteenth century on, torture as a means of extracting a confession was more and more widely used in central Europe, less so in England.[166]

At first an unsystematic practice, torture was developed in the fourteenth and fifteenth centuries along with the inquisitorial trial. In 1215, Pope Innocent IV authorized its use in inquisitorial trials, and the initially unregulated and arbitrary practice of torture was put down in writing along with the procedures of the inquisition. Under the influence of Roman law and Italian jurists, a regulatory system was developed in which torture was permitted, among other things, only for crimes subject to corporal or capital punishment, and only if there was strong circumstantial evidence or eyewitnesses whose testimony gave grounds for strong suspicion. In Germany it was initially the "Bambergische Halsgerichtsordnung" (Bamberg Rules of the Criminal Court) (1508), and later, and with far more influence, the *Constitutio Criminalis Carolina* (1532), issued under Emperor Charles V, which tried to translate the traditional demands for moderation in torture into usable procedures. The legal codes did not change local practice immediately; instead, they were subsidiary, that is to say, they gradually acquired the force of law where other regulations were absent or were reformed or adapted. According to paragraph 58 of the *Carolina,* "torture ("die peinlich frag") should be applied, depending on the suspicious nature of the person, frequently, often, or little, harshly or mildly in accord with the judgment of a good and reasonable judge." Therefore, as all later commentators point out, judges were to take into account the severity of the crime, the weight of the circumstantial evidence, and the physical condition of the delinquent when ordering the use of torture. This was no easy task: on the one hand torture could not be applied twice unless there was new evidence; on the other hand judges were criminally liable if the use of torture was excessive. One way of compelling moderation, checking abuses, and making it possible to control both was the quantitative regulation of torture according to degrees of severity, whereby the commentators usually distinguished either three or five degrees. In the first degree, the "territio," the accused were threatened with the use of torture and then shown the torture chamber and the instruments. In the second degree they were supposed to feel the pain, for example, by being briefly pulled up by their hands which were tied behind their backs. In the third and fourth degree these procedures were intensified by striking the rope, adding weights, and using fire and cold water. Such intensification was permissible only against those suspected of extraordinary crimes, such as treacherous murder, theft of Church property, the killing of ecclesiastical dignitaries and secular princes, incitement to rebellion, and heresy. The goal was to extract a confession; if possible the accused should suffer no permanent injury, for they were supposed to repeat their confession after the torture in a clear state of mind, and only after this would they be sentenced and punished.

The second approach to moderation that was widely discussed since the beginning of the sixteenth century was the quantitative regulation of torture according to time. Its origins, presumably in Italy, are not entirely clear. Paulus Grillandus recommended one Paternoster, one Hail Mary, or one Miserere as the time measure for the brief hoisting up in the second degree; in the third degree the hoisting should last two Miserere or somewhat longer; in the fourth degree it should last one-third, one-quarter, one-half, two-thirds of an hour, or one hour.[167] The new method of time limitation was given a boost by a bull of Pope Paul III (1548), in which he prohibited the wardens in the Roman prisons from torturing for more than one hour in normal cases. Pietro Follerio, who was commenting on Neopolitan statutes, invoked Bal-

62. Torture scene by Alessandro Magnasco (1710). Frankfurt am Main, Städelsches Kunstinstitut. Photo: Ursual Edelmann.

dus de Ubaldis (died 1400) when he recommended the use of a sandglass ("ampollecta") for the observance of these limits. In the writings of the jurists we can subsequently trace the gradual refinement of this procedure. Hortensio Cavalcani demanded the setting up of a sandglass ("horologium pulverinum") before the eyes of the judges in order to control the quality and quantity of torture; before the eyes but not within sight of the delinquents, who, according to Tranquillo Ambrosini, had begun to ask: "Hey! How much longer for the clock to run?" In 1706, Johann Valentin Kirchgeßner, the Most Princely Chancery Registrar of Würzburg, also desired to see torture limited to one hour per day and have this limit "accurately observed" with a sandglass, and he suggests, moreover, invoking authoritative writers, that judges be suspended in the torture for no less than fifteen minutes so they might feel what it was like.

Word of the disadvantages of objective time controls gradually spread. Some complained, for example, that thieves preferred to be tortured in Saxony, where it was done according to the time. The short-lived *Constitutio Criminalis Theresiana* (1769–1787) therefore stipulated that one should make it impossible to endure the dark hours of torture by not informing the accused beforehand of the intended torture procedure and thus of its length. Against "strong people," though, and against "obstinate Jews accustomed to denying things," it was permissible to extend the torture over several days.[168] However, those who were tortured the most were the least protected by the procedures for moderation through time limits. Witchcraft was regarded as an extraordinary crime, where torture was allowed to exceed all stipulated measures. Protocols of witchcraft trials reveal everywhere that the learned commentaries regarding the degrees as well as the duration of torture were not worth the paper they were written on. We don't have to search long for accounts by notaries about cruelties that violated the rules and about tortures that went on for hours and days.[169] After the witch craze subsided, this cruelty did not by any means wane everywhere. In the Netherlands, in the Spanish parts of which the *Carolina* had no force, several hundred victims were tortured to death in the eighteenth century in extremely long sessions; during the final decades the notaries furnished the protocols of torment with indications in minutes. In other cases, however, delinquents survived even twenty-four hours under torture and thus saved their necks.[170] The topic of the limitation of torture, as well as the practice of torture itself, was rendered obsolete in the late eighteenth century through the advance of the concept of the discretionary assessment of the evidence by the judge and also through the humanitarian criticism of the enlighteners.

As early as the fifteenth century we also find the first reports that clock-time punctuality at tournaments and shooting contests was demanded not only of the participants; a new, objective time measure was imparted to the individual rounds of a contest by means of sandglasses. We hear this, for example, from Olivier de la Marche in his account of the tournament games on the occasion of the marriage of Margaret of York and Charles the Bold of Burgundy in Bruges in the year 1468. Sport as the contest for speed and the quest for time records began much later at the horse races in England, where in the seventeenth century running times were recorded by minutes and minute fractions. Reports of stop watches and times taken to the half second reach us once again from England from the middle of the eighteenth century. Only the technological leap in the building of precision clocks has, since then, turned record lists into eternal records of seemingly unlimited acceleration.[171]

Temporal orders of a special kind are also the numerous court and chancery statutes issued in the sixteenth and seventeenth centuries. As attempts to set a "house" in "order" in a comprehensive way, they are certainly comparable to the ecclesiastical constitutions and territorial administrative statutes, and in part the regulatory material overlaps. They differ from the state of affairs I have called the urban temporal order primarily in that they did not arise gradually in many different areas, but sought to organize court and chancery according to a uniform design. In this they resemble the school statutes, though in respect to the daily routine of household members and employees the approach was usually far more thorough. As registers of offices, court and chancery statutes regulated not only who had to do what, but also when everybody had to work, eat, or simply be present.[172]

During the first century of the diffusion of public clocks, most modern time regulations in the urban sphere can be assigned to two types: "fixing the time" and "setting time limits." Even where modern time fixations merely specified old daytime regulations with greater precision, with no observable objective need for more precision, the status of the regulations changed imperceptibly. When venerable old fixations—in part natural ones, in part connected to bells—were given clock time, they appeared as social fixations. Progress in precision is also progress in the possibilities of change and conflict. Wherever clock-time regulations appear as the results of conflicts over temporal fixations, they produce unequivocal organizational forms that were largely removed from the arbitrariness of particular interests and were also capable of being changed.[173]

Greater possibilities of changing regulations encounter limits, however, when regulations arise that involve "coordination," regulations that have become part of complicated timing patterns. To be sure, the potentially endlessly complex temporal orders of the nineteenth and twentieth centuries, made possible in principle by clock-time regulations, were still a long time away, and we cannot yet speak of an "avalanche effect of temporal fixations" (N. Luhmann). Still, early stages of processes of temporal coordination, which led to time commitments that could no longer be easily revised, can certainly be detected in the realms of the church and the schools. Many different indirect repercussions for the urban temporal order came, for example, from regulations that set the time of the main religious service apart by prohibiting many activities in the city during this time: selling, games, tavern visits, later also taking walks and driving carriages. Added to this were the requests not to schedule any committee meetings during this period, if possible, and not to ask for compulsory labor services. If the regulations also called for the closing of the city gates during the time of the main service on Sunday, all possibilities of an individual disposition of time were suspended for a while. In the larger cities the great number of services in the various churches necessitated careful time-coordination. This involved problems that were in part quite old. Parish churches had to defend themselves against competing mass-times especially from the churches of the mendicant orders. Questions of jurisdiction and rank merged with quarrels over income and donations. Temporal order thus also meant the balancing of financial interest. Another perennial topic, which played a large role also in Protestant ecclesiastical constitutions, was the time of weekday services, which were to be located such as not to interfere with the work-time of city dwellers.[174] To the extent that after the Reformation church spaces had to be used by both confessions, additional temporal fixations were necessary. Here, too, it could happen that the confessional parties accused each other of manipulating the clock.[175]

A new practice was the temporal fixation of baptisms, weddings, and funerals, with the Protestant constitutions preceding the Catholic ones only chronologically. Two different goals were being pursued by this move. As was the case with the festal and sumptuary regulations, decency dictated that parishioners came to these ceremonies in a sober state and did not disturb the other schedule of services with requests for baptisms and funerals. In addition, these times had to be coordinated with school times. One issue was the fact that in the smaller cities the same people often held several posts. Priest, cantor, and sacristan were also teachers at the municipal school; teachers and sacristans were also town clerks and were, in addition, frequently charged with maintaining the public clock. "To the schoolmaster 20 gr. for setting the clock"—such notices are found with great frequency, for example, in the visitation reports since the first years of the Reformation. Complaints about this onerous additional task that was held in low esteem are heard until the end of the nineteenth century.[176] Another reason why times for services and school had to be coordinated was that the students provided the church choir that was indispensable for the main services and funerals. "In order that the students do not miss any classes," the ecclesiastical constitutions constantly urged that services end punctually and that funerals be held in the early afternoon.

In spite of such examples, temporal regulations that are explicitly coordinating in nature remain relatively rare in late medieval and early modern cities. The potentially endless differentiation of time structures "internally" that clocks made possible in principle was realized only in rudimentary form.

But clocks also expanded the physical space in which a temporal order was applicable beyond the zone delineated by acoustic or optical time indication: they made possible the coordination of temporal fixations independent of the time signal. Virtually none of this was put into practice. All the small steps of abstraction and innovations in time organization I have described remained literally "under the bell tower." Only in the eighteenth century do we begin to note that people were becoming consciously aware of the possibilities of coordination also beyond the boundaries of the "urban monads." For example, in his *Historie der Kirchenceremonien* (1772), Christian Gerber discussed the possibility of detaching the temporal order from the accustomed system of urban signals: "Those who are far away, say in villages, . . . cannot, after all, hear the bells, especially when the wind carries the sound in the other direction; they must therefore go by the clock and the hour to know when mass usually starts."[177] Adherence to a clock-time regulation outside of the reach of a certain public time signal became a part of everyday conduct only in the age of the train, which was at the same time an age of the private ownership of clocks. Up until the beginning of the nineteenth century, life by the clock remained simultaneously a life under the urban bells.

Time-measurement and "scientia experimentalis"

The transition to the continuous indication of equal hours was a transition to a system of reference that was more abstract with respect to the period of daylight though not yet completely detached from it. The equal hour as a time period was one step in the direction toward abstract, measurable, and—to that extent—more objective time measures. Of course it is necessary to point out the boundaries of these innovations that were not yet crossed in the Late Middle Ages. The "time of the cities" was not yet the abstract or mathematical time of the sciences.

Misunderstandings in the historical literature arise when one fails to realize that up until the end of the sixteenth century, the clock-hour was the twenty-fourth part of the full day or the twelfth part of the day or the night at the two equinoxes. In everyday life this hour was divided into halves, thirds, quarters, sometimes into twelve parts, but it was not divided by sixty or understood as the period of sixty minutes.[178]

The great scholarly syntheses from Lewis Mumford to Alistair C. Crombie continuously repeat—invoking the work of Lynn Thorndike—that the division of the hour into sixty minutes and of the minute into sixty seconds had become customary as the framework of thought and action as early as 1345.[179] However, at least until the end of the sixteenth century one simply cannot talk of a customary use of minutes, let alone seconds. Minute and second indications—and this is also the only way in which Thorndike's evidence can be understood—are found only in theoretical discussions and in astronomical or astrological time indications. They were ancient and theoretical but not measurable time units. Clocks that indicated minutes might have existed at the end of the fifteenth century, though we have no reliable reports about this.

If we were talking only about everyday linguistic usage in the Late Middle Ages, this correction would be of interest only to specialists. But such small misunderstandings become misjudgments of considerable magnitude when it comes to the history of ideas and consciousness. The conventional thesis, as formulated by Crombie (who is drawing on Mumford) in his book *Augustine to Galileo,* reads as follows: "The invention at the end of the 13th century of the mechanical clock, in which the hands translated time into units of space on the dial, completed the replacement of 'organic', growing, irreversible time as experienced by the abstract mathematical time of units on a scale belonging to the world of science."[180]

This thesis has problems that are typical of many discussions of the change in time consciousness: with a view toward the final outcome it severely condenses the actual historical process, and it treats the experience of time and the scientific conception of time—in modern parlance: the subjective sensation of time and "objective," physical concepts of time—on one and the same level. Moreover, one gets the impression that linear conceptions of time—here infelicitously called "organic"—were replaced by discrete conceptions. For example, Daniel C. Boorstin, in his book *The Discoverers* (a popular synthesis), explains that once clocks had become common, "people would think of time no longer as a flowing stream but as the accumulation of discrete measured moments."[181] While the ticking clock makes such ideas plausible, they remain far removed from the historical developments of consciousness at that time.

In an essay on the problem of quantification in medieval physics, Crombie formulated his theses more precisely and modified them: beginning in the High Middle Ages, efforts toward quantification were stimulated, on the one hand, by the biblical phrase that God "has ordered all things by measure and number and weight" (Wisdom 11.20),[182] and, on the other, by the concept, borrowed from Platonic philosophy, that the universe was rationally created by God mathematically or geometrically, and that it was therefore comprehensible according to mathematical or geometric rules.[183] For Crombie this poses two different levels of investigation for a modern inquiry: that of quantitative theories and models, and that of quantifying procedures and measuring units and scales, that is, the level of counting, measuring, and weighing.

Medieval physics was primarily concerned with the theoretical explanation of the world and the translation of qualitative descriptions into quantitative relationships. On the level of theory, however, measures, numbers, and weights were of interest only as relationships such as bigger-smaller, slower-faster, harmonious-unharmonious, or proportional-nonproportional. It is from this perspective that we must understand the quick reception of the mechanical clock into the arsenal of the metaphors and models of the natural scientists and theologians. The demiurge-God became a clockmaker-God, and the "machina mundi" became the clockwork-like universe, and the starry heavens, the ideal clock. In these texts the clock stands for a fixed order, for impulse-free, even movement, for the harmonious coordination of seemingly contrary movements or the balancing of movements of varying speed, for an elaborate artisanal contrivance or in general for an ingenious machine.[184] In addition, clocks served as the drive for heavenly models that illustrated astronomical theories. In the fourteenth century, the clock, as an ideal timekeeper, was used by late scholastic discussions on the concept of time as an example, but it never played a role as a measuring instrument. The early clocks were simply not suitable for such a use. Measurements or experimental verifications of theoretical statements were of little interest to the sciences that were pursued at the universities.

Measurements were made—largely without any connection to theories—only where there was a demonstrated practical need, for example, in astronomy for calendric calculations, in navigation, in surveying, and in mining. According to Crombie, it was thus not the universities but practical requirements which promoted the quantification of space and time in the experienced world. This hypothesis, on the whole plausible and supported by subsequent studies,[185] has only one minor flaw: with respect to the quantification of time it bases itself once again on the legendary ordinance of Charles V.[186] The sources on clock-use and abstract time measurements, by their very scarcity alone, substantiate Crombie's thesis of the minor importance of practical time-measurement in the late medieval sciences. In only a single instance from the middle of the fifteenth century do we hear about an attempt to draw water samples at two different thermal springs at a specific hour of the day which was determined with the help of a clock.[187] From the end of the century a report exists on the use of a mechanical clock for temporally limited astronomical observations.[188] Suggestions for using mechanical clocks to determine degrees of longitude at sea date only from the sixteenth century;[189] however, a solution to this particular problem was not found until the eighteenth century. At the level of scientific concepts, too, clocks and equal hours did not yet play a role in the sciences of the fifteenth century. I shall illustrate this with an example.

In the middle of the fifteenth century, physicians picked up a suggestion from Greek physicians in antiquity to supplement the qualitative descriptions of the pulse, which was based on the prevailing teachings of Galen, with quantitative descriptions, that is, to count the pulse. Temporal categories like "fast" or "slow" were also known in the classical doctrines on the pulse, but they were only subjective and empirical categories like "faster than normal based on experience," or they expressed relationships such as "fast movement" as against "short pauses." Around 1450, Michele Savanarola, professor of medicine in Padua and Ferrara, proposed in a tract on the pulse to determine the difference between the pulse of a healthy person and that of a sick person as the difference in the pulse rate. He evidently did not know of a suitable time measure for the pulse rate and thus recommended instruction by musicians in conductor's tempo. He was confident that this could be learned in eight hours (!).[190] However, at this time mensural music did not yet know an absolute time measure; instead, it used a system of tempo levels that was built up proportionally and learned by practice. Still, it seemed to Savanarola a sufficiently "objective" time measure to determine the pulse rate. At this time Nicholas of Cusa also recommended counting the pulse and breaths of healthy and sick people for making diagnoses and prognoses. Following classical suggestions, a clepsydra was to be used for this purpose. For the duration of one hundred breaths or pulses (not pulse beats!) each, the narrow outflow spout of a clepsydra should be opened and the water collected in a basin. While the pulse was counted here, it was not counted for a given time unit but up to an arbitrary number. The unit of measurement was not a time measure but a volume measure or a weight, for Nicholas suggested comparing the collected amounts of water by weighing them.[191] Scale and water clock were to be used also for determining the weight of air through falling experiments, the depth of unknown bodies of water through the rising time of floats, the speed of ships, the running speed of people and animals, and astronomical data such as the relationship between the lengths of orbits and the diameters of planets (see chap. 2, Clepsydrae and scientific measurements).[192] The "experimentalis scientia" projected by Nicholas of Cusa was to expand human knowledge through individual, continuously recorded experimental steps of measurements and comparison. But the suggested experiments do not fall outside the bounds of the armchair experiments of classical and Islamic textual tradition. For example, the procedure for measuring unknown depths of water, which is also suggested by Leon Battista Alberti and Giovanni Fontana, is found in the MS Ripoll 225 from the eleventh century as well as in other astrolabe tracts. How far these experiments were from actual practice is revealed by the fact that the problems of the outflow calibration of simple water clocks, well known in antiquity as well as in China, seem to have been simply forgotten. These suggestions also overlook that such measurements were feasible at best in concrete situations and were not suitable for documentation. Even more important to our questions is this: what comes into view are only relations and proportions, and not abstract time measures that were made possible by the use of mechanical clocks or sandglasses and were thus documentable in the long run. Although clocks and modern hours were certainly known to these authors, they do not appear in their scientific discussions. Only Giovanni Fontana seems to have worked on the further development of both types of clocks and on the construction of a stopwatch needed for such measurements.[193] But whether time measurements were carried out on physical phenomena in the fifteenth century seems very doubtful also in this case.

If we follow the history of pulse measurement further, it becomes clear by what circuitous and protracted paths modern time-measurement gained access to scientific theory and practice. At the end of the fifteenth century, music theorists began to search for an obligatory time measure for notes, suggesting first the pulse, and then, from about 1530 on, mechanical clocks.[194] Girolamo Cardano used the pulse in 1570 for astronomical time measurements and established a relationship between the length of the moon's orbit during one hour and the distance it traversed between two beats. As the basis for his calculations he took the fairly accurate value of four thousand beats per hour.[195] In 1583, Galileo checked the regularity of the swings of a thread pendulum at first by the pulse. Only later did he arrive at the reverse concept, that of determining the rate of the pulse with the help of the swings of a pendulum.[196] As late as 1604, Kepler, too, explained a problem of determining an astronomical time by the duration of a pulse. Not until 1618 does he report accurate pulse measurements in healthy and sick, old and young people.[197] Only from this time did it become customary to take pulse measurements per time units and, with the appearance of minute-showing clocks, use them in medical practice.

This short look at one slice of the history of science shows that technical limitations as well as weighty conceptual reasons explain why the innovations we have discussed gained entry into the realm of the sciences only with a considerable delay, unlike what took place in the sphere of public urban life. Theses that establish a direct connection between the development of physical conceptions of time and the diffusion of mechanical clocks in the Late Middle Ages are thus not tenable. Norbert Elias has also pointed to this late "emergence" of the concept of "physical time."[198] Only since the "Scientific Revolution" in the middle of the seventeenth century can one speak of experimentally quantifying scientific procedures and of conceptions of time as a scaled continuum of discrete moments.[199]

9

Work Time and Hourly Wage

Work time has been one of the great themes of social conflict since the beginning of industrialization. At first the struggle was over the length of a working day, which, under the dictate of the "economy of time," had been inhumanely lengthened in many places when the age of the machine began; only after fierce conflicts at the end of the nineteenth century was it gradually reduced to the "normal workday." However, the new concept of the "economy of time" that arose along with workshops and factories did not concern primarily the length of work time, but above all its regularity and intensity, and thus the preconditions for its efficient economic use. These aspects have been widely discussed in the last decades under the catchphrases "social disciplining" and "the loss of individual control of time."

A broad consensus has emerged that work in the factories made entirely new demands on the rural population, which was used to an uneven volume of work and uneven workloads. Part of the problem was supposedly also that money wages alone were not enough to motivate this population to continuous paid work over and above what they subjectively determined to be the necessities of life. Only over the course of several generations, by means of compulsion and then also through internalized norms of productivity, did the adjustment to these new, more strongly regulated, and more intensely demanding work rhythms succeed. This recent European development has become, unawares, the model case for the path from an agrarian to an industrial society, and the accompanying problems appear as the typical problems also of countries that still have the transition to our type of industrial society ahead of them. This whole concept is so plausible that objections—be they against its view of modern European history, be they against the interpretation of the problems of the "world in transition"—are rarely voiced.[1] After N. McKendrick, S. Pollard, and K. Thomas described the factory-industrial disciplining also, and above all, as a process of disciplining in the everyday handling of time, E. P. Thompson, in a famous essay, raised the question of to what extent the changes in the perception and apprehension of time which preceded industrialization prepared for the adaptation to the new demands to such an extent that they could become, in spite of all initial resistance, the "second nature" of the inhabitants of Western Europe.[2] His answer was that the revaluation of time into a commodity was the product of a marriage of convenience between the ethic of Puritanism and industrial capitalism. The key for the discussion of the change in the conception of time was also for Thompson the social use of timekeeping. Jacques Le Goff's essay on "Church's time" and "Merchant's time," in which the communal clock is described as "a distant precursor of Taylorism," extended this perspective back into the Late Middle Ages.[3] Le Goff addressed the problem of working time in another essay, describing the victory of modern time-reckoning as the adjustment to an altered economic situation, or more precisely: to the changed conditions of urban wage labor.[4]

Time for sale: The working day

For the large majority of workers in the Middle Ages, working time was not limited or measured, but was determined either by the demands of the peasant economy or by custom and authority. Within the boundaries drawn by tradition, the work rhythm and work intensity of the manorial lord, the master, or the master of the house determined directly the duration and intensity of the work of all subordinates. Work outside the house was usually farmed out as job orders ("ad tascam"), sometimes with a delivery or completion time attached, though not with working time regulations. Codified working time regulations are thus found only with the organized guilds and for the initially small group of wage laborers who were almost always employed on a daily basis. The customary word was "dieta"; derived from a measure of distance, it described in the fourteenth century, apart from a daytrip, a rural day's work or in general a daily quota or a day's wage.

The selling of "one's own time" as a period to be used by others was considered theologically acceptable in the Middle Ages, in contrast to the selling of "time belonging to all," for example, in expectation of interest profit.[5] Work time that was for sale or owed because of an unfree status could be measured in years, weeks, days, or half days. With respect to the intensity or result of work, labor for a day's wage was considered "unmeasured" right up to the eighteenth century.[6]

Work assigned on the basis of a day's wage is a quantity that is difficult to calculate. In part this has something to do with the number of yearly working days, a number that is important for the estimate of average incomes but one that can hardly be generalized. Assuming continuous employment—a fictitious assumption since it rarely applied—most studies posit about two hundred and sixty working days per year. On average this yields a five-day week, whereby we must also take into consideration that feastdays and half feastdays were very unevenly distributed across the year.[7] A large number of pertinent statutes attest unequivocally that from the thirteenth century, at the latest, the problem of what constituted the time of a day's work led to conflicts in many cities. And here we can distinguish two types of conflicts and two forms of regulation. There were, first of all, conflicts among guild members involving guarantees of working times of equal length and thus of equal earning opportunities. The resulting regulations of working time were part of the arsenal of guild measures to reduce competition, measures which also included, for example, controls on weights, measures, and quality, and restrictions on the number of apprentices or looms. They further included regulations about the minimum time that had to be spent for the production of one unit of a certain product, or about the maximum number that could be produced during a certain time period. More frequent, however, were conflicts and regulations in connection with the working time of laborers who received time wages. Since the wages were often paid out on Saturdays, these laborers were in fact weekly wage earners; but since the rates were generally calculated by the workday, it makes sense to call them day laborers. These working time regulations were usually connected with limits on the maximum wage permitted, limits which could hardly become the subject of conflicts or negotiations.

By contrast, the workday was, within narrow boundaries, a comparatively elastic quantity. Conflicts arose over the length of breaks and the end of work. If the regulations stipulated that work ended with daylight, it was open to debate, for example, whether it was permissible or in keeping with ancient custom to let workers return from construction sites or vineyards near the city while it was still light out. In many cities the end of work coincided with the peal of certain bells, and custom determined exactly how close to sunset they were sounded. Attempts to push the end of work into the late afternoon were not attempts to add "spare time" to "working time"—this modern distinction was unknown at that time, both conceptually as well as terminologically[8]—but efforts to gain time for personal work, and to that extent questions of time, were, especially for day laborers, in those days also unequivocally questions of money.

Opinions diverge on the percentage that day laborers accounted for in the labor market. The majority of work was awarded as job orders or piecework. Day laborers were a small minority—though a noticeably growing one in the Late Middle Ages. Despite the fact that wage labor went through a period of crisis—especially in the years 1336–1343, 1349–1360, 1415–1430[9]—many authors speak of a golden age of wage labor, a result of the shortage of labor in the wake of the plague epidemics.[10] A flood of national, regional, and local ordinances fixing maximum wage rates were directed against wages that were showing a tendency to at least a nominal rise. Traceable since the thirteenth century, these ordinances appeared with increasing frequency in the decades following the Black Death, and a great many of them contain regulations of working time.[11]

For the day laborers affected the trend was clear: the work regulations issued by various authorities pursued a restrictive or restorationist policy. They consistently stipulate the full period of daylight as the standard measure of daily work, valid since time immemorial, and seek to enforce it with new controls and new penalties.[12] These measures reflected the distribution of power, since the urban employers everywhere also controlled city government and could at any time have such steps confirmed by the territorial lords.

However, if we follow the technical details of organization, we find innovations whose way was paved precisely by the incorporation of clock-measured times, innovations which in the long run turned working time into a negotiable quantity. The new fixations of working time with the help of clocks spread so rapidly because the vagueness of the old time indications and the limits of what the old signal systems could do had for some time been recognized as a problem. Hence—and this, too, Jacques Le Goff has correctly perceived—new solutions to this problem were sought from the thirteenth century on.

Since nearly all work was done collectively and in spatially tight-knit communities, daylight was a sufficiently accurate measure of daily working time. But in working time regulations since the thirteenth century, we also find time indications of greater specificity that supplemented naturalistic indications for times of diffuse light with experiential values. In Paris, for example, fullers were to begin work in the winter "as soon as one can recognize a person in the street in daylight."[13] According to a statute from the year 1324, tawers were to use the possibility of distinguishing two coins as the criterion whether there was sufficient light to work.[14] In the 1375 Hamburg statute on smiths, work ended in the autumn when "the sun turns golden," in the winter "when day gives way to night."[15]

In the cities, working time was determined in part by daylight, in part by the ringing of the Hours in various churches, in part by civic time signals (for example, the wine or fire bell). A look at the collections of older working time regulations reveals that working time signals, too, developed over the course of time into quite complicated signal ensembles, at least in the very large cities. In the process problems evidently arose with the reach as well as the differentiation of signals. The working time regulations for about one hundred guilds in the *Livre des Métiers,* which was drawn up in 1268 at the instigation of Etienne Boileau, prefect of Paris, contain, for example, three types of bell signals: Hour ringing from churches in quarters where certain trades were located; for changing work sites ringing "wherever it can be heard at the time"; and the ringing of Notre Dame especially for the end of work on evenings preceding major feast days.[16]

Calculations in which projects were computed in terms of their expected time or costs as the sum of "day works" were complicated but not impossible to carry out. There were experiential values on how large a field could be plowed in one day or how large a piece of cloth could be woven. At the end of the thirteenth century, the English steward Walter of Henley reckoned the working year at "44 weeks woorkable."[17] The limits of the workday were determined by the length of daylight, that is, work usually began at sunrise and ended at sunset. This practice is neither especially medieval nor especially urban. The workday was thus shorter in the winter than in the summer, and wages fluctuated accordingly. For the purpose of setting wages, the year was divided into two, four, or six periods. Winter wages were on average one-quarter to one-third less than summer wages. The intensity of work, which varied strongly seasonally and with local feast day customs, could be raised from case to case by giving general permission for work in the evening or at night, which was usually prohibited.[18] Typical was the suspension of the prohibition against night work prior to markets and fairs or to allow orders from high-ranking individuals to be carried out. Construction projects that were commissioned by high-ranking persons or projects that were technically difficult—for example, in underground engineering, where groundwater posed a threat—also justified the lifting of the prohibition on night work and even work on feast days.

Church bells and working times

It is hardly possible to say to what extent the wealth of church time signals in the working time statutes of the thirteenth and fourteenth centuries led to effective time regulations. However, a famous example, the conflicts over working times in the vineyards outside of Sens and Auxerre in Burgundy, reveals that the ringing of the canonical hours was increasingly less suitable for regulating working times in such a way that they were immune from conflict.[19]

At the beginning of the year 1383, the nobility, clerics, and citizens of Sens obtained a royal ordinance which newly regulated the working time of the day laborers in the vineyards surrounding the city. The initiators of the ordinance had complained that the workers were demanding excessive wages, and that for these wages they were also not working the usual time until sundown but left the vineyards between midday and None, in any case long before sundown. Moreover, they were easy on themselves so that they could work in their own vineyards after working hours. The royal ordinance put an end to this: it decreed that men and women had to appear before sunrise at the place where day laborers were hired, and subsequently had to work continuously and loyally in the vineyards until sundown, with reasonable ("raisonable") breaks. At the same time a maximum wage was set. This statute evidently met with resistance, for two poor winegrowers appealed to the parliament of Paris because the royal official in the city had imposed severe penalties on them after they gathered to defend their customary working hours. Invoking previous, similar decisions, the parliament rejected the appeal; however, since the appellants were merely "ignorant people," their act of bringing the appeal, after payment of a fine, was forgiven without any further consequences.

Ten years later the conflict was repeated in almost identical form in Auxerre, about a two days' journey from Sens. The wines of this region were precious, and in good times the shortened day's work had apparently become a tolerated custom. Since 1359, however, the situation of the vineyard owners had worsened because of war-related devastation, and they now sought to reclaim the old, long working time. At first, in 1392, they received a royal ordinance for Auxerre and the other winegrowing towns of the region; this ordinance was an almost verbatim copy of the one granted to Sens in 1383. Subsequently conflicts erupted in the city, which led to the arrest of workers who refused to follow the new regulations. A second complaint from the burgher owners to the parliament led to the suspension of the pending dispute. Two delegates came to Auxerre and heard both sides. The case put forth by the owners is contained in a second, unrevised draft. The decision was the same as in Sens—work during the entire period of daylight—and in itself unremarkable. What is striking, however, is the trouble both parties had in defining the contested end of working time.

The wage laborers raised the following objections: the ordinance was unjust because the custom of ending work in the afternoon was so old that nobody could remember it ever having been different. The soil of this region was particularly heavy, and ending work at None had always been the case in Auxerre as well as in other areas. A further reason was that None in Auxerre was presently being rung later than in previous times, between the fourth and fifth hour of the afternoon instead of at the third. The shifting of None, generally completed by 1300, was thus not yet common in Auxerre, and the workers expressed their objection with the help of a new system of reference, clocktime, which had long since become established in Auxerre. The only open question is thus how they knew that in the past None had been rung much earlier.

The vineyard-owning burghers complained of poor yields and high costs. They denied that the abbreviated workday was customary and lamented the ineffectiveness of the royal ordinance. The workers, "fat, insolent, and obstinate like Pharaoh," had carried on as before, denouncing, moreover, the commissioner as a "bloody rogue" and the burghers as "misers." They had tried to incite a revolt like the peasant uprisings and tax revolt in Paris (in 1372)—a very serious charge, since both of those bloody events were still fresh in everyone's memory.[20]

According to the text of the "arrêt," the burghers drew a clear picture of the purposeful reduction of working time that the workers had pursued. At first they had worked to the end of the ringing for None. Then they had left at the beginning of the ringing. Currently they were ceasing work with the "cliquet," the pre-signal to the actual ringing.[21] And when the ringing came late in the judgment of the workers, they decided the end of work freely according to the position of the sun. All told they were now working little more than half a day. The historical argumentation by the burghers who had brought the complaint has flaws. As shown by the draft of the complaint, they were not at all agreed when the workers did in fact cease working in the afternoon: we hear "two hours," "two and a half hours," "three and a half hours" (sometimes with "approximately" added), and finally long ("grant espace de temps") before sundown. In situations of conflict, the traditional system of reference for the times of day proved to be inadequate. In Sens and Auxerre, however, the formal side of the conflict was not yet fully settled, since the king once again tied the end of work to sundown.

After the Hundred Years' War, the conflict flared up one more time in Auxerre. The parliament of Paris was once again brought in. Its decision did not amount to any fundamental change for the workers; it merely protected them from objections like the ones mentioned above and tied working time to communal clock-time. According to an "arrêt" from the year 1447, in the summer work was to go on—with a three-hour break—until the last stroke of the seven o'clock ringing.[22]

Work bells ("Werkglocken")

The great importance that attached to the regulation of the length of daily work is borne out not only by the growing flood of working time regulations, but also by the spread, since the thirteenth century, of a new urban time signal. With the help of the so-called "Werkglocken" ("work bells"), the time of day work was actually and symbolically detached from the intra-urban temporal order and separated in terms of signaling technique. If we trace the history of the work bells into the fourteenth century, it becomes clear that the problems of working time had given rise to some sort of need for greater precision, which then manifestly promoted the diffusion of clocks.

It is not entirely clear where working time was given a separate signal for the first time. The podesta code of Orfinus Laudensis that was mentioned earlier gives the impression that a "campana laboris" for the urban trades ("artes") was customary in northern Italy by the middle of the thirteenth century.[23] Such bells, which in this area were often called "Marangona" (derived from the word for a carpenter's ax), are attested, apart from Venice and Florence, in Ferrara (1287, "Marangona"), Novara (1295, "campana paraticorum"), Padua (1308, "Marangona" for the "laboratores ad precium"), Verona (1314–1315, "Marangona"), Parma (1317–1318, for the "laboratores manuales et diurni et muratores et magistri manariae"), and Treviso (1314–1315, "Marangona" for the "laborantes de quacumque arte sive opere").[24]

Work bells are mentioned under many different names around the same time also in the textile regions of northwestern Europe. A "clocque des teliers" can be found in Douai around 1250.[25] In the fourteenth century in Therouanne, the bishop was divested of the right of ringing the "campana operariorum," a right that went back to an "arrêt" issued by the parliament in Paris in 1261.[26] Bruges had a "weversclocke" ("weavers' bell") in 1269. In Provins in Champagne, the textile workers had rebelled after the burgomaster had moved the bell signal for the end of work into the evening. Since the unrest, in which the burgomaster was killed, had been rung in with the communal bell, that bell was destroyed during a punitive expedition against the town. The amnesty issued in 1282 by Edward I, King of England and Count of Champagne, explicitly permitted the ringing of the new bells also for the working hours of day laborers.[27] Such bells can be attested at the beginning of the fourteenth century in many other cities of this region,[28] and judging from contemporary phrase books, they appear to have been common everywhere.

The names of the bells and the circles to whom they were addressed bear out quite clearly that the spread of work bells must be seen in close connection with the growing importance of wage labor. At the same time, however, they also symbolized the political importance and the financial muscle of the guilds who were the major urban employers. The "Arte della lana" in Florence was not only the mainstay of the cathedral workshop, it also paid for the largest bell in the campanile of the cathedral. In working time regulations that bell was referred to simply as "campana dicte artis."[29] Work bells were frequently taken as seriously as communal bells. They were described as an attribute of a "bonne ville,"[30] and the penalties for the deliberate misuse of the work bells were often no less severe than for the communal bells.[31] Like the communal bells, the work bells marked out a sphere of jurisdiction. In Bruges in 1284, the weavers were described as "belonging under their bell"; they also had to pay for the signal themselves.[32] Non-guild members, for example, the Beguines, were placed under the weavers' work time signal, though not so readily other guilds.[33] The ritual of how punishment was exacted was also taken over from the communal bells. Following a rebellion in Bruges in 1380, the weavers were humiliated by having their bell removed "in perpetuity" and being subjected to the working time of the other trades.[34]

In the cities of northwestern Europe we can observe, following the great plague, a wave of newly acquired work bells, newly granted permissions, and new directives for them, a wave that lasted for decades. The best-known case is the granting of a work bell for Aire-sur-la-Lys in 1355, since it was stated that the city was governed by the cloth trade and needed a bell so that the workers, but also the communal committees, could keep "certain hours."[35] One cannot infer from this, however, that this form of working time regulation was typical for cities that were strongly dependent on the cloth trades.[36] Work bells are also found outside of the textile regions, for example, for rural workers in Riga, for vineyard workers in Würzburg ("Heckerglocke"), or for rural wage laborers in Sisteron in Provence.[37]

A large number of work bells regulated working times at the great construction sites: 1354 at the Tower in London, "to ring the hours for the workmen"; 1356–1357 Windsor Castle, "campana pro operarii"; 1365 Moor End (residence), "to 'excittand' the masons, carpenters and other workmen to hasten their work"; 1370 at the York Minster, "campana in logio cementariorum"; 1365 at the cathedral in Florence, "campana per operarios"; 1390 at the cathedral in Milan, "campanile fabricae"; 1396 at the Certosa in Pavia, "campanella posita super laboreriis."[38] Working time problems were no less here than they were at weaving looms or in vineyards.

Increasing precision

When work bells are spoken of in the sources since the beginning of the fourteenth century, we often also hear about "certain hours" that are desirable. Leaving aside the de-coupling of a special time signal for wage laborers, what changes did the work bells bring about in the temporal reorganization of working times?

Descriptions or regulations of the ringing of work bells are much rarer than reports about their installation. But the few reports we do have reveal that precisely the ringing was a problem for which there was no easy solution. A famous intervention by the French crown into urban labor law was the permission for a new work bell in Amiens in the year 1335.[39] At the initiative of the linen weavers, cloth fullers, wool workers, and cloth shearers, the guilds in Amiens had asked for a stronger enforcement of older working time regulations which had lost their effectiveness. The representatives of the guilds believed the reason for this was that the assessors had not imposed and collected the appropriate fines. The assessors, for their part, argued that workers were able to come and go as they pleased because the old regulations were inadequate in making it clear that workers were obligated to come, take breaks, and leave at certain hours. The king gave permission for the installation of a new bell that was to be different from the existing bells in the belfry. It was to be rung four times a day: in the morning, in the evening, and before and after lunch break. As for the details of the signal regulations and the determination of the penalties for violating them, the city should make an effort to obtain comparable regulations from Abbéville, Douai, Montreuil-sur-Mer, and St. Omer. Requests to those cities to that effect apparently did not receive a satisfactory response, and so the assessors, protected by an anticipatory authorization from the king, drew up such regulations on their own.[40]

For the signaling sequence they used two kinds of distance-time measures, one of which can hardly be reconstructed today but was undoubtedly quite common at the time. The other distance-time measure was local; it made sense only in Amiens. In the summer the work bell was to ring beginning at sunrise for a period of one mile ("une lieue de terre"). The break should last "2 lieues," whereby the signal for the return to work should be sounded long enough to allow a person to return to the workplace from any point in town. The signal of the end of working time was to be given around Vespers in such a way that it was still possible to walk "une lieue de soleil" afterwards. In the winter, working time was demarcated by daylight and the pause was limited to "une bonne demie lieue."

When it comes to concrete situations, defining or converting distance-measures used at the beginning of the fourteenth century is no less difficult than it is in the case of time measures. The editor of the texts from Amiens estimates—plausibly, in my view—that the distance-time for a walk through the city was a quarter of an hour. But how long was a "lieue de terre" in Amiens in the year 1335? The Gallic "leuca" that was customary in England corresponded to a mile of about 2,400 meters. At the end of the century Geoffrey Chaucer reckoned it at twenty modern minutes, "three Mileway maken an howre."[41] The French "leuca" is considered to be twice as long as a "mileway" and is given as about 4,000 meters. The editor reckons it at forty-five modern minutes. The summertime work break would thus have lasted ninety minutes, or perhaps only forty minutes. Comparable time indications are rare and also do not reveal much. At the construction site of the cathedral in York in 1352, the duration of the summertime breakfast break was set for the time of half a mile ("spacium dimidiae leucae"). In the redaction of this regulation in 1370, the breaks were given, among other things, as whole or half "mileways," the end of working time in the summer as "a mileway byfore ye sone sette."[42] Such experiential time indications were no doubt suitable for the particular purposes and to that extent precise enough. But they would hardly have been able to prevent the sort of conflicts we have seen in Sens and Auxerre. It therefore comes as no surprise that in the course of the general development toward greater precision in working times, the possibilities of the new clocks were seized upon quickly and in many places.

New methods of determining working time appeared first in 1358 in Valenciennes. According to the last entry in the final account for the new work bell in this important cloth city in Hainaut, the smith Jehan Biaulieu received a considerable sum for the purchase of "uns orloges" for the belfry, where the new bell was being installed.[43] The city had ordered the clock to regulate the four signals of the bell. It was in all likelihood a mechanical clock, probably without a striking work. In later accounts, to distinguish it from the large communal clock, it was described as "les petis arloges" in the keeper's room. We do not know how the temporal regulation of the four signals was accomplished. It does not appear to have been an innovation, however, for the city dispatched a messenger into nearby Tournai to obtain work bell regulations. In Tournai and Valenciennes, working time—in regard to beginning, breaks, and end—was thus being regulated with the help of a clock. However, a work time statute from Tournai shows that in 1365 the signals were not yet described as clock times.[44]

In the subsequent years, the use of clocks and sandglasses continued to spread quickly in cities and on construction sites. At the site of the cathedral in Orvieto, a "temperator" for the "arlogium" was employed from 1365.[45] At the quarries in Candoglia, where marble was quarried for the cathedral in Milan, two half-hour clocks appear in the accounts in 1392; presumably they were used to time breaks. In 1418 a mechanical clock was procured.[46] In Hamburg in 1375, the end of work for the smiths in summertime was limited to the time "when the clock has struck five"; during the other seasons, indications according to natural daylight (mentioned in the statute) remained in use. In Frankfurt, clock time indications appear in 1377 in various work statutes. In Paris, the work time of the cloth shearers was newly regulated in 1384: journeymen working by the day began their work in the winter at twelve o'clock at night. The work lasted until sunrise, interrupted by two one-hour breaks and one half-hour break. In summertime work went from sunrise to sunset. The main break began one hour after midday and lasted two hours on "long days," one hour on "normal days." However, journeymen who lived with their masters were excluded from this regulation. Modern hours appear in Cologne in 1377, when it was stipulated that the break for wage laborers could last at most "one hour."[47]

Breaks were usually timed with sandglasses. In a statute from the first half of the fifteenth century for the caulkers in the shipyards outside of Genoa, the length of the break has been given the name of the timekeeper used to measure it. For those engaged in repair work, the midday break lasted one "clock" ("unum horologium" = 1/2 hour), vesper break lasted a quarter of an hour or "half a clock" ("medium horologium"). When new ships were being built, the midday break was allowed to last one full hour ("una hora integra"), and vesper a half hour ("unum horologium sive mediam horam").[48]

The story of the public clock of Beauvais, procured in 1390 (that is, relatively late), shows once again how easily the traditional time indications could become embroiled in conflicts, what sort of solutions to such conflicts were envisaged at the time, and how important public clocks were as modern organizational aids. Our source is, as it was for Sens and Auxerre, a decision by the parliament of Paris, which reports in the preamble the arguments presented by the contending parties.[49]

Great conflicts, which had been brewing for decades, had erupted between the mighty guild of weavers on the one side and the guilds of the wool shearers and fullers on the other. The main problem was that the division of labor between the three guilds no longer functioned. The weavers had begun to have some of the preparatory work, which the other two guilds used to perform, done in house. In addition, because they considered their craft to be more difficult than the other two, they had increasingly closed off their guild through longer periods of apprenticeship and higher admission fees. The journeymen and apprentices of the other two guilds could no longer switch over; more and more skilled workers became unemployed. Added to this was the quarrel over the traditionally different working times.

The wool shearers and fullers maintained that their crafts had each formed an independent guild with its own statutes since time immemorial. The physical strain on the fullers was much greater than on the weavers. With a helper a weaver could easily work an entire day ("dies integer") from sunup to sundown. By contrast, a fuller, if he wanted to preserve his health, could put in only a shorter day's work. Invoking "antiqua statuta" that were at least fifty years old or even older, in any case far too old for anyone to remember anything to the contrary, the fullers gave their version of the old working time regulations. According to the fullers, the differentiation among the guilds should begin at the market of the day laborers. At sunrise the weavers were the first who had to appear at the market and hire their people. Only then ("platea textorum finita") did the fullers, and after them the shearers, come to the market. Thereafter the fullers had a short breakfast break ("per modicum spacium temporis"), about the time of the sacrament at the high mass at St. Pierre. The main lunch break began for them "around None" and lasted until the striking of None at St. Pierre. In the winter this time was so short that the bell sounded before the workers had finished eating. Contrary to all customs, the weavers arrived late at work, namely, not before the early mass, and some had also talked them into taking breaks as long as those of the other guilds.

The weavers denied the special regulations for the other guilds. They accused them of extending the first break to a full hour ("una hora") and of using all breaks and the time in the early morning and in part even the night to produce inferior products in violation of guild regulations. They also claimed that all had to start work at sunrise and take a short break around the time of Prime. The duration of the midday break in Beauvais, as in the other cities (Arras, Hesdin, St. Omer), was fixed in the summer at one "leuca terre," in the winter at half. The weavers suggested, "for the evident benefit of the entire city," that Beauvais follow the example of the other textile cities in Picardy and install a work bell with a corresponding regulation of working time. Apart from the simultaneous market for day laborers, the break-time regulations were to be maintained as described.

The decision of the parliament reflected the usual restrictive policy and followed the suggestion of the weavers in this matter. In the process, however, it stipulated the installation of a modern instrument of control, a reliable clock, and at the same time also changed the much more unequivocal time indications it made possible.[50] The market for day laborers should be held for all guilds at the same time, but different locations should be agreed upon for the various guilds. In the summer the market was to begin between "the fifth and the sixth hour" in order that work could begin at six o'clock. In the summer the first break should last one hour, the second one hour and a half. In the winter the working time should last from five to seven o'clock, and the breaks half an hour and one hour, respectively.

The modernization of the municipal work time regulations in Beauvais that was decreed from Paris is characterized, in regard to the time indications (which actually render a work bell superfluous), by an unusual degree of consistency. In other cities the communal clock only gradually became the normal time giver for the regulations of working time. When the communal clock appears in the statutes of the Paris cloth shearers in 1415 ("orloge du palais"), in the statute of the fullers in Orléans in 1406, in that of the smiths in Troyes in 1412, and that of the stonemasons in Regensburg in 1440, these time indications are usually isolated cases amidst a sea of time indications of the old type.[51] Frequently only the end of work in the summer or the duration of a break were tied to a clock, while the other time regulations, as, for example, the beginning of work at sunrise, were not. There were still clear limits to the emancipation (in terms of temporal organization) of working time from daylight. Occasionally we also sense distrust of the reliability of the devices, for example, when certain times are given in two different ways. In Ulm in 1420, for example, the end of work for the threshers was fixed both by the canonical hours and in the modern way: they were to work "until the bell strikes three or Vespers is rung." In Rouen, the canons of the cathedral had lent a clock to the masons so they could follow it with respect to their working times. In the middle of the fifteenth century, the break for the masons was then regulated by the communal clock. But in case the communal clock broke down, it was stipulated that the break should last as long as the great mass in the cathedral.[52]

Regardless of whether it was structured by a work bell with clock-regulated "certain hours," directly by the main communal clock, or with the help of sandglasses, the length of daywork was at first not changed by these innovations.[53] New was the fact that working times were signaled in a way that was transparent to the various groups concerned and allowed these times to be checked. Only very slowly and in very scattered instances can we observe that the use of clocks made it possible to shorten working times without involving any major changes in regulatory technique. In Valencia in the fourteenth century, a statute had decreed that no worker was permitted to leave the slopes prior to the stroke of the evening bell in the cathedral. In the sixteenth century, the period of daylight ("de sol a sol") was reiterated as the normal workday. For particularly difficult kinds of work, however, working time was de-coupled from the old signal and, with the help of the clock signal, shortened in two steps. For digging and chopping, the statute on rural laborers of 1537 stipulated the time from seven in the morning to five in the evening, and in 1555 this time was further shortened to eight in the morning to five at night.[54]

With respect to working time, too, abstract clock time opened up possibilities of negotiating working time and regulating it anew or in such a way that it was immune to conflict. That this was an essential prerequisite for all reforms in the organization of working time was certainly understood at the time. The second book of Thomas More's account of the happy island of Utopia was written at the beginning of the sixteenth century, presumably in Antwerp, one of the busiest commercial cities of late medieval Europe. Here we read that the Utopians divided night and day into twenty-four hours of equal length and assigned only six to work. This was possible because the socially necessary work could be done in a much shorter time by involving all idlers.[55] More, the English jurist and later Lord Chancellor, combined his critique of the unmeasured workday with clear references to the prerequisite — not yet self-evident at the time — for the new organization of labor: modern hour-reckoning.

Working times "underground"

Already during the High Middle Ages there had existed special forms of the normal working day in underground mining. The sources, sparse with respect to working time regulations, show that the working time of miners had long since been independent of daylight and had been organized independent of the urban bell system. It likewise seems doubtful whether the unequal or canonical hours were ever used in mining for the regulation of working times. The oldest known working time regulation for miners, the "Jus regale montanorum," was an attempt, initiated by King Wenceslaus II, to standardize the mining law of Bohemia-Moravia. A Magister Goczius from Orvieto is listed as the editor of the redaction that was put together shortly before 1300. This mining law was also valid for the important mining cities of Iglau and Kuttenberg and is thus also called "Kuttenberger Bergordnung" ("Kuttenberg Mining Code"). Johann von Geilnhausen translated it into German at the end of the fourteenth century. The section on the shifts of the miners is based on an abstract division of the day, though without clocks. "Hora," the miners' shift, is in this text both a quarter of a day as well as an (equal?) hour: "[From the hour-caller] . . . one shall know that all hours [of the day and night in the mines] . . . should be divided into four hours . . . And the caller . . . shall call out the beginning . . ."[56]

The interpretation of this code is difficult, since it remains unclear what time-keeping technique was used to divide the working day of the miners. In the German version the passage goes on to say that these shift signals were unsuitable for fixing inspections "at a regular hour" because they "changed unevenly with the lengthening and shortening of the day," and consequently the time for inspections was determined by the times of divine services.[57]

The next code with usable information about shift times, the Mining Code of the County Judge Johann von Üsenberg in the Breisgau (1372), already takes modern hour-reckoning for granted. We read that the smelters and workers in the mines, the ore mill, and the smelting mill "work eight hours a day, four hours before noon and four hours after noon, that is the working situation in the mountains for all miners."[58]

Miners' shifts were subsequently fixed at three times eight hours or four times six hours. The descending times that were included in these regulations, the prohibition of working two shifts in a row, and the questions whether it was permissible to dig for oneself in the "free time" outside of working time provided endless sources of conflict, though at the same time they also made people aware of the problem of a measured normal working day.[59] The history of the regulation of working time "underground," too, allows us to trace the growing importance of modern time-measuring techniques. From the Late Middle Ages on, the methods of time indication and time control in mining were improved. The first great account of mining and metallurgy by Georg Agricola (1556), city physician in Chemnitz, mentions the widely used special work bells for shift times (whose signals were passed on into the mines through the banging of the tools), but it also speaks of tallow lamps as basic time-keepers.[60] Where the miners tried to bring the time indication under their control with the help of clocks, the costs for these timepieces were frequently imposed on them.[61] Occasionally the attempt was made to compensate for the imperfections of the mechanical technology through large sandglasses. For example, in Altenberg in the Erz Mountains, a "sand clock of nine hours" was procured because the striking clock was not functioning properly.[62] A considerable modernizing impulse in respect to the organization of time thus also came from the mining industry.

Time measurement and production

Regulations of working time and breaks made no sense where the rhythm and duration of work were determined directly by work tasks or production methods. In the smelting of ores, the casting of bells and cannons, and in work at glass melting furnaces, complicated production methods dictated the measure of work without regard for day and night, Sundays or feast days. But even in these areas we can observe changes brought about by the new methods of time measurement. There are, to begin with, the manufacturing directions and recipes handed down in writing (still very rare in the Late Middle Ages). Time indications appear naturally where certain production results were not immediately obvious but were known from experience to occur only after a certain minimum or maximum time had elapsed. The famous *Diversarum Artium Schedula* of Theophilus Presbyter, a collection of instructions in the artistic crafts from the beginning of the twelfth century, supplies a few daytime and hour indications. For example, gold used in the illustration of books should be ground in the mill "for two to three hours"; ash and sand in glassmaking should be melted for one day and one night. By melting certain mixtures until Terce or None, one could produce yellow or crimson glass. The experience of craftsmen was described and passed on by using the vague system of reference of the canonical hours. There is only one passage which mentions what is in all likelihood an estimate rather than a measurement: we read that one should wait for "one half hour" when cooling a coating of enamel.[63]

Only from the beginning of the fifteenth century on do we encounter such descriptions with greater frequency in the writings of gunsmiths and blasters. Modern hour-indications appear with recipes in the redaction of Marcus Graecus's "Book of Fireworks" by Master Achilles Thabor, which is probably an assumed name of the writer Johannes Hartlieb (died 1468). Subsequently the *Feuerwerkbuch* ("Book of Fireworks") from the beginning of the fifteenth century, a famous and much copied compilation of knowledge concerning artillery and pyrotechnics, provided not only recipes—for example, how to make strong gunpowder from sulfur, saltpeter, and ammonium chloride by stirring the mixture over a fire for half an hour—but also instructions on how fuses could be measured off with the stroke of the bell.[64] The Tower in London has a copy of the *Feuerwerkbuch* with the depiction of a powder mill in which it is not entirely clear whether the illustrator, in using the sandglass, was referring to the pounding time needed to produce the fine granulation of the powder or to the work shifts of the mill workers (fig. 63). Gunsmiths also had other uses for modern time-measurement, for example when they report of alarm clocks that set off a fire or a shot at a certain time, or when they set to work constructing time fuses ("feuerschloss auff stunden und tag gerichtet wie man will") (fig. 64).[65] Technical manuals of the sixteenth century, like Piccolpasso's *Arte del Vasaio* and Biringuccio's *Pirotechnia,* are full of instructions in which old experiential values were frequently translated into modern time indications. When it comes to the timing of work processes and work shifts, since we simply do not know enough about production processes (for example, in work at the printing presses) (fig. 65), we cannot always tell if our texts are talking about the supervision of working time, the guiding of a work rhythm, or adherence to production-related time limits.[66]

Time controls and hourly wage

Mathematically the hourly wage of a medieval day laborer could be determined even if the wage and working time regulations on which it was based contained no or only very few modern hour indications. With the help of tables for the time of sunrise and sunset in the various latitudes, of festal calendars for specific locations, and of data on the liturgy and thus the approximate length of masses, one could, first of all, determine the duration of the working day in hours and then, with the help of a wage table for the appropriate season, also the hourly

63. Work in a powder mill. Illustration from manuscript of the *Feuerwerkbuch* (around 1450). Royal Armories, HM Tower of London. Crown Copyright, The Board of Trustees of the Royal Armories, London.

wage. After adjusting for the influence of fluctuations in monetary value and of non-monetary wage components, one would be left with an arithmetical hourly wage that would be usable for comparative studies on standards of living and economic cycles. Like historical calculations of caloric consumption, such studies presuppose a series of complicated considerations that operate with abstract measuring units, considerations that are conceivable for the Late Middle Ages only in rudimentary beginnings. Modern hour-reckoning is one prerequisite for such concepts, and the process of increasing precision that began in the Late Middle Ages and transformed "hora" as a vague time pe-

64. Time fuse in *Newe vuerfarne treffenliche vortheile zu allerhand Kriegsübungen im veld und bevestungen durch Veitt Wolffen von Senfftenberg aus Österreich Itzo der von Dantzig Czeugmeistern fürgegeben Anno 1568.* Dresden, Sächsische Landesbibliothek: "Item, in cities, castles, and hamlets which one must abandon after a long siege, one can with such hidden, buried explosives throw such an obstacle in the face of the enemy so that within two or three days he will come to regret this conquest: namely by placing buried into the earth one great fireball or several in more places in the chambers and rooms, here and there also in the stables, with a running clock attached with a fire lock, everything properly arranged as best one can. A number of such hidden explosives can be delayed as long as one wishes, and set at such hour as one desires. Indeed, with such a setup or in such a brief time many marvelous things can be done afterwards and with the cocked fire lock, it is more than one can relate."

65. The rhythm of work and the setting of time limits in a printing shop. Copperplate engraving by Abraham von Weerdt, 1666. Photo: Deutsches Museum, Munich.

riod into the modern hour led to advances in abstraction also in the sphere of time-money relationships, advances that have so far received no attention.

In order to stabilize—for the purpose of calculation—the quantity "time" in time-money relationships, one needs a time measure such as year, week, day, hour, or minute. The advances made in this connection concern also the transition—made possible by the new time-measuring technique—from day to hour or the specifying of the working day as the sum of equal hours.

That the period of daylight had twelve hours was a familiar notion in the High Middle Ages. If we leave aside the complications caused by the varying length of daylight, there was thus nothing to prevent a corresponding division of the day wage. The vineyard laborer of the biblical parable who appears at work in the eleventh hour of the day and still receives a full day's wage (Mt. 20.6–12) did not unduly concern medieval commentators. Here and there at the beginning of the fourteenth century, the parable of the vineyard was used as an example of the disadvantages of time wages as against piece wages, but discussions of the just division of the day's wage or appropriately calculated parts of the wage are not found in the theologians.[67] This is a bit surprising, since division of the work day and division of the day's wage were quite common. On Saturdays and days preceding feast days the work time was shorter and in most cases only half the wage was paid. Deductions of thirds and fourths, at the end of the fifteenth century even of sixths and twelfths, were common at construction sites, for example, as deductions for bad weather.[68] But not even the division into twelve parts led directly to the idea of paying or deducting an abstract time measure such as the modern hour. From the middle of the fourteenth century it became apparent, however, that the customary divisions of the day and the day's wage were no longer adequate. New forms of dividing the day and new wage demands appeared in the cloth industry and on an even larger scale in the construction industry.

Bruno Dini has found a whole series of noticeable time measures that differentiate the time of day in the account books of a large Florentine woolen cloth company in the years 1355–1370. In every case the issue revolved around the "overtime" (a later term) of the time-wage workers.[69] During the winter period, there was the "notte," more rarely called "veglia," for work in the evening until midnight. Both periods of overtime were remunerated with half a day's wage in summer and winter. Summertime work in the evening, "lunata" or "sera," would earn nearly a quarter of the day's wage. The wage lists had periods even for pre-dawn work, though without separate names. For example, while "sera" was by the end of the thirteenth century the name for a quarter-day or the corresponding wage, the quarter days that were also calculated and called "ore" were completely new.[70] These references to division of the day differentiated in this way are, however, a rarity and they appear in the bookkeeping of a company in which work was evidently carried on for unusually long periods of time in summer as well as winter. We cannot yet decide in this instance whether the "ore di straordinario" (B. Dini) can already be understood as modern hours or whether we are still dealing with differentiations within the old timekeeping, which would mean that "ore" were merely vague experiential values.[71]

Another new — though equally short-lived — time measure appeared in the middle of the fourteenth century. The above-mentioned supervisor at the construction site of the cathedral in Orvieto in 1364–1366 had not only the task of maintaining the clock of the site, he also had to record completed working times. Those times were called "doctae magistrorum et operariorum."[72] "Docta/dotta" is attested on one prior occasion in wage deductions in the accounts for public buildings in Siena (1341).[73] The term appears once more in later statutes of the cathedral workshop in Orvieto from the year 1421, among other things in the expression "doctae pro horis." According to the major dictionaries, "dotta" is a short time period but nowhere a temporally specific one.[74] According to our sources, "dotta" could stand for a kind of work shift, possibly for a still-vague hour, or for the wage that corresponded to this period. We do not know, however, how long such a "dotta" lasted.

The construction site of the cathedral in Milan, begun in 1387, was in many respects organized in a far more modern way.[75] While old divisions of the day were occasionally still used in making deductions for bad weather, modern hour-reckoning was largely familiar when it came to supervising working time and breaks. For the quarries in Candoglia, two half-hour clocks appear in the accounts in 1392, presumably used to supervise break times. In addition, the daily work time was recorded in hours. A mechanical clock was procured for the quarries in 1418 "in order to be certain about the hours when the masters and workers leave their work on account of rain and when they take a break to eat." The construction site in the city, too, had its own clock with a bell and after 1433 also with a dial.[76] And at this time the new hour reckoning was already used not only for—still sporadic—calculations of working times, but also for modern cost calculations. In 1392, a man with two donkeys was to bring iron parts and foodstuffs from the valley to the quarries so that the masters and workers would have no pretext for absenting themselves from the work site. In the justification of this measure, the loss to the cathedral workshop from such absences was calculated by work hours. In another instance in 1411, the written recording of daily working hours was explained in part by the need to determine the unit costs for individual blocks of marble.[77] With these early steps of calculating an abstract measure of work as a cost factor we have reached the limits of the mathematical abstraction possible in the fourteenth and—as far as I can see—also the fifteenth century. Even though all extant sources are not accessible at this time, we can state further that such calculations were extraordinarily rare. Only for payment by the hour do we find, around 1400, scattered references also outside of Milan.[78] The above-mentioned decision of the parliament for the guilds in Beauvais (1390) stipulated, for example, that the wool fullers did not necessarily have to hire themselves out by the day wage ("ad dietam"), but could also do so by the weight of the product ("ad libram et pondus lanarum") or by the hour ("ad certum numerum horarum"). In the statute on fullers in Orléans (1406), too, hiring oneself out "a heure" was provided for as a possibility next to "a terme."[79]

The subsequent development of this new form of time wage took an indirect course. It was not the case that day wages were divided with growing frequency into equal parts (in correspondence with the equal hours). Rather, hourly wages emerged much more often from payment for overtime or deductions for missed hours. We can trace this development above all at the construction sites from the middle of the fifteenth century.

At the cathedral building site in Orvieto, the statutes of the cathedral workshop regulated not only the operation of the construction site clock but also the recording of missed hours ("hora obmissae").[80] In Nuremberg a comparatively complicated form of modern hour-reckoning was in use, which is why the records of the city architects

Lutz Steinlinger (1452) and Endres Tucher (1464–1465) contain thorough lists concerning the working times during the various seasons of the workers employed by the city.[81] Endres Tucher also reports on a regulation by which a certain amount was taken from the pay envelopes (handed over to the workers on Saturdays) for hours missed or added for additional hours worked. These amounts corresponded to about one-tenth of the relevant day wage, and as far as I know this is the first time they are called "hour money" ("stundgelt").[82] In the accounts for the construction of the choir of the Church of St. Lorenz (1445–1449, 1462–1467) we still find hardly a trace of these kinds of wages.[83] Later, in the accounts for the work on the towers of the same church, which began in 1481, there is mention of both overtime payment for working through breaks ("under iren stunden") as well as overtime work at fixed rates ("je von ainer stund 2 d.").[84] Here we are clearly looking at hourly wages, since their amount, in contrast to day wages, no longer varies with the time of year.

The system of controlling breaks and of punitive wage deductions for missed working time presents itself as more severe and more precisely regulated in the statutes for Calais (1474). This city was the English bridgehead in France, and the workers were in a quasi-military situation of subordination. Working time at "the king's works" was regulated with great precision also because work on the harbor and fortifications had to be performed quickly on account of the tides. When it came to work during the day or night that was dependent on the tides, all rights to breaks, for example, were suspended. Normal working times were regulated in part by daylight, in part by the public clock, and when supervisors and artisans were penalized for being late, differentiations were made by the hour and half hour.[85] In the building accounts for royal residences, churches, and colleges, too, the wages for extra hours—which were above all paid breaks ("hour times and drinking times")—show themselves as precursors of hourly wages. These hours, as well, were in the vast majority of cases timed by sandglasses—in this regard England was no different from France or Poland.[86]

The hourly wage that was developed in the fifteenth century was the rare exception in the Late Middle Ages, and this is even more so the case with the corresponding cost calculations. Only punitive deductions for missed hours are frequently encountered. At least until the

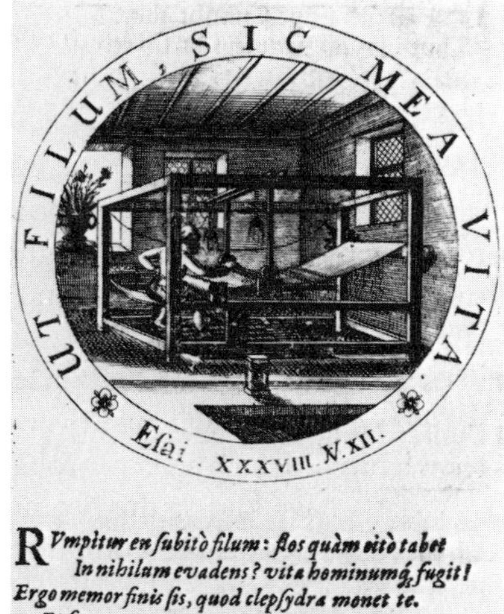

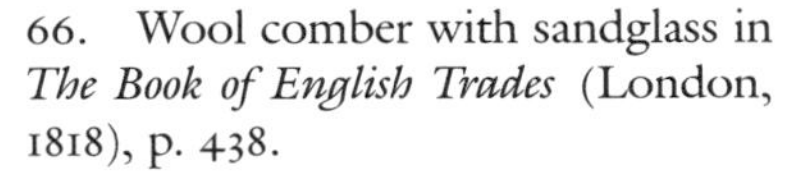
66. Wool comber with sandglass in *The Book of English Trades* (London, 1818), p. 438.

67. Weaver with sandglass in Johann Manich's *Sacra Emblemata* (Nuremberg, 1624). Wolfenbüttel, Herzog-August-Bibliothek.

end of the eighteenth century, the day wage remained the normal form of wage for time-wage workers. In Paris, for example, the concept and the word for the hourly wage were unknown until the beginning of the eighteenth century.[87] In this area, too, the period developed only possibilities whose realization was still a long way off.

Standard clock and factory clock

The linkage of working times to the hour stroke of the standard municipal clock, and the timing of work by an instrument—the sandglass—that was largely immune to manipulation, (fig. 66) reduced the sphere of freedom of those concerned. At the same time, however, it was also protection against the arbitrariness that was at least conceivable behind the vagueness of the old time indications. The employers had committed themselves with both devices. At least this was the case as long as the public clocks were not removed from public control.

The control of the clocks became, first and sporadically in the countryside, then frequently and massively in the factories, a problem whose practical importance may have been slight, but whose considerable psychological repercussions can be traced down to this day. The prerequisites for this were, on the one hand, the rapidly improving accuracy of the clocks after the middle of the seventeenth century, and, on the other, the widespread private ownership of clocks in the seventeenth century among the burgher class and from the eighteenth century also among the middle and lower classes. The monopoly of the communal public clock was preserved only in the countryside, and even there not everywhere. Since almost all appointed times and time periods had been switched over by now, the important question was less having a clock than controlling it.

What I have said at the beginning of this chapter about urban day laborers applies also to compulsory labor in the countryside as late as the eighteenth century — to the extent that it was temporally fixed, that is, in terms of days. The daily corvée of the rural subjects was unmeasured work the extent of which was limited only by the principle of fairness and the breaks that were occasionally prescribed, two hours in the winter and one hour in the summer. Labor-owing peasants, however, could no longer make sure their breaks were kept or demand them if the manorial lord had the clock changed or the sundial taken down, as is reported to have happened in Wormlage in Niederlausitz in the year 1724.[88] But the exploitive despotism of tyrants (known from folk tales) who falsified clocks[89] could be counteracted from the eighteenth century to a certain extent by private timepieces. Occasionally the state supported the afflicted subjects in their struggle. The *Reglement der Hoff-Dienste* ("Regulation of Manor Services") which the Demesne Chamber (*Domänenkammer*) of Strelitz issued in 1724 for the district ("Amt") of Stargard in Pomerania, sought to put a stop to the exploitation of the peasants by the leaseholders. Various services are described as precisely as possible, summer work-time and breaks by the clock are listed with the following comment: "and here the subjects are free to take a proper hour-glass into the field with them and to take their cues from it."[90] The same technical device was used by the rural cottage workers who — euphemistically speaking — controlled their own work time. An older depiction (fig. 67) makes an emblematic connection between the timekeeper that was undoubtedly often used in weavers' cottages—though surely no longer called "clepsydra"—and the passing of the time of life.

With the transition to industrial production from the end of the eighteenth century, the struggles over the organization of time in the factories attracted public attention. The factory codes combined authoritarian-patriarchal elements of urban and domestic time systems with new goals in the economy of time. Hatred for the factory bell and the experience that factory owners "falsified" clocks can be traced from the early factory reporting, which Marx and Engels summarized in detail, to the memoirs of workers in our century.[91] Already at the very inception of sociocritical fiction, the new time systems—the social conditions of agrarian society had been turned into an idyll—were described in drastic terms:

> [H]ere, too, the lark had trilled its evening song: except that nobody had time to pay any attention to it!
>
> Not until the great, hoarse factory clock chimed, when the gates of the immense working halls swung open on creaking hinges, and the supervisors, through clouds of dust and steam and sweat, announced the end of working hours with raspy voices, did the day come to an end also here. (Robert Prutz, *Das Engelchen,* novel, 1851)[92]

Time measurement and time indication appear here as a prerogative that was abused on a large scale and against which people learned to defend themselves only slowly and with effort. According to older English and Swiss factory laws, factory clocks had to be set by the public clock. Occasionally factory owners prohibited the workers from bringing private clocks into the factories. From as late as 1907 comes a report that rural workers in Apulia pooled their money to buy a clock as a countercheck on the prescribed working times.[93] The dual character of the clock, as an instrument of anachronistic lordship and simultaneously a means of overcoming it, is revealed in an account (contained in one of Marx's footnotes) of a conflict in a power-loom spinning mill in Wiltshire in 1863. Here the cloth manufacturer had the working time announced by the blowing of a horn from the "time watcher." Following a strike, the workers refused to return to work until a clock was procured and a new scale of fines was introduced.[94]

Occasionally, however, the clock was physically attacked by the workers as an instrument and symbol of the tyranny of industrial forms of work.[95]

With regard to external aspects of the organization of time, the factory of the nineteenth century adopted practical and symbolic elements of the time organization of the late medieval city, with the difference, however, that clocktime, in the minds of those affected, had become largely alienated time dominated by the powerful. Down to the present, the architecture of factories has also reflected the new work ethic, first with small clock towers on the ridge of the roof in the manner of the medieval predicant orders, then with massive free-standing bell and clock towers.[96]

Precursors to Taylorism

Profound changes became visible in the eighteenth century with respect to the precision (and the shortening it made possible) of the units of calculation that were used in paying for work and assessing fines for missed time. In the "Law Book" (drawn up before 1700) for the "Crowley Ironworks," at the time the largest European ironworks, we find a detailed time regulation of almost pathological, hardly realizable perfectionism. The working time of "office workers" should be supervised from the office of a "monitor" and recorded to the hour and the minute in a "time sheet," with the numbers written out in words. The office had a special clock, "minute dial," which was to be the only authoritative one. Its function as the standard clock was underlined by the fact that only the monitor was allowed to operate the work bell. To accommodate the many individual work activities, the grounds of the works were expansive, and not all activities of the "office workers" could be supervised from one office. For that reason—and this, too, is quite new—these workers were to keep protocols of where they went and what they did, whereby they themselves were to record time not spent for the benefit of the works ("loitering hours"). The time that was recorded in the office ("Monitors Neat Time") and the time spent outside minus breaks ("Corrected Extra Time") should then be added and paid by the hours and minutes.[97]

With respect to time controls, this painstakingly precise law book goes far beyond what was common in the eighteenth century. In vari-

68. Manufacturing of razors in the English manner. Iron plate engraving (1783). Photo: Deusches Museum, Munich.

ous "model factories," which were probably atypical but much written about because of the new experience of highly divided labor, considerable efforts were made in the control of working time also with the help of " time clocks," a device which at first served to supervise night watchmen in the cities. In the porcelain factories of the English entrepreneur Josiah Wedgwood, which the workers called "bell-works" after the introduction of the factory bell, primitive forms of time clock cards were used. At the end of the century, the factories hired a man who had to promise to make sure not to get caught observing the work activities with the help of a stopwatch.[98]

L'Art de l'Épinglier, the account of the manufacturing of pins reworked by M. Perronet in 1762, not only describes the highly divided production steps but also indicates their times.[99] Even more modern procedures of time recording are found in the *Architectura Hydraulica* of the French fortifications engineer Bernard Forest de Belidor, where an astonishingly subtle form of time recording was used in calculating the time and costs for driving in pilings for breakwaters and harbor walls.[100] For the operation of the pile driver, Belidor calculated the time needed for hoisting and dropping the ram in relationship to the various soil layers and took into account corrections when the pile deviated from a straight line. Next he estimated the output of the workers, which declined because of fatigue, by reckoning for twenty working hours a real value of eighteen hours, thirty-two minutes, and twenty-four seconds. Because of the high cost of drawing water, he recommended having the work proceed day and night. When drawing up a contract with the contractor, each piling that was driven in should be estimated, inclusive of incidental expenses, at six livres, ten sous, eight deniers. There followed other recommendations for the timesaving use of the driver, "especially since in hydraulic constructions everything must revolve around this." This regulation was not so much motivated by economics as compelled by the natural circumstances. Tides created time pressure. The result was very modern, and since fatigue was taken into account it is in fact part of the prehistory of Taylorism and the rationalization movement. At the end of the eighteenth century, people had begun to calculate not only time spent in the workplace but also productive capacities. The Municipal Archive of Cologne has a list, written in French and dating from the last years of the eighteenth century, in which the performance of humans, horses, and simple machines is given in meters and kilograms per second.[101]

In the workshops and factories of the early nineteenth century, it was often natural circumstances (for example, the uneven operation of water-driven machines) as well as economic problems (for example, unevenly distributed workloads) that prevented work from becoming regularized and intensified. The steam engine and favorable economic circumstances subsequently made possible the headlong industrial surge and at the same time the new forms of factory-industrial working times that, for a small segment of the workforce, eliminated the distinction between day and night, summer and winter. Since then, working time has been one of the dominant themes of economic and social policy, in two respects. For employers, the rational use of working time, "the economy of time," has been paramount, though its external control has remained a major topic with great symbolic significance. When Alfred Krupp admonished that "the time and the regulation of the clock" be observed in his factories because "each lost minute now costs 100 Sgr. in wages without production," the "horrible differences of the clocks" was for him the symbolic expression of impending chaos. To prevent it he had a central chronometer installed in the villa Hügel: "We must never tolerate a difference of one minute; only then will I be satisfied and reassured, only then do we have an orderly factory system (*Fabrikordnung*)."[102]

69. Ad for a punch-clock system from the first issue of the *PRITEG-Nachrichten,* the company paper of the Fuld & Co. Telephone and Telegraph Works, 1923.

Workers, too, discovered the topic of working time and in the nineteenth century made the workday into the object of political struggle. The often-described road to the shortening of working time led, at a different pace under varying political constellations and in various countries, to the normal workday and accordingly to leisure time as a measured and contested quantity.[103] With the new aspects of vacation and the shortening of lifetime work, working time has remained a prominent theme down to the present, and discussions about flexible working hours are keeping people aware of the significance of measured time.

10

Coordination and Acceleration: Timekeeping and Transportation and Communications up to the Introduction of "World Time" Conventions

The terms "universal time" and "world time" came into common use from the end of the nineteenth century in connection with the discussions and conventions on standardizing and coordinating all local and regional customs for determining the time of day. Uniform practices in the chronology of years and in dating had been widely established in the Late Middle Ages and the early modern period; it became apparent from the end of the eighteenth century that the acceleration and growing density of transportation and communications also necessitated the synchronization of local and regional times. Until then, local times or the various "burgher times," as they were called in the nineteenth century after their origin with the city burghers, had remained the times of unconnected "urban monads."

While mechanical clocks and sandglasses had made it possible to compare day times and abstract periods of time since the Late Middle Ages, the conventions of universal time at the end of the nineteenth century now made it possible to relate them to one another with respect to their temporal location. The international Meridian Conference in 1884 initiated the final disappearance of the "urban monads."

In the Middle Ages, too, people had been aware not only of the connection between geographic latitude and the length of the day, but also of the difference between local times. At least the educated, for example, had been familiar with the fact that in the summer daylight in England was much longer than in Italy, and that the sun rose later in Paris than it did in Regensburg. These differences were taken into consideration for astronomical observations; however, given the slow speeds in communications and transportation, this kind of knowledge made no difference on a practical level in everyday life.

At first sight only one epochal threshold becomes apparent: the development of railroad and telegraph technology. Prior to that, terrain, weather, and the strength and fatigue of animals and humans set immutable natural limits to all speeds of travel and transport. These speeds could be raised to a certain extent only through a system of "postal" transport that was known already to the high cultures of antiquity.

"Postal" refers to a technique of transport which eliminates the natural factor of fatigue with the help of messenger relays, horse changes, and the maintenance of changing stations or relays (posts).[1] In this way the short-term maximum effort attainable by animals and humans could be turned into average, continuously maintained speeds for transports over long distances, as well.

Ancient authors report quite astonishing speeds and distances in the transmission of signals and messages via chains of fire, smoke signals, and calling signals. Despite the fact that recent scholarship has debunked many legends about ancient communication techniques, it remains uncontested that from the fifth century B.C., at the latest, effective signal relays for governmental and military purposes were known and used. Deserving of our interest among these reports are schemes by ancient authors that sought to use contemporary time-keeping technology to move from the transmission of single signals to the transmission of messages. Polybius reports the following suggestion by Aeneas Tacticus ("the Tactician"), a Greek writer of military treatises in the fourth century B.C.: two calibrated outflow vessels exactly equal in size should be equipped with a float and a rod divided into fields to which various messages could be assigned ("cavalry have entered the country," "hoplites," etc.) (x.44). Once a fire signal has been given, both "water clocks" should be opened. When the message field on the sinking rod of the "sender" reached the rim of the vessel, a second fire signal stopped the "transmission," and the "receiver" could read the message from his own rod. In a collection of war stratagems written in the second century, Polyainos, a jurist living in Rome, attributed to the Carthaginians the development of an improved "clock telegraph" (vi.16). The dials of two clepsydrae of the Vitruvian type (see chap.

2, Continuous time-indication and automata mechanisms) should be inscribed with the kind of supplies that had to be taken from Africa to Sicily. A starting signal began the message; a second fire signal stopped it and would allow the message to be read from the clepsydra, which would have to be operated after sunset. Leon "the Mathematician" is said to have suggested a similar procedure with two synchronous water clocks to emperor Theophilus in the ninth century for setting up a message link between Cilicia, where attacks from the Saracens were expected, and Constantinople. In the nineteenth century, all these projects, unworkable for many different reasons, were made into precursors of optical telegraphy, and they contributed to an overestimation of ancient signaling technology that has in many instances persisted to this day.[2]

No new transmission and transportation technologies were developed in the old European world. In his handbook of postal courses, Ottavio Codogno explained around 1600 that with respect to the speed of transport, the contemporary courier system was inferior to that of Roman imperial times.[3] If we compare the individual reports that are extant, we find that the post had in fact not gotten any faster. During the Roman Republic, as during the French Revolution, a letter could be taken from Rome to Paris in ten days, at best. The difference lay in the fact that until the beginning of the modern period, such rapid connections were in every case maintained only for brief periods for privileged users at great economic and political-administrative cost. During the French Revolution, by contrast, everybody who paid the postage could use these links and, moreover, expect that the forwarding time would be met and connections to other cities would be ensured. The early modern postal system had turned what had earlier been unusual transmission speeds into the rule.

The history of communication and transport speeds and of the postal system in the various countries has been well researched, and we can concentrate here on the question of what role the new technologies of timekeeping played in the gradual coordination and acceleration of communication and transportation links. In the process I shall try to shed more light on the "creeping" evolution — so to speak — of acceleration processes within the transportation systems that were tied to natural conditions. Our starting point is the observation that the time-organizational means of acceleration appear to have been due to a process of "literary reception," and that the first known practical uses of the new time-measuring techniques can be observed once again in Milan under the Visconti.

Reports about the postal system in Asia

The word "post" for stations of a message system operated on a regular basis was used in medieval Europe for the first time in Marco Polo's account of his travels in Asia, which was written around 1298. This famous traveler reported that the Great Khan had the roads radiating into the country from Peking furnished with stations where his messengers and envoys could change horses. The distance between these posts was between twenty-five and forty miles, depending on the terrain. Between the posts were small hamlets, about three miles apart, that were relay stations for the unmounted couriers. This part of the postal system worked as follows:

> They [the runners] wear large belts, set all around with bells, so that when they run they are audible at a great distance . . . And at the next station three miles away, where the noise they make gives due notice of their approach, another courier is waiting in readiness. As soon as the first man arrives, the new one takes what he is carrying and also a little note given to him by the clerk . . . And I can assure you that by means of this service of unmounted couriers, the Great Khan receives news over a ten day's journey in a day and a night . . . So in ten days they can transmit news over a journey of a hundred days . . . I assure you that the [mounted] messengers ride 200 miles in a day, sometimes even 250.[4]

The postal system, like paper money, was one of the standard topics in medieval travel accounts of Asia. In the thirteenth century the Mongols had set up a postal system with horse relays and express messenger relays on the old Persian model, and after conquering China they restructured the similarly organized Chinese postal system.[5] These posts, set up for military and administrative purposes and organized in a correspondingly military and bureaucratic manner, struck travelers from the West with admiration. Aside from the relays that were posted in regular intervals independent of settlements, the travel accounts express amazement, especially about speeds. Marco Polo tells us that the

Chinese express messengers covered a distance of a normal ten days' journey in one full day. The account by Odoric of Pordenone (ca. 1325) speaks of the distance of a thirty-days' journey, a number which some manuscripts corrected downward to three.[6] The Dominican friar Johannes von Cara reports (ca. 1330) that news from a distance of a three-months' journey were transmitted within fifteen days,[7] and the widely read *Mandevilles Travels* (1356), though entirely compiled from second hand, gives in the original French version once again a ratio of one to three.[8] There is not much one can say about these relative numbers; hence the information in Marco Polo's account, which on the whole is considered to be reliable, about the distance covered in one day by mounted messengers also causes us some difficulty. If we take as our base value the Venetian mile (1 miglio = about 1.7 kilometers), relatively short compared to all the miles in use at the time, we arrive at a daily distance of between 340 and 425 kilometers. Even these figures are still far above the speeds of the posts as ordered by Gazan, the Mongol ruler of Iran (1295–1304). Here the distance between the postal stations was to be 3 parasangs (16.5 kilometers). The official messengers were to cover 30 to 40 parasangs within twenty-four hours (165 to 220 kilometers), the express messengers as much as 60 parasangs (330 kilometers).[9]

Following Persian and (indirectly) Egyptian models, an expansively organized postal system ("cursus publicus")—serving exclusively the purposes of government and using relay stations and relay riders—had been organized in Roman imperial times. However, as with the Asian postal systems, it is unclear whether it was continuously operated over longer periods of time. The maximum daily distance in this system is given as 165 to 220 kilometers, the average distance as 80 kilometers.[10] In late antiquity, however, the cursus publicus collapsed along with the political structure. Reasons for its demise were the high costs and the problems of requisitioning horses, carts, quarters, and supportive services from the populace. Beginning in the thirteenth century, these postal systems were rebuilt—on the Mongol model—in Egypt and Syria. Horse relays, as well as a well-organized pigeon post, linked the strategically important lines from Cairo to Damascus (reachable in four days), Beirut, Aleppo, Tripoli, and Baalbek.[11]

Medieval Europe no longer had large-scale political and administrative structures that could have maintained such postal systems. Thus the speed of messages reported from Asia had to cause astonishment. Although the maintenance of relay horses and messenger services are attested as part of feudal tenures, a state-organized postal and messenger system no longer existed in the early and high Middle Ages. The papal curia and great territorial lords, such as the kings of Aragon, maintained their own messengers or used traveling clerics and merchants as the need arose. Cities, monasteries, and universities also employed messengers (mostly on foot), who were in part organized into messenger establishments.[12] The city of Strasbourg, for example, maintained a major establishment with dozens of permanently employed messengers. According to the messenger codes of 1443 and 1484, they were paid a fixed rate for each mile they covered. If they had to run, ride, or drive day and night, extra pay was due.[13]

Regular messengers were also maintained from the thirteenth century by the Italian merchant houses. Contacts with the merchants' agents (the factors) at the cities of the great fairs in Champagne and with the branch offices in France, Flanders, England, and Spain led to permanently organized communication links with collective transports of letters on fixed routes and dispatch times that were coordinated with the dates of the fairs. In the regulations concerning the "cursores" in the statutes of the guild of clothmakers in Florence (before 1301), the couriers were obligated by oath to complete the journey to and from a place within fixed periods, without detours and without interruptions.[14] The institution of the fair couriers disappeared in the second half of the fourteenth century. However, after this time we continue to find letter services maintained by merchants. F. Melis has reconstructed the routes and conveyance times within this network, which linked Europe with the Mediterranean region, from the letters—more than 100,000—in the Datini archive in Prato.[15] Melis points not only to the extraordinary frequency of dispatched letters, but also to what appear to have been fixed mail departure times for individual routes. Thus, in Florence around 1400, mail was dispatched every Saturday, at the least, to Genoa, Milan, Pisa, and Rome.

The speeds of communications and transport

By now, numerous studies on the speed of the transmission of messages and the transport of passengers and goods within old world Europe have yielded, on the one hand, remarkably consistent average values, and, on the other, extreme variations even with transports and transmissions that are well suited for comparison.[16] A day's journey is pegged at an average of thirty kilometers, fifty kilometers at the most. The actual distance frequently fell short of average values; high values could hardly be attained several days in succession. Even on identical routes, the seasons, the weather, and the conditions of the road (which varied a good deal) led to travel times that fluctuated to an extraordinary degree.

The usual means of transporting news was the messenger on foot, whose daily performance probably did not exceed the normal speed of travel. But far greater speeds could be reached in the transporting of messages, letters, and light goods. Here we must distinguish between sensational, exceptional performances, standardized speeds, and average values calculated on the basis of extensive data. A skilled horseman could cover between seventy and one hundred and fifty kilometers a day, a pace that could be maintained over several days only with a continuous change of horses and couriers. That required organizational prerequisites, horse stations, lodgings, and considerable financial means. Courier relays were therefore set up only on rare occasions and then only for limited periods. For example, King Alfons of Naples in 1444 established relay lines—some of them with daily connections—to Calabria (Cosenza and Atri).[17]

There is no lack of examples from ancient times on of extraordinary courier speeds with and without change of horses. In order to send a message that concerned a cathedral provostship in Würzburg from Nuremberg to Venice in the year 1494, Duke Albrecht of Saxony promised the Nuremberg resident Jakob Krause eighty-four Rhenish gulden—twice a worker's yearly wage or a small pension—on the condition that he complete the distance in four days. For each additional hour two gulden would be subtracted. Jacob Krause's receipt reveals that he set out from Nuremberg on Friday evening, February 14, "at twenty-three hours as it strikes in Venice," and arrived at his destination Tuesday night "at one quarter before ten o'clock at night."[18] For the ten and three-quarters hours he was over, his wages were therefore reduced by twenty-one and a half gulden. This example—chosen because of the apparently common conversion of Venetian into Nuremberg time in the fairly precise hour indications—shows that speed commanded a high price. Jakob Krause's ride, however, was no pursuit of a record, and his payment was in keeping with normal fees. In the papers of the Augsburg company Paumgärtner we find after 1490 a list of costs for express messengers. The "common wage" for the distance from Venice to Nuremberg amounted to twenty-five gulden for the normal delivery time of six days; it rose as high as eighty gulden for a delivery time of four days.[19]

The delivery time for a normal letter from Pisa to Champagne amounted in the thirteenth century to twenty to twenty-two days. According to F. Melis's analysis of the Datini letters, a letter on the comparable route from Florence to Paris around 1400 took twenty-one days on average, in the best case ten days. The same time is given in the section on letter delivery times in the *Practica della mercatura* (around 1450) of Giovanni Antonio di Uzzano from Florence.[20]

Fast delivery times (in parentheses) and average delivery times in days on routes mentioned with greater frequency:

	Datini letters	*Practica*
Florence-Rome	(2) 5	5–6
Florence-Milan	(2) 6	10–12
Florence-Paris	(10) 21	20–22
Rome-Paris	(13) 26	
Rome-Bruges	(20) 31	
Paris-Brussels/Bruges	(2) 4	

The section on delivery times ("Termini di Chorieri di andare da luogo a luogo") in the *Practica* and the empirical times extracted from the letters resemble the standardized times rather accurately, of course with the exception of the extremely long time given by Giovanni Antonio for the route from Florence to Milan. In the middle of the fifteenth century, the general traffic in messages thus did not function any faster than it had in the thirteenth century, though in all likelihood far more regularly and reliably.

Hour passes

Connections that were calculable and punctual to within a few days were sufficient for merchants to transact their normal business in goods and money. Commercial practices and the legal framework took the efficiency of the means of transport and conveyance into account. Obvious need for greater speed in transport and communications links existed above all in mercantile speculation and in the military sphere. News about harvest sizes and prices, about cargoes and shipwrecks were as prized and — figuratively speaking — highly traded among merchants as reports about the actions of the enemy among generals, for whom a quick shifting of troops could decide the outcome of a campaign.[21] In both cases there existed an ardent interest in the fastest possible and exclusive conveyance of information. News of speculative importance lost value in the degree to which it became widely accessible, and movements a military enemy knew about or could follow became useless.

In cases of deaths, proclamations of rulers, or declarations of war, the speed of transmission could become a factor of political or military relevance. News of the election of the Borgia Pope Alexander VI on August 11, 1492, had spread in Florence within about twelve hours, but it reached a Venetian delegation in Swabia only nine days after the event.[22] "Sensational" speeds are reported especially for news of this type. Much more rarely do we hear about crucial delays, about the political consequences of the time difference in the transmission of rumor and authorized news, or about the effects of transmission speeds that were far below average.[23]

Then, at the end of the fourteenth century, attempts were made to secure or even accelerate the standard delivery times of news by setting up rates in a specific way. In 1390, the managers of the cathedral workshop in Milan and their messenger agreed that he would receive ten florins if he brought the letters to Rome in eight days. If it took him nine days he would receive only nine florins; if it took him ten days, only eight florins.[24] F. Melis mentions a messenger contract from the year 1394 that stipulated not only deductions from the wage for exceeding delivery times but also rewards for coming in under them.[25] Until the end of the fourteenth century, however, delivery times continued to be measured only in days.[26]

With the help of clocks, a transition to smaller time units was made at the end of the fourteenth century. New methods of time control appeared first in the military-political sphere, and from the end of the fifteenth century they were permanently established in the emerging European postal systems.

Marco Polo's travel account was a great literary success. The original version in Old French was soon followed by translations into Latin and Tuscan, in which the text was in part shortened, in part purged, but in part also expanded with explanatory additions. The first printed version, which most older translations followed, was published by Gian Battista Ramusio in Venice in 1559. It presumably followed a Latin manuscript that has now been lost, and in which the text quoted above was expanded by an interesting passage. According to the addition, an official at the Chinese postal stations recorded the day and hour of the arrival and departure of the messengers.[27]

The other contemporary accounts by travelers to Asia contain no comparable information. Time controls, however, were quite common in the Asian states. In Iran, courier passes were marked with a sign for on-time or late arrival.[28] In China, too, passes were marked with the times of arrival and departure using the Chinese hours (unequal double hours). It is unclear, however, what sort of timekeeping devices were employed.[29] These certificates served the purpose of checking up on the messengers, who were supposed to cover three stations per day. But they were also intended to keep the messengers from wrecking the horses. Whether reports of these practices reached Europe, whether they might go back to Marco Polo himself, or whether they were only added in the fourteenth century is not something we can determine. It is striking, though, that they circulated at a time when two comparable techniques of control came into common use in two places in Europe.[30]

Among the approximately 30,000 extant pieces in the letter archive of the Teutonic Order there are about 1,800 "express letters," that is, for the most part letters directed to the Grand Master at whatever place he happened to be and marked with formulaic notations of urgency. About 1,500 of these "express letters" contain in addition "praesentata notices," list-like notations with hour indications of the places of dispatch and transit.[31]

> 1420 Apr. 5, To the honorable Grand Master with all reverence day and night . . . Left from Königsberg on Friday before Easter hora IX before midday. Left from Brandenburg on the same day hora XII at midday. Left from Balga on the same day hora sexta.[32]

Striking clocks were consistently used for these notations:

> 1409 Aug. 6, Left . . . when then clock struck IX before midday.—Left from Papau when the clock struck I after

midday.—Left from Grudenz when the clock struck V after midday on the same day.[33]

Indications such as "when the clock was about to strike V" or "between IIII and V after midday" show the limits of more precise time indications. It is also striking that the praesentata notices with hour indications begin only on the territory of the Order, which means that they were still not common outside of this area.[34] The commanderies, "Vogteien," and "Pflegeämter" were linked together into a network of relay stations for the Order's communications system, whose beginnings lie presumably in the years around 1400, from which time come the first reports about a tower clock at the Order's headquarters, the Marienburg. However, many of the other houses of the Order—convents, military strongholds, and centers of landed lordships—were also equipped with mechanical tower clocks as early as the beginning of the fifteenth century. Moreover, thousands of individual notices show that the installation of clocks was already very dense. This is confirmed not only by the large number of hour notations in letters, but also by the time indications in the records of the diets from the sphere of the Order's jurisdiction. Dials also dominate the town views of the former branches of the Order in Christoph Hartknoch's *Altes und Neues Preußen* (1648) (fig. 70).[35]

To be sure, the Teutonic Order did not play a pioneering role in the diffusion of modern time-keeping at the beginning of the fourteenth century.[36] In the fifteenth century, though, it knew how to make quick use of the new technology of time-keeping for its undeniably modern administrative organization.

In looking at the delivery times on the most frequented express letter routes in the territory of the Order, we notice remarkably high speeds in individual cases, on the one hand, and extreme variations in the speeds, on the other. It took anywhere from one hour to fourteen hours to cover the route from Königsberg to Brandenburg. The purpose of the praesentata notices was evidently to check on the messengers and to pinpoint the causes of delays. This communications system spanned an enormous territory. The Baltic was reached via Königsberg and Memel. From Thorn, messengers went to Poland, Hungary, Silesia, Austria, and Bohemia. The branches of the Order served as horse and messenger relays and received in return subsidies from the treasury of the Order. Although it was exclusively in the service of military and

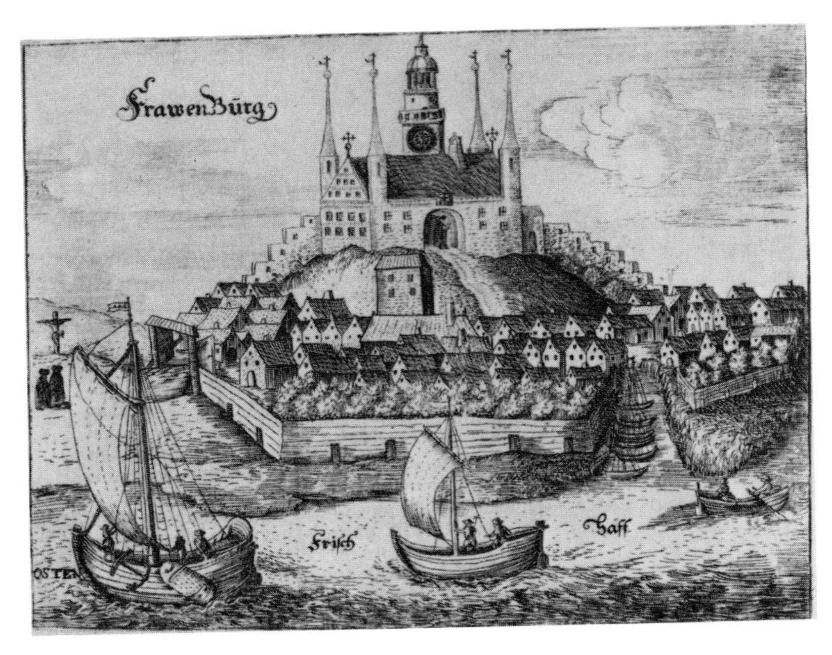

70. View of Frauenburg from Christian Hartknoch's *Altes-und Neues Preußen*, 1648.

administrative purposes (merchant letters were transported considerably more slowly in the same region),[37] this communications system still displays—despite the irregular delivery time—typical features of a more modern postal transportation technique. Whether the Order, in systematically setting up permanent relay lines, drew inspiration from Asian models remains a speculative question. However, when it came to the technique of controlling delivery time, there were also models from northern Italy.

In Milan of the Visconti at the end of the fourteenth century, relay lines (called "poste") were maintained at least in times of military tensions, and here too we find the formalities of what were later called "hour passes." From the small town of Trezzo sull'Ada, Barnabò Visconti informed Lodovico Gonzaga in Mantua on June 4, 1373, that his enemy, the Signore of Ferrara, who was allied to the pope and the Duke of Savoy, was moving toward Bologna. On the back the letter bears the posting mark with a modern hour indication of a messenger who set out on horseback from Brescia. A little later (1387–1389) we find continuous markings with the departure times for individual stations and references to chains of couriers organized like relays. Later examples then show what was for hour passes a typical combination of the demand for lightning-fast conveyance with the threat of horrible punishments. For example, we read the following on a dispatch of Filippo Maria Visconti, dated November 27, 1427: "On penalty of a thousand gallows it shall be carried by our messenger day and night, not fast but like lightning, fast, fast, fast, fast, fast, fast, fast; posted in Milan in the twenty-second hour."[38] From the beginning of the fifteenth century, relay courses after this model — which naturally presupposes the use of modern clocks — were set up in Italy. A postal code of Ottoboni Terzi (Lord of Reggio Emilia, Parma, and Piacenza) from the year 1408 combined modern controls of letter delivery times with regulations on the normal speed of messenger traffic. His officials were to pay heed that the mounted couriers covered five miles per hour in the summer and four in the winter, the foot couriers a mile less. Every mile under the norm was to be punished with one blow from a cane.[39]

With the help of relays of "posted" couriers and horses, the natural limit to the acceleration of news had been reached, or reached again, around 1400, at least for certain stretches. The early hour passes, in a sense an early form of the postmark, were not an indispensable but certainly an effective and relatively modern means of control for this technique of transmission. At the same time, modern hour-reckoning changed the general perception of routes and distances. In the correspondence of the dukes of Milan, the postal distance Rome-Naples was given as thirty hours, that from Rome to Milan as eighty and a half hours.[40] No longer the day but the hour had become the measure of the speed of travel.

Already around 1400, all organizational means for accelerating the conveyance of mail that would be available until the beginning of the nineteenth century had been exhausted. But only the postal services that arose at the end of the fifteenth century combined methods of transport and time control into permanent and large-scale organizations.

The "invention" of the postal service and the Taxis family

To legitimate its claim to the heritable position of postmaster in the German Reich, the Taxis family always maintained that in the time of

Emperor Frederick III it had invented, through posts and couriers, the means of imparting the requisite speed to the news service of the princes.[41] Members of the Taxis family were "the first inventors and originators of this postal service" and thus had claim to the imperial fief.[42] In the letter-happy century of the Enlightenment, the invention of the postal service was placed in the same category as the discovery of America.[43] However, the typical features of the postal service were in fact not revolutionary innovations; new in the postal system of the Taxis family, even though this has been occasionally denied, was only its setup as a "sub-contracted" state enterprise and its gradual internationalization.[44]

The rise of the Hapsburgs to a European power was the occasion and prerequisite for the development of a stable postal system. Through his marriage to Mary of Burgundy, Maximilian I, elected king of the Romans in 1486, had become heir to the highly developed Netherlandish part of the Burgundian kingdom. In 1489 he also took over the government in the Duchy of Tyrol, which he inherited in 1496. Through his son Philip the Fair the Kingdom of Spain was subsequently linked with the house of Hapsburg. In order to secure communications between the widely dispersed parts of this realm, Maximilian engaged members of the family of Taxis, which hailed from northern Italy and had previously been active in Venetian and papal courier services.

The first documents of this new, international postal system are five hundred years old. Chronicles of the city of Memmingen report its establishment in the year 1490. The term "postal system" (German: "Post") refers to a setup in which nearly all known organizational means of conveying information were combined to give rise to the early modern postal system: post stations at regular intervals, changes of horses and couriers, fixed delivery times, acceleration through continuous transport, and—implicitly—the organizational framework of the state. It was this combination which appeared as a significant innovation.

> In this year they began to establish the posts at the order of Maximilian I the Roman king, from Austria all the way to the Netherlands, France, and Rome. Everywhere one post was separated from another by five miles [38 kilometers] . . .

> (and everywhere one courier had to wait for the other, and as soon as the other came riding to him he blew into his horn; the courier lying in the lodgings heard this and had to rise immediately) . . . one courier had to ride one mile each hour, that is two hours far, or deductions were made from his wage, and they had to ride day and night. In this way a letter often made it from here to Rome in five days.[45]

The resemblance to Marco Polo's account is striking but surely coincidental. Although it based itself in part on contemporary reports, the chronicle undoubtedly gives too short a time for the delivery of a letter to Rome. The distance between the posts and the speed of one mile an hour, though, whereby one geographic mile was set at the distance of a two-hour journey, correspond approximately to the conveyance activities we can ascertain from the hour passes.

For France, too, it is reported that individual courier relay lines were set up at the end of the fifteenth century for the royal message traffic. However, the edict of Louis XI (dated June 19, 1464), according to which the main roads of the kingdom were to be furnished with stations for relay couriers at intervals of four lieues and letter packages were to be signed with the day and hour of transit, is presumably a seventeenth-century forgery.[46]

According to the hour passes of the Taxis postal service that have survived from 1494 on (at the time called "lista de posta" or "Post-Zedl"), the postal routes under Maximilian initially linked Innsbruck with Milan, Vienna, Augsburg, Worms, Mecheln, and Brussels.[47] The wording of the passes clung closely to the Milanese models. As an example I shall give an hour pass for a dispatch that ran between the seventh and eleventh of February from Milan via Bormio to the king, whose whereabouts were not known to the postmaster in Milan. On the outside: "Lista. Per postas Cito Cito Cito Cito. Expeditum." Next to it once again a drawing of a gallows. The formulaic threat of the gallows remained in use in the Taxis postal system until the middle of the sixteenth century, whereby it was gradually stylized into a symbol: XOX.[48]

> 1495 Milan Friday February 6 [7], at the eighth hour [= 1 o'clock A.M.] couriers of the post to Bormio and onwards

> to where the most gracious Roman king is! Transmit this pouch ["bolgieta"] quickly, quickly, quickly, quickly, flying day and night, without losing any time, because these are matters of the greatest importance . . . by signing this slip from post to post.[49]

There follow the various transit stations with transit times—according to the Italian hour-reckoning as far as the mountain pass, thereafter according to central European reckoning. In the earliest examples the notices of the change of courier on the Italian side are always signed off by a high local official, the procurator or referendarius. The verification of transit times by political authorities still points to the original governmental character of the organization of relay lines.

As with the Asian post and the couriers in the territory of the Teutonic Order, the routes were not links between cities, because the latter did not lie along the shortest route and because they would have delayed the activities and made them more expensive with their cumbersome gate-closing regulations. The conflict between the time requirements of postal conveyance and the urban time system that was still largely oriented to daylight was a permanent topic for the Taxis postal system. The routes, instead, linked the administrative seats of the court with the respective whereabouts of the temporary residence of the king and thus had to be frequently newly "laid." Cities that were reached by the imperial post via branch routes continued to run their own messenger services.

The analysis of the extant hour passes confirms the information in the chronicle of Memmingen. The intervals between the stations were between four and five (geographic) miles, and the average speed of transport was not quite one mile per hour, that is to say, between one hundred twenty and one hundred eighty kilometers per day.[50] According to an hour pass filled in on March 31, 1506, by Franz von Taxis, a package of letters went from Mecheln to Innsbruck in five and one-half days. One hundred and thirty-two hours were needed for one hundred and three miles (164 kilometers). On one segment of the route the shipment also included private postal matter, which was delivered separately to an urban recipient.

> There is a package in this bag, it belongs in Augsburg into the hands of Anthony Welser, and you will find a letter with

> it and XII plapart [half groschen] inside; with it you can send a courier immediately to Augsburg.[51]

As security against the always precarious financing by the emperor, the quasi-official Taxis postal system had opened itself to private customers who could use it for a fee.

In January of 1505, Philip I the Fair, King of Spain and Archduke of Austria, drew up a contract with Franz von Taxis, who had been named chief postmaster in 1501. According to the terms of the contract, von Taxis was to set up postal routes from Brussels to the Spanish-Dutch residence of the governor, to the king in Germany, to the French court, and to the sometime residences of the king in Aragon, Castille, and Granada.[52] In return for 12,000 livres per year, Franz von Taxis was to maintain, at his own expense, mounted posts and transport the official correspondence at no additional charge. In times of war he had to ensure the fastest possible shifting of threatened routes and maintain contact with the commander in the field by means of posts that were spaced four miles apart. The delivery times were also fixed in the contract: from Brussels to Innsbruck 5 1/2 days in the summer (6 1/2 in winter), to Paris 44 (54) hours, to Granada 15 (18) days. In a new contract, Charles I of Spain—after 1519 Emperor Charles V—had the routes expanded, among other destinations to Rome and Naples. Rome should be reached from Brussels in 10 1/2 (12) days, the route Brussels-Paris was to be shortened to 36 (40) hours.[53] Even if these delivery times could not always be consistently maintained, we can see that the stable organization of the Taxis postal system cut the conveyance times of letters and messages approximately in half compared to those of Italian merchant letters.[54]

Adherence to the contractually guaranteed conveyance times in the "Extraposten" ("special dispatches or conveyances") was checked in all European postal systems by hour passes, which from this time on were customary in some systems, and obligatory in others.[55]

Although the contracts stipulated that the postal lines of the Taxis were to be run or established only for royal letters and business, the transport of private postal matter was not explicitly excluded. This ensured the Taxis a profitable monopoly, though in the years to come it had to be protected with much effort, by means of numerous privileges and patents, against competition from older, local courier and postal institutions.

"Postal reform": the coordination of routes

The expansive political structure of Hapsburg rule had been favorable to the creation, in the German Empire under the leadership of the Taxis family, of a postal system that spanned large areas. Inevitably the postal system was also affected by the political crises of the Hapsburg empire. The division of the Hapsburgs into a Spanish-Netherlandish and an Austro-Bohemian line, the strengthening of the territorial lordships within the Empire in the course of the confessional conflict, the revolt of the Netherlands, and the bankruptcy of the Spanish monarchy: all this threatened the financial and administrative up-front services as preconditions for the running of the postal system and cast the various branches of the Taxis family into conflicts of loyalty and financial difficulties. The Protestant estates of the Empire, in particular, mistrusted the Netherlandish postmaster general. Competing postal installations, particularly urban ones, flourished again. Around 1570, the imperial postal system had largely collapsed. The end of the century saw the beginning of the "postal reform."[56]

Following the stabilization of conditions in the Catholic provinces of the Netherlands, Leonard von Taxis was appointed chief postmaster general of the German Reich in 1595, and once the debt questions had been settled the reorganization of the postal routes was undertaken. The conveyance of private letters was now intended to finance the conveyance of official correspondence and the maintenance of the relay routes. The interest of merchants in *reliable* connections emerges as a noticeable motive alongside the governmental-administrative interest in *fast* connections. The first step was the elimination of the "secondary courier systems." In 1597, Rudolph II elevated the postal system to an imperial prerogative and made its unauthorized exercise punishable. Only the Imperial Post of the Taxis was allowed to transport letters "postally," that is, with horses posted at intervals and with courier relays. City couriers were permitted on the condition that they transported their parcels "without any change" of couriers.[57]

The European postal routes were subsequently linked together through numerous agreements. Punctual transportation, that is above all adherence to fixed delivery times, was now important not only for the guaranteeing of minimum speeds, but also for the intermeshing of routes, their coordination. Thus the postmaster of Cologne, Jakob

Henot, as the plenipotentiary of the Taxis Post, explained to the emperor (in a report of May 8, 1596) the importance of the punctual processing of the mail from Prague to Augsburg by pointing out that the imperial mail would be idled in Augsburg for one week if it did not reach the main route from Augsburg to Speyer, Cologne, and the Netherlands, on the one hand, and to Innsbruck and Italy, on the other, punctually on the hour.[58]

Hans Fugger, in his capacity as imperial commissioner for postal reform, had written to Leonard von Taxis: "everything depends on reliability; the merchants, too, are especially concerned with this." The delegates of the imperial postmaster tried to organize the system of routes around this maxim of reliability. Examples are the arrangements which Jakob Henot concluded in the spring of 1598 in northern Italy. First he entered into an agreement with the postmaster of the dukes of Mantua by which the latter was to serve the route from Mantua to Triest with couriers within twenty-four hours, whereby a separate register of the hours of arrival and departure had to be kept. On the same day Henot concluded in Verona an agreement with the Venetian postmaster by which the mail that accumulated in Verona had to be taken on Saturdays to Volargne to the postal route from Mantua to Trieste. In order to better serve the merchants and traders, the mail coming from the north should be brought from Volargne to Verona in four hours via an express courier service.[59] A few days later, Henot came to an agreement with the royal Spanish postmaster general in the Duchy of Milan to the effect that the mail from Rome, Genoa, Florence, and Lombardy was to be in Mantua Fridays at the midday hour, and that the mail coming from the north was to be delivered to the recipients in northern Italy by express couriers ("con cavalcata espressa").[60] Numerous similar postal contracts and postal regulations made the careless filling in of the hour passes or delays on the various segments of the routes punishable.

To be sure, the development of the postal network and the adherence to specific transportation speeds served directly to accelerate urgent news or to transport letters for certain senders or recipients.[61] But it had also become a means of coordinating the postal routes, and the system of these routes, according to the wording of the sources, now served primarily the commercial exchange of news.

Increasing precision in the eighteenth century

After the Thirty Years' War state postal services were deliberately set up, especially in the Protestant states, and those that set them up had recourse to a (legally controversial) territorial postal privilege. Such systems were set up in part because of a mistrust of the imperial postal service, in part because of the obvious profitability of such institutions. What distinguished the territorial postal services from the imperial postal service was their legal form as pure state institutions, a network of routes that covered their territories more densely, and the linking together of the transport of letters, passengers, and small goods. The organization and technique of transportation did not change.

However, postal laws and regulations of the eighteenth century reveal the attempt to use the existing possibilities to stabilize the speed of transportation and to arrange the coordination of the routes with greater precision. Demands for punctuality are reflected in the constant admonitions to mail coach drivers and sub-postmasters to adhere to the hours "precisely" and to keep the "hour-cards" or "hour-passes" (which were now printed as forms) accurately and without any "bias" "so as to speed up the post stations." Enforcing these norms was not easy, since mail coach drivers were supposed to be not only sober but also literate. The prescribed speed for ordinary postal riders of the regular service as well as for relay couriers of the special dispatches was at least one mile per hour; mail coaches and carriage post should cover a mile, depending on the time of the year and road conditions, in one hour and a quarter to two hours, a norm that combined the desire for punctuality with the desire to go easy on the horses.[62]

Room for acceleration seems to have existed with dispatch and changing times, and consequently these were strictly limited. The carriage post could be delayed for half an hour at the most, the mounted post only one quarter of an hour. While the Saxon Postal Code of 1713 still stipulated that relays had to be dispatched "with the loss of only a few minutes," but at the very most within a quarter of an hour, in Prussia during the second half of the century this time was set at five minutes in small stations.[63] Moreover, the scheduled running of the mail service was to be disturbed as little as possible by the affairs of the lords. Thus the Prince Elector of Saxony issued a patent in 1693 according to which the posts leaving from Dresden should wait at most one hour in affairs of the territorial ruler, so that "the orderly running of their posts is not impeded to the disadvantage of the correspondence, in that the other posts, who are coordinated with them, must also wait long and in excess of the time." The Saxon Postal Code (1713) adopted this regulation "in order that the connection with the other posts is not severed and the entire route upset," and for official purposes it called for special relays that were supposed to catch up with the mail.[64] Finally, control and penal regulations were toughened and drawn more narrowly in temporal terms. According to the Code of the Electorate of Saxony, every quarter-hour delay in the expedition of the relays was to be fined half a gulden. The renewed Prussian Postal Code of 1732 set a fine of one Reichstaler for each half-hour delay with the carriage post, and for each quarter-hour delay with the mounted post.[65]

However, the prerequisite—by no means self-evident in the eighteenth century—for such increased precision was a reliable public indication of the time. The Prussian Codes (1710–1712) no longer accepted the unequal running of clocks as an excuse for time lost at the stations and demanded that the postmasters urge the authorities and magistrates to set the public clocks by the sun. Tax commissioners ("Steuerräte") of the Electorate of the Mark of Brandenburg were asked in 1719 to urge the magistrates of the cities subject to their inspection to set the clocks properly, on the grounds that this was "a necessary part of good administration."[66]

These normative texts undoubtedly create an unrealistic image of the postal stations, describing how they were supposed to function but never in fact did. They should thus be understood as concepts for systematically organized transportation and communication links, which, if we trace them over a longer period, reveal the growing importance of the techniques of time measurement and time control.

"Transport revolution": The acceleration and increasing density of traffic

Mercantilist policy, which was aimed at creating national markets with the freest possible internal circulation, included the development of trunk roads and canal systems. The goal of this policy was to develop the national territory as completely as possible and to free domestic traffic from the impediments of urban as well as feudal forms of economic interchange. The result of the development of the national transportation systems in the century prior to the age of the railroad was the so-called transport revolution, the precondition for and later part of the process of the industrial revolution.[67]

The repercussions of the development of land routes and waterways manifested themselves quickly in England, more slowly in France, as Arthur Young has observed with astonishment, though his verdict is undoubtedly correct.[68] The most noticeable changes were an expansion in the range of the profitable transport of heavy goods and an increase in the volume of all areas of traffic, including the transportation of letters. The new postal systems were also preconditions for the letter culture of the Age of Enlightenment. The intensity of correspondence is closely related to the frequency with which letters were conveyed.[69] Through improved and standardized road construction techniques, the straightest possible alignment of routes, and new carriage designs, wheeled traffic on the trunk roads of the eighteenth century was gradually brought to the speed of mounted messengers and couriers of earlier times. Judging by its schedule, the express coach ("diligence") from Paris to Brussels in 1775 reached quite precisely the speed of the old express couriers (two days). Moreover, new roads and macadams connected to the network of routes cities which had previously been hard to reach and thus far away in terms of time—for example, those located in central France or Wales. Comparatively remarkable speeds were attained on such routes. By contrast, the general tempo of the transmission of news hardly changed between 1500 and 1775. The maps published by F. Braudel, on which the speed of news is indicated with isochronic lines, indicate, for example, an average speed of two weeks for the distance from Paris to Venice during this entire period.[70]

Between 1830 and 1850, during the time when the first railroad and steamship routes were being established, the conventional means of road transport reached their peak capacity in terms of volume and speed. The cumulative effects of tighter organization, better roads, and better carriages produced a doubling of the average speed to about 10 kilometers per hour in the transportation of letters and passengers. The Prussian express posts for mail and passenger transport, established in 1821, was to cover 1 mile on macadamized roads in 35 minutes (13 kilometers/hour); the continuing "courier and relay service" was to do it in 30 minutes (15 kilometers/hour). In France the average speed of coaches rose to 9.5 kilometers per hour prior to 1848. In printed service regulations, the English "Mail Guards" were admonished in February of 1800 to spend every possible minute on the road. In 1812, the Post Office considered 5 minutes changing time for the coaches sufficient on the line from London to Brighton. The same time applied to the "express coaches" of the Prussian express post.[71] According to P. Bagwell, in England after 1820 the traveling speed of the coaches exceeded that of riders.[72]

These accelerations, which seem very modest today, struck contemporaries as notable progress and by no means an ordinary fact of life. In his *Monographie der deutschen Postschnecke,* Ludwig Börne in 1821 mocked the transportation style of the Turn and Taxis carriage post. After having spent forty-six hours on the Frankfurt-Stuttgart route he provides statistical data ("doctrine of standstill") according to which the alcohol-tinged breaks of the postal carriage amounted to a total of fifteen hours. For Börne the personnel of the postal service was a reflection of the German burgher: not dangerous or malicious, but calm and slow, not inclined to undertake anything daring.[73]

Transports were a profitable business, because the users hardly had to bear the costs for the infrastructure, and from the turn of the century there is a growing number of clues indicating that rapid and reliable transport was not only a functionally necessary organizational task, but that punctuality and speed had also become values of intrinsic worth.

The demand for punctuality constantly moved its guarantors, clocks, into the center of attention. Initially the question was which public clocks should be binding for the posts. When the Prussian Postal Inspector F. Chasté announced the first Rhenish express post in 1821, he still identified one of the city clocks in each location as the standard clock: "The departure in Cologne, Koblenz, and Düsseldorf shall occur with the bell-stroke of the clock, that is in Cologne the Cathedral clock, in Düsseldorf the clock on the church of the Franciscans, and in Koblenz the city clock closest to the postal station shall serve as the guide."[74]

The new, more precise time of traffic and the cult that was made of valuable minutes had been made possible by advances in the technology of precision measurement.[75] Public clocks participated at least in part in these advances. Beginning in the middle of the seventeenth century they had been gradually converted to pendulum drive and thereby moved into a range of accuracy of a few minutes per day. The development of the clock spring as the regulator for transportable chronometers and portable watches made the reckoning in minutes accessible to everybody and made possible the construction of watches that were off by only a few seconds a month. Chronometers gradually discredited the public clocks, for with their help anybody could easily determine that the sun was only a moderately accurate time standard. Owing to the inclination of the earth's axis and its elliptical orbit, the duration of the only thing known up to that time, the "true local day," that is, the time between the passages of the sun's center through the north-south line of a locality, varied. Once this had been recognized it was seen as troublesome, and in the large cities (Geneva 1780, London 1792, Berlin 1810, Paris 1816), one began to set the public clocks according to the "solar mean time." The difference between "solar true time" and "solar mean time" (between −14.3 and +16.4 minutes) was published in tables with the "equation of time."[76] The civic day was now no longer an object of direct experience. To ease the transition, in Berlin in 1811, for example, the sundial, now superfluous, was removed from the authoritative clock of the Academy.

What had begun as a bow by the large cities to the scientific spirit of the age was turned into common practice by the needs of transportation. In Prussia, for example, the transition to solar mean time after 1825 was ordered by the Ministry of the Interior on the urging of the General Post Office, because the "correspondence of the public clocks . . . is indispensable for all transportation, particularly for the postal system which is now so well ordered." At the same time the city magistrates were admonished to have one clock, usually that of the post office, regulated by solar mean time and thereby turn it into the standard clock.[77] Clocks in the church towers and town hall were to be set by it; they had lost their central function of ordering time. In sealed "course-clocks," the postilions of the mounted and express posts, in England, the "mail-coaches," brought the time of day also to places where the maintenance of a standard clock or a "regulator" was too expensive.[78] Already before the beginning of the age of the railroad, the "time of transport" had rendered the "time of the cities" obsolete.

Railroad time — hour-zone time — world time

With the considerable increase in the tempo of carriage traffic in the first decades of the nineteenth century, the limits of the physical capacity of animal power, the horses, had been reached. It was clear to contemporary observers that even with an optimal organization of the relays and with the most advanced carriage designs, a continuous pace of two miles per hour (ca. fifteen kilometers/hour) was possible only at the price of exploiting the natural resources — horses and humans.[79]

By contrast, the speeds on the first railroad lines — very moderate in hindsight — meant already at the time a doubling or tripling of the tempo, which could be maintained along the entire stretch. In describing the new experience of speed, the first travelers reached for the old image of the ultimate in speed, the flight of birds, and measured the distances in hours of marching or riding time. "One sits on the train and slides in thirty to thirty-five minutes to the five-hour-distant Potsdam." Like Jacob Burckhardt, David Friedrich Strauß also spoke of the "magical flying" on the occasion of a train trip from Heidelberg to Mannheim, which he completed "in half an hour instead of five hours." The Ludwigs Railroad Company in 1850 advertised a trip from Nuremberg to Fürth with the promise to travel "one and a half hours in ten minutes."[80]

What stands out in the metaphors of the travel accounts of "railroad time," in the flood of evidence for the experiential change triggered by trains, is the identification of the historical change with the progressive "Zeitgeist" and the persistent topoi "emancipation from nature," "destruction of space and time," and "conquest of time."[81] "Emancipation from nature" meant the liberation of the distance surmounted "from all resistance, variation, and adventure." Contemporaries noted that natural impediments to travel were not experienced because the machine ensemble of vehicle and road, of locomotive and rail, cut through mountains and spanned valleys. The image of the "landscapes flying by" that was common from the age of carriages became intensified into a landscape panorama that was unrolled in front of the compartment window.[82] The even, reliable, and tireless functioning of the steam engine drive was also experienced as an emancipation from nature. Even if traveling speeds were occasionally described as "still modest at present," in the imagination they were already being infinitely accelerated.[83]

The actual traveling speed of trains rose, and the railroad networks of the various railways grew together with increasing density. For the purpose of setting up their schedules, which accustomed the traveling public to everyday dealings with minutes, the railways had each taken the local time of their headquarters as the basis and had made this time binding for the so-called internal operating service.[84] For the dealings with the public at train stations and in the schedules, the so-called external service, the respective mean local time was at first retained. In southern Germany, however, the "internal" railroad time was also valid for the external service, and in many cases it became in this way the local time. State authorities supported this synchronization of distant places with the local time of the railway's administrative seat. No longer the postal service, but the railroad mediated the capital city's time of day into the regions, and it was precisely this which prompted the absentminded mailman Bornike to utter the malapropism: "It is high train, the time arrived three hours ago."[85]

In Bavaria time was reckoned after Munich time south of the Danube, after Ludwigshafen time north of the Danube. In Württemberg it was reckoned after Stuttgart time, in Baden after Karlsruhe time. As late as 1874, the mean times of Berlin, Cologne, Königsberg, Lübeck, Oldenburg, Elmshorn, Gießen, Frankfurt am Main, Leipzig, Munich, Stuttgart, and so on, were used for time reckoning in Germany. That same year Berlin time was prescribed for internal service, at least for the North German railways. Railways worked like "great national clocks."[86] In other countries, especially those with not much of an East-West spread, the transition to railway time as nationwide local time had been made far more quickly and with less fuss. In Germany and Austria the public discussion over the standardization of time reckoning began in the 1870s. The plans to synchronize all local times into one world time or "standard time" came from America, which had seventy-one railway times in 1873. Sandford Fleming, chief engineer of the Canadian Pacific Railway, had suggested in a number of memoranda to select 24 normal meridians at intervals of 15 degrees—that is, at one hour apart—from the 360 latitudes of the circumference of the earth. Localities belonging to different normal meridians would thus have exactly one hour time difference but the same minutes and seconds. Support was sought for these proposals at a series of conferences, and the quarrels revolved primarily around the question whether Greenwich, Jerusalem, or a fictional place in the ocean should be the prime meridian.

Various countries followed this concept—some quickly, some hesitantly—by assigning their national times to a time zone, for example, the central European zone. The Reichs Law "concerning the introduction of uniform time reckoning" decreed the following on April 1, 1893: "The legal time in Germany is the solar mean time of the fifteenth degree latitude east of Greenwich." France at first made Paris time instead of railway time binding (1891). Not until 1911 did it join Greenwich time, though no mention was made of the English observatory: legal time in France and Algeria was now mean Paris time minus nine minutes and twenty-one seconds (= Greenwich time). In the German Empire the "time reform" was not without controversy. But the defenders of the "dear old local time" had no chance against the advocates of reform.

In the Reichstag, Field Marshall von Moltke, too, spoke up as an advocate of a uniform national time. To be sure, the introduction of a "world time," demanded by the scholars at the observatories, was not a necessity, he maintained. However, the difference between railway time and local times was a remnant from the time of political disunity, "which now, after we have become one Reich, should rightly be done away with." A uniform national time was demanded not only by the operational needs of the railways but also by an important group of passengers: "The noblest travelers, gentlemen, are the troops." Working time regulations of the factories could be easily changed; "And as far as the rural population is concerned—well, gentlemen, the rural worker doesn't look at the clock much, for the most part he doesn't have one; he looks around to see whether it is already light, then he knows that he will soon be summoned to work by the court bell ("Hofglocke"). However, if the court clock is wrong, which is usually the case (laughter) . . . he arrives a quarter of an hour early . . . , except he will also be let go by the same clock a quarter of an hour earlier: the working time stays the same." In his last Reichstag speech, Moltke then drew a picture of bourgeois daily life at the end of the nineteenth century in which minutes and the synchronicity of all clocks were still of no great concern:

> Gentlemen, practical life rarely demands a punctuality reckoned in minutes. In many places it is customary to set the school clock back ten minutes so that the children will be present when the teacher arrives. Even the court clock is frequently set back so that the parties will assemble before the proceedings begin. Conversely, in villages that lie close to the railroad, the clock is usually set ahead by a few minutes so that the people will not miss the train. Indeed, gentlemen, even this honorable house keeps an academic quarter hour, which sometimes also gets to be a little longer.[87]

This picture was no longer realistic. Not only the railways, but the needs of factories, schools, and public transportation in large cities for punctuality and coordination, too, made the manner, technique, and quality of public time indication into a perennial topic. Professor Förster of the Berlin observatory declared in 1890 that urban time indication that did not guarantee everywhere punctuality to the minute was a device that was downright contemptous of people.[88] Centrally controlled public clocks were installed at an exceedingly rapid pace. From the beginning of the twentieth century, under the influence of the railway route books, a new kind of hour-reckoning—continuous counting from midnight—slowly gained in importance, and the introduction of daylight saving time has brought to our awareness once again the fact that the relationship between daylight and hour of the day is established by society.

The universal time promoted by the railroad system—"world time" was never really able to catch on terminologically —was still astronomically determined and to that extent a natural time. Transmitted telegraphically at first, then via the radio, the synchronicity of all local times—in a world in which small local times under city clocks and court bells were still in operation—opened new horizons for the sense of time. But the observable universe was not suitable as a time standard where the utmost in precision was called for, and since the 1960s, time has been determined physically, in accordance with the 9,192,631,770 oscillations per second of the cesium atom.

ABBREVIATIONS

AH — *Antiquarian Horology and the Proceedings of the Antiquarian Horological Society.* Ticehurst, 1953–.

APT — *Archiv für Post und Telegraphie.* Berlin, 1873–.

AU — *Alte Uhren. Zeitmeßgeräte, wissenchaftliche Instrumente und Automaten.* Munich, 1977–.

BN — Bibliothèque Nationale

CC Cont. Med. — *Corpus Christianorum continuatio medievalis.* Torhout, 1966–.

CCM — *Corpus Consuetudinum Monasticarum.* Siegburg, 1963–.

CSEL — *Corpu scriptorum ecclesiasticorum Latinorum.* Vienna, 1866–.

MEW — *Karl Marx, Friedrich Engels. Werke.* Berlin, 1961–1974.

MGH — *Monumenta Germaniae Historica inde ab a. C. 500 usque a. 1500.* Hannover et alia. 1826–.

Migne PL — *Patrologia cursus completus, series latina.* Edited by J. P. Migne. Paris, 1844–.

Muratori — *Rerum Italicarum scriptores . . .* Edited by L. A. Muratori. Milan, 1723–1751.

ORF — *Ordonnances des roys de France de la troisième race.* Paris, 1723–1849. Repr. Franborough, 1967–1968.

RE — *Realencyclopädie der classischen Altertumswissenschaften.* Stuttgart and Munich, 1893–.

R.I.S. — *Rerum Italicarum scriptores.* New edition. Edited by G.

Carducci and F. Fiorini. Città di Castello-Bologna, 1900–.

Rer. Brit. M.A. script. *Rerum Brittanicarum medii aevi scriptores.* London, 1858–1896.

SFAU *Schriften des historisch-wissenschaftlichen Fachkreises "Freunde Alter Uhren" in der deutschen Gesellschaft für Chronometrie.* Ulm, 1960–.

UB Urkundenbuch

NOTES

Chapter 1

1. *Dives et Pauper*, ed. by P. Heath Barnum, Early English Text Society 275 (Oxford, 1976), 119ff.

2. R. Koselleck, "Gibt es eine Beschleunigung in der Geschichte?" MS (Bielefeld).

3. Saint Augustine, *Confessions*, transl. by R. S. Pine-Coffin (Penguin Books, 1975), 269.

4. R. Glasser, "Die humanistische Wertschätzung der Zeit," *Die Welt als Geschichte* 7 (1941): 165–180; E. Stürzl, *Der Zeitbegriff in der Elisabethanischen Literatur. The Lackey of Eternity* (Vienna, 1965); compare Quinones, *Discovery;* E. Panofsky, "Father Time," in *Studies in Iconology* (New York, 1939).

5. Pliny, *Natural History* VII.191 ff., ed. and transl. by H. Rackham (Cambridge: Harvard University Press, 1942). On this literary genre see Copenhaver, *Historiography of Discovery.*

6. L. White, Jr., "Cultural Climates and Technological Advance in the Middle Ages," *Viator* 2 (1971): 171–201, here 173, 174.

7. R. Tuve, "Notes on Virtues and Vices," *Warburg Journal* 26 (1963): 264–303; 27 (1964): 42–72.

8. Uncertain date: the frescoes were completed in 1337–1338; Lorenzetti probably fell victim to the plague in 1348; the section in question was restored prior to 1360 after a fire; see C. Brandi, "Chiaramenti sul 'Buon Governo' di Ambrogio Lorenzetti," *Bolletino d'Arte* 40 (1955): 119–123. On the interpretation see N. Rubinstein, "Political Ideas in Sienese Art: The Frescoes by Ambrogio Lorenzetti and Taddeo di Bartolo in the Palazzo Publico," *Warburg Journal* 21 (1958): 179–207.

9. The other examples, very similar: Chantilly, Musée Conde, MS fr. 491, fol. 240 verso; Paris, BN MS fr. 9186 fol. 304; Oxford Bodl. Lib. MS Laud.

Misc. 570, fol. 16 (ca. 1450); London, BL Add. MS 6797, fol. 276 recto (with clock and spectacles); Rouen, Bibl. mun., MS 12, fol. 17 verso (1454); Examen de conscience, Rouen (1492?), fol. 191 v.; P. Breughel, Temperantia (copperplate engraving, 1557); with clock and spectacles on two tapestries (made in Brussels in 1520) of the series "Los Honores," Segovia, San Ildefonso; on the interpretation see C. C. Willard, "Christine de Pisan's Clock of Temperance," *L'Ésprit Créateur* 2 (1962): 149–154; R. Tuve (note 7).

10. L. White, Jr., "Iconography."

11. A. G. Keller, "A Renaissance Humanist Looks at 'New' Inventions: The Article 'Horologium' in Giovanni Tortelli's De orthographia," *Technology and Culture* 11 (1970): 345–365.

12. The new things also include sugar, tallow candles, and the blowpipe; ibid. 363.

13. For example, Bartolus Lucanus, *Oratio metrica ad Innocentium VIII* (Rome: E. Silber, 1486), from H. Diels, "Über die von Prokop beschriebene Kunstuhr in Gaza" (text, translation, and commentary), in *Abhandlungen der königlich-preußischen Akademie der Wissenschaften, Phil.-hist. Klasse,* no. 7 (1917): 1–41, 23.

14. See M. Luther, *Tischreden* (Weimar Edition, 1919), V, 23 (no. 5241).

15. Karl Marx to Friedrich Engels, Jan. 28, 1863, in *The Letters of Karl Marx,* selected and transl. by Saul K. Padover (Englewood Cliffs, New Jersey, 1979), 168.

16. H. D. Kittsteiner, "Reflections on the Constitution of Historical Time in Karl Marx," in *History and Memory* (Tel Aviv, 1992).

17. F. Engels, *The Condition of the Working-Class in England* (Moscow, 1980), esp. 190 ff.

18. Ibid., 41.

19. Compare M. Bloch, *Feudal Society,* transl. by L. A. Manyon (Chicago, 1974), II, 74. Objections by P. Dubuis, "Les paysans médiévaux et le temps. Remarques sur quelques idées reçues," *Études de lettres. Rev. de la Fac. de Lettres de l'Université de Lausanne* (1987, 2–3): 3–10.

20. Max Weber, *General Economic History,* transl. by Frank H. Knight (Glencoe: Illinois, 1927), 365; *The Protestant Ethic and the Spirit of Capitalism,* transl. by Talcott Parsons (New York, 1958), chap. V: "Asceticism and the Spirit of Capitalism."

21. W. Sombart, *Der moderne Kapitalismus,* 2nd ed. (Munich, 1921), vol. 2, 126 f., and vol. 1, 505 f.

22. Mumford, *Technics,* 12–18, 470.

23. M. Bloch, "Un manuel d'histoire des techniques," *Annales d'histoire économique et sociale* 3 (1931): 278 f.

24. M. Bloch, "L'heure et l'horloge," *Annales d'histoire économique et sociale* 9 (1937): 217 f.

25. Y. Renouard, *Les Hommes d'affaires,* new ed. (Paris, 1968), 240 ff.

26. J. Le Goff, "Merchant's Time and Church's Time in the Middle Ages,"

in *Time, Work, and Culture in the Middle Ages,* transl. by Arthur Goldhammer (Chicago, 1980), 35.

27. W. Paravicini, in F. Seibt and W. Eberhard eds., *Europa um 1400. Die Krise des Spätmittelalters* (Stuttgart, 1984), 219.

28. A. J. Gurjevich, *Categories of Medieval Culture* (London, 1985), 147. Transl. by G. L. Campbell.

29. By now there are some skeptical voices: P. Rück, "Die Dynamik mittelalterlicher Zeitmaße und die mechanische Uhr," in H. Möbius and J. J. Berns, eds., *Die Mechanik in den Künsten* (Marburg, 1990), 30.

30. P. Wolff, "Le temps et sa mesure au moyen-âge," *Annales d'histoire économique et sociale* 17 (1962): 1141–1145.

31. The sources for the history of clocks, clockmakers, and modern hour-reckoning up to about the year 1500 have been collected in an inventory, which is being prepared for duplication.

Chapter 2

1. English translation by Robert J. White (Noyes Press: Park Ridge, New Jersey, 1975), 175. On the interpretation of dreams in antiquity see S. R. F. Price, "The Future of Dreams: From Freud to Artemidorus," *Past and Present* 113 (1986): 3–37.

2. For this chapter see by way of introduction: Sontheimer, "Tageszeiten," in RE XXVIII (1932): cols. 2011–2023; A. Rehm, "Horologium," in RE VIII (1913): cols. 2416–2433; G. Bilfinger, "Die Zeitmesser der antiken Völker," in *Festschrift/Programm des Eberhard-Ludwigs-Gymnasiums in Stuttgart* (1886); G. Bilfinger, *Die antiken Stundenangaben* (Stuttgart, 1888); H. Diels, *Antike Technik,* 2nd ed. (Leipzig, 1920); W. Kubitschek, *Grundriß der antiken Zeitrechnung,* Handbuch der Altertumswissenschaft I,7 (Munich, 1928): despite many important references this is a chaotic text; J. Marquardt, *Das Privatleben der Römer,* 2nd ed. (Leipzig, 1886; reprint 1964), 250 ff. and 788 ff. These works contain a wealth of references, only a few of which have been cited.

3. Gellius, *Noctes atticae* III.3.5, ed. and transl. by J. C. Rolfe (Cambridge: Harvard University Press, 1927; 1984), 246 f.; Seneca, *Epistolae morales ad Luclium.* Transl. by E. Phillips Barker (Oxford, 1932); Suetonius, *De vita caesarum* XIII.16.2, ed. and transl. by J. C. Rolfe (Cambridge: Harvard University Press, 1914; 1970); Martial, *Epigrammata* IV.8, ed. and transl. by W. C. A. Ker, revised ed. (Cambridge: Harvard University Press, 1968); Cassiodorus on the occasion of the sending of two clocks to the Burgundian king, *Cassiodor Senatoris Variae,* ed. by T. Mommsen, MGH Auct. Antiq. XII (Berlin, 1894), p. 42; compare p. 62.

4. Herodotus, *The Histories* II.109, ed. and transl. by A. D. Godley (Cambridge: Harvard University Press, 1920).

5. Pliny, *Natural History* VII.212–215, ed. and transl. by H. Rackham (Cambridge: Harvard University Press, 1942).

6. On the Babylonian hours see F. Thureau-Dangin, "Sketch of the History of the Sexagesimal System," *Osiris* 7 (1939): 95–141, esp. 111 ff.

7. After Bede, *De temporum ratione* c. VII (Bedae Opera de Temporibus), ed. by C. W. Jones (Cambridge, Mass., 1943), 195; on the ancient models see J. Marquardt, *Privatleben* (note 2), 254.

8. According to O. Neugebauer and H. B. Braun van Hoesen, *Greek Horoscopes* (Philadelphia, 1959), 169, the majority of horoscopes since the time of Christ's birth also gave hour indications in temporal hours, diverging in this from astronomical usage.

9. Aristophanes, *Ecclesiazusae* 651–652, ed. and transl. by Benjamin Rogers (Cambridge: Harvard University Press, 1924); the references are collected in Bilfinger, "Zeitmesser" and *Stundenangaben* (see note 2).

10. The fundamental work is that of J. Drecker, *Die Theorie der Sonnenuhren* (Berlin, 1925).

11. L. Borchardt, *Die altägyptische Zeitmessung* (*Die Geschichte der Zeitmessung und der Uhren,* vol. I, part B; Berlin-Leipzig, 1920), 26 ff; R. W. Slowley, "Primitive Methods of Measuring Time with Special Reference to Egypt," *Journal of Egyptian Archeology* 17 (1931): 166–178.

12. The extant pieces are catalogued in S. L. Gibbs, *Greek and Roman Sundials* (New Haven and London, 1976).

13. See E. Buchner, *Die Sonnenuhr des Augustus* (Mainz, 1982; repr. from *Römische Mitt.* 83 (1976): 319–365; 87 (1980): 355–373, with references and plates).

14. J. Marquardt, *Privatleben* (note 2), 788 ff.

15. Climate: for example, Pliny, *Natural History* II.186 (see chapter 1, note 5); E. Buchner, "Antike Reiseuhren," *Cheiron* 1 (1971): 457–482 (and plates); this essay by Buchner, and his work cited in note 13, stand out for their terminological clarity and systematics.

16. The best introduction to water clocks is by Turner, *Water-Clocks,* 1–44. The older references are assembled in M. P. Schmidt, *Kulturhistorische Beiträge zur Kenntnis des griechischen und römischen Altertums,* No. 2: *Die Entstehung der antiken Wasseruhr* (Leipzig, 1912).

17. On India see J. F. Fleet, "The Ancient Indian Water Clock," *Journal of the Royal Asiatic Society* (1915): 213–230; on China see below chap. 4, The "Heavenly Clockwork."

18. O. Neugebauer, "Studies in Ancient Astronomy VIII: The Water-Clock in Babylonian Astronomy," *Isis* 37 (1947): 37–43.

19. L. Borchardt, "Altägyptische Zeitmessung"; A. Pogo, "Egyptian Water-Clocks," *Isis* 25 (1936): 403–425.

20. L. v. Mackensen, "Neue Erkenntnisse zur ägyptischen Zeitmessung. Die Inbetriebnahme und Berechnung der ältesten erhaltenen Wasseruhr," in *AU* 1 (1978): 13–18.

21. S. West, "Cultural Interchange over a Water-Clock," *Classical Quarterly* 23 (1973): 61–64.

22. See the account (supplementary to M. P. Schmidt, note 16) accom-

panying the presentation of an archeological find by S. Young, "An Athenian Clepsydra," *Hesperia* 8 (1939): 274–282.

23. Tacitus, *Dialogus,* ed. and transl. by W. Peterson and M. Winterbottom (Cambridge: Harvard University Press, rev. ed. 1970).

24. Pliny the Younger, *Epistolae* II.ii.14: "Dixi horas paene quinque. Nam duodecim clepsydris, quas spatiosissimas acceperam, sunt additae quattuor," ed. by R. A. B. Mynors (Oxford, 1968), 49; compare IV.9.9, 107.

25. Caesar, *De Bello Gallico* V.13.4, ed. and transl. by H. J. Edwards (Cambridge: Harvard University Press, 1917), 251.

26. T. F. Glick, "Medieval Irrigation Clocks," *Technology and Culture* 10 (1969): 424–428.

27. On the sinking-bowl clock see O. Kurz, *European Clocks,* 2–4.

28. See, by way of introduction, E. Weiß, "Der Rechtschutz der Römischen Wasserleitung," *ZRG Rom. Abt.* 45 (1925): 87–116.

29. Bilfinger, *Stundenangaben* (note 2), 103–109.

30. The account appears in the Roman physician Marcellius (second century A.D.); text: H. Schöne, in *Festschrift zur 49. Versammlung Deutscher Philologen und Schulmänner in Basel* (1907), 448–472; here cited from M. P. Schmidt (note 16), no. XXXVI.

31. Cleomedes describes the method using the "hydrologion" as one that was invented by the Alexandrians. With reference to a lost work of Heron of Alexandria, Pappus and Proclus also report such measurements with the help of water clocks: see W. Schmidt, *Herons von Alexandria, Druckwerke und Automatentheater* (Leipzig, 1899), 456 f. and 506 f.; Ptolemy, *Almagest* V.14, transl. by G. J. Toomer (New York, 1984), 252; Sextus Empiricus, *Against the Astrologers* V.24, 25, 75, ed. by G. Bury (London, 1987 [1949]), 332 ff.; O. Neugebauer, *A History of Ancient Mathematical Astronomy* (Berlin and New York, 1975), part 2, 657 f.; handed down to the Middle Ages in Macrobius, *Commentarii in Somnium Scipionis* I, 21.9–22: "duobus igitur vasis aeneis praeparatis, quorum alteri fundus erat in modum clepsydrae foratus," ed. J. Willis (Leipzig, 1970), 86 ff.; also in Martianus Capella, *De nuptiis* VIII.847: "multiplici enim clepsydrarum appositione monstratum omnia signa [signs of the Zodiac] paria spatia continere"; compare VI.860, ed. J. Willis (Leipzig, 1983), 320.

32. See D. R. Hill, *Arabic Water Clocks,* 8.

33. J. E. Armstrong and J. McK. Camp II, "Notes on a Water Clock in the Athenian Agora," *Hesperia* 46 (1977): 147–161 (here also the comparison with the clock in Oropos).

34. J. V. Noble and D. J. Price, "The Water Clock in the Tower of the Winds," *American Journal of Archeology* 72 (1968): 345–355, and plates 111–118.

35. See O. Benndorf et al., *Jahresheft des Österreichischen archäologischen Instituts* 6 (1903): 32–49; D. J. Price, in C. Singer, *History of Technology* III (Oxford, 1957), 617 f.; A. Rehm, "Neue Beiträge zur Kenntnis der antiken Wasseruhren," *Sitzungsbericht der Bayerischen Akademie der Wissenschaften, Phil.-hist.-Klasse,* 1917–1920 (1921).

36. D. J. Price, *Origins of Clockwork,* 92–94; *idem* in J. T. Fraser, *Study of Time*

II, 374–378; *idem, Gears from the Greeks;* A. G. Bromley, "Observations of the Antikythera Mechanism," *AH* 18 (1990): 641–652.

37. John Lydos, *De magistratibus* II.16 and III.35, ed. and transl. by A. C. Bandy (Philadelphia, 1983), 109 f. and 188 f.; the other topographical references in R. Janin, *Constantinople Byzantine. Développement urbain et répertoire topographique,* 2nd ed. (Paris, 1964).

38. See chapter 1, note 13.

39. J. Marquart, *Osteuropäische und ostasiatische Streifzüge* (Leipzig, 1903), 221 f.; M. Izzedin, "Un prisonnier arabe à Byzance au IXe siècle," *Revue des Études Islamiques* 15/1 (Paris, 1941–1946): 59.

Chapter 3

1. E. S. Kennedy, ed., *The Exhaustive Treatise on Shadows by al Biruni,* 2 vols (Aleppo, 1976); E. Wiedemann and J. Frank, "The Gebetszeiten im Islam" (1926–1927), in E. Wiedemann, *Aufsätze zur arabischen Wissenschaftsgeschichte,* vol. 2 (Hildesheim and New York, 1970), 757–788.

2. *Gisleberti Chronicon Hanoniense,* ed. G. H. Pertz (MGH script. rer. Germ. 29), 188–191; the reference in M. Bloch, *Feudal Society,* transl. by L. A. Manyon, vol. 1 (Chicago:University of Chicago Press, 1961), 74.

3. Bilfinger, *Horen,* 68 ff.; Bilfinger's findings are confirmed by Lehner, *Tagesteilung,* passim; M. Fallet-Scheurer, "Zeitmessung," 254 ff.; B. Ejder, *Tider och Maltider* (Lund, 1969).

4. E. M. Casalini, "Condizioni economiche a Firenze nel Duecento," *Studi storici dell'Ordine dei servi di Maria* 9 (1959): 3 f.

5. A. Lecocq, *Histoire de cloître de N.-D. de Chartres* (Chartres, 1867), 127 ff.

6. A. Chéruel, *Normanniae nova Chronica* (Caen, 1851), 31 f.

7. L. Mumford, *The Myth of the Machine* (New York, 1967), 264; this is also where he uses the expression the "etherealized and moralized megamachine."

8. Mumford, *Technics and Civilization* (New York, 1963), 12–18, 470.

9. H. E. Hallam, "The Medieval Social Picture," in E. Kamenka and R. S. Neale, *Feudalism, Capitalism, and Beyond* (London, 1975), 29–49.

10. E. Zerubavel, "Benedictine Ethic," and *Hidden Rhythms: Schedules and Calendars in Social Life* (Chicago, 1981), 31–69.

11. The importance of the day as the basic unit is underscored by the parable-like statement that the brother who entered the monastery in the second hour of the day will be junior in rank to one who entered in the first hour, regardless of age and former position (The Rule of St. Benedict, chap. 63).

12. Detailed and in part schematic descriptions of the daily monastic routine in: J. Biarne, "Le Temps du Moine d'après les premières règles monastiques d'Occident (IVe–VIe siècles)," in *Temps Chrétien,* 99–128; P. Buddenborg, "Zur Tagesordnung in der Benediktinerregel," *Benediktinische Monatsschrift* 18 (1936): 98; C. Butler, *Benedictine Monasticism* (1921, repr. 1964), 275–290; D. Knowles, *The Monastic Order in England* (Cambridge, 1963), 714 f.; idem, "The Monastic Horarium 970–1120," *Downside Review* 51 (1933): 706–726; L. Thomas, *La Journée Monastique* (Paris, 1982); B. H. Rosenwein, "Feudal

War and Monastic Peace: Cluniac Liturgy as Ritual Aggression," *Viator* 2 (1974): 134 f.

13. Aptly described by P. Mesnage, in M. Daumas, ed., *Histoire Générale des techniques* II (Paris, 1962), 290: "La succession des offices était réglée suivant la durée du jour et de la nuit et n'imposait pas la mesure d'intervalles réguliers du temps." Erroneous is E. Zerubavel's interpretation that "the duration of most daily activities and events was not intrinsic to them"; banal are his observations that the sequence of offices was "purely artificial" and that their temporal location rested on "social convention alone"; see his *Hidden Rhythms* (note 10), 41. What would a "natural" sequence of offices look like?

14. Graphic examples for an intrinsic timing can be found in the Carthusian Consuetudines of Chartreuse (1128): from the middle of September to the beginning of February, Vigils (= Nocturn) was to last long enough to allow fifty psalms to be chanted without haste. From the beginning of February until Easter, they would be gradually shortened to a period that was sufficient for praying the "matutin sanctae mariae." The pause after Nocturn was to last seven penitential psalms. The interval between Sext and None during winter should be as along as the one between Terce and Sext in the summer, or like a "regularis hora" and two "sanctae Mariae," and so on; *Coutumes de Chartreuse* (Paris, 1984), 226 ff. Landes's description of the Hours in *Revolution,* 61, is off: "As the very names indicate most of these were designated and set in terms of clock hours." Landes makes this statement even though he sees the distinction between the name of the Hour and the time of its observance; see 404, note 22.

15. On the time punishments see O. Zimmermann, *Ordensleben und Lebensstandard. Die Cura corporis in den Ordensvorschriften* (Münster, 1973), 446 f.

16. Borst, *Astrolab,* 62 f.

17. H. E. Hallam (note 9); on the relationship between asceticism and profit see Schreiner, MS 39 f.

18. Compare, in contrast, once more E. Zerubavel, *Hidden Rhythms* (note 10), 62: "The Benedictines were clearly the first ones to have established a rigid temporal regularity that was not directly geared in any way to nature."

19. The Rule of St. Benedict, chap. 8: "Hiemis tempore . . . iuxta considerationem rationis octaba hora noctis surgendum est, ut modice amplius de media nocte pausetur et iam digesti surgant," ed. R. Hanslik (CSEL 75) (1977), 58; H. M. Hanssens, *Aux origines de la prière liturgique: Nature et genèse de l'office des Matines* (Analecta Gregoriana 57) (Rome, 1952).

20. "Locus sane intricatus et obscurus," Benedikt van Haeften, *S. Benedictus illustratus sive monasticarum disquisitionum,* Part II (Antwerp, 1644), lib. VII, disq. IV, p. 776; "non consentiunt commentatores," E. Martène (1690), in Migne, PL 66, col. 410; P. Delatte, *Commentarie sur la règle de St. Benoit* (Paris, 1913; repr. 1969), on chaps. 8 and 47.

21. On the early history see M. Trumpf-Lyritzaki, "Glocke," in *Reallexikon für Antike und Christentum* 11 (1981): 164 ff.

22. G. Durandus, *Rationale divinorum officiorum* (1284), I, 4.8 (no date,

Nuremberg edition, 1494); on the urban chronicles see, for example, *Cronica di Giovanni Villani* (Florence), ed. F. Gherardi Dragomanni, part 2 (Florence, 1845; repr. 1969), 324.

23. M. A. Schroll, *Benedictine Monasticism as reflected in the Warnfried-Hildemar Commentaries on the Rule* (New York, 1941), 107 ff.

24. Durandus (see note 22).

25. In a commentary to the *Dictionarius* of Johannes de Garlandi, ed. by H. Géraud, *Paris sous Philippe-le-Bel* (Paris, 1837; repr. Tübingen, 1991), 590.

26. There is no serviceable account of this genre and its development; an introduction is provided by C. W. Jones (chapt. 2, note 7), 3–113; J. Flamant, "Temps sacre et comput astronomique," in *Le Temps Chrétien,* 31–44; see also L. Thorndike, "Computus," *Speculum* 29 (1954): 223–238. A vivid account of the connection of chronological, liturgical, and historical questions can be found in A. Borst, "Ein Forschungsbericht Hermanns des Lahmen," *Deutsches Archiv für Erforschung des Mittelalters* 40 (1984): 379–477; recently published in an expanded and revised version under the title *Computus* (1990).

27. T. Tally, "Liturgical Time in the Ancient Church: The State of Research," *Studia Liturgica* 14/2–4 (1982): 34–51.

28. On the history of the Paschal controversy see C. W. Jones (chap. 2, note 7).

29. Ed. by C. W. Jones (chap. 2, note 7); new edition, C. W. Jones, ed., *Corpus Christianorum, Series Latina* 123 B (Turnhout, 1977).

30. K. Harrison, "Easter cycles and the equinox in the British Isles," *Anglo-Saxon England* 7 (1978): 5; differently, Borst, *Computus,* 33.

31. J. Wiesenbach, "Wilhelm von Hirsau. Astrolab und Astronomie im 11. Jahrhundert," in K. Schreiner, ed., *Festschrift Hirsau, St. Peter und Paul 1091–1991,* part 2: Zur Geschichte eines Reformklosters (Stuttgart, 1991), 109–156, here 119 ff.

32. See chap. 2, note 31; Johannes Scotus, *Annotationes ad Martianum,* ed. by C. M. Lutz (Cambridge, Mass., 1939), 138 f.; Remigius of Auxerre, *Kommentar zu Martianus Capella,* ed. by C. M. Lutz (Leiden, 1965), 138 f., 277.

33. According to Isidor, *Etymologiarum* V.29: "Momentum est minimum atque agustissimum tempus, a motu siderum dictum," ed. W. M. Lindsay (Oxford, 1962).

34. Bede, *De temporum ratione,* chap. 3: "De minutissimis temporum spatiis" (chap. 2, note 7).

35. Scattered references on divisions in the High Middle Ages in L. Thorndike, *History* II, 419, and III, 123 f., 266 f., 290, 298, 318. An early example of the use of hour-minutes appears in the chronicle of Robert de Monte in reference to the duration of a solar eclipse in the year 1181: "spatium unius hore equalis et triginta octo minutorum," ed. G. H. Pertz, MGH SS VI (1844), 532. A later example is the observation of an eclipse in Augsburg on January 7, 1415: "do was ain vinsternus zwischen 7 und 8 stund vor mittag und werot 15 minut," *Chronik des Hektor Mülich,* ed. by K. Hegel, Chroniken deutscher Städte 22, Augsburg 3 (Leipzig, 1892), 59.

36. P. Ruppert, *Die Chroniken der Stadt Konstanz* (Constance, 1891), 35; other, similar, examples can also be found there.

37. In G. Rucellai, *Zibaldone,* ed. by A. Perosa (London, 1960), 58; compare 56; compare the account of the same earthquake in the Chronicle of Hektor Mülich (note 35), 289 f.: "geworet hat als lang bis das man sprechen hat mügen den psalm miserere."

38. Bede, *De temporum ratione,* chap. 2, 182; a similar distinction is also made by Roger Bacon in the thirteenth century; see Borst, *Computus* (note 26), 67.

39. Nicholas of Cusa, *Philosophisch-theologische Schriften,* ed. by L. Gabriel (Lat.-German), vol. 3 (Vienna, 1967), 324.

Chapter 4

1. Dom Alexandre, *Ausführliche Abhandlung von den Uhren* (Lemgo, 1738; repr. Osnabrück, 1981), 16.

2. North, "Monasticism"; Landes, *Revolution,* 53–66.

3. See the bibliographic references in note 109.

4. Jean Froissart, *Li Orloge Amoureus,* ed. by A. Scheler, Oeuvres, Poésies vol. 1 (Brussels, 1870; repr. 1977), 53 ff.

5. In J. Spöring, *Die Uhr im Zytturn uff Musegk zue Luzern 1385–1535* (Lucerne, 1975), 17. The expression "woman's temperament" was not a casual phrase used in jest. It still appears a hundred years later in an account of the city of Solothurn (1487); see H. Morgenthaler, in *Anzeiger für Schweizerische Altertumskunde,* Neue Folge *22* (1920): 137.

6. Sketches: Milan, Bibl. Ambrosiana C. 221: "compositio . . . multiformis et communis est," *Johannis de Dondis Paduani Civis, Astrarium* (versio A), ed. by E. Poulle (Padua and Paris, 1988), 8.

7. Ibid., 105: "Quod si vellocior motus fuerit, aut ponderi detrahas secundum bonam extimationem, considerata vellocitatis quantitate, aut alliquid ponderis appendas ad frenum, quoniam illud alliqualiter tardat motum. Si vero tardior fuerit vel alliquid addas ponderi vel freno detrahas . . ." It follows from this, in my view, that there were two ways of regulating the movement through changes in weight, unless "frenum" referred also to the entire clockwork including the weights, as is shown in the illustration. In the *Almanus Manuscript,* ed. by J. H. Leopold, 46–49, a sketch of the "corona" of the second clock shows notches for weights on the horizontal beam.

8. Text: North I, 472 ff., commentary: North II, 330–334. Latin "stroba" is attested for the crossbow since the fifteenth century. On the basis of the formal and functional similarity between the "semicirculus" and the mechanism that holds the string of the crossbow, A. Lantink-Ferguson, *Nature* 330 (1987): 615, argues for a joint technological development, which, he maintains, is made more likely still by the fact that crossbow makers appear as early clockmakers. One difficulty with this hypothesis arises in the fact that only very few crossbow makers can be found among the early clockmakers.

9. North II, 332; D. J. Price, "Clockwork before the Clock," *Horological Journal* (Jan., 1956): 34, 35; there is evidence for the claim that the Leonardo

sketches are copies of plans originally prepared in the late fourteenth century for a royal palace clock in Paris; see North II, 331 f. Clues to the diffusion of this escapement type could come from an unedited manuscript referred to by North II, 333. Apparently it also mentions an escapement with a double crown wheel. "Nota quod hoc retardativum possit verti cum uniformitante, quia uniformitans possit esse rota duplicata et habere circumferenciis se respicientes in medio dentes . . ." This is a manuscript of German provenance from the end of the fourteenth or the beginning of the fifteenth century, today in Cracow: Bibl. Jagel. 551, fol. 44 v–491 r, "In composicione horalogii praemittenda sunt prius . . ."; E. Zinner, *Verzeichnis der astronomischen Handschriften des deutschen Kulturgebiets,* gives 1388 as the date and Erfurt as the place of origin, though without making clear the reasons for these attributions; see North III, Addenda, 276. The two manuscripts mentioned by L. Thorndike and P. Kibre, *A Catalogue of Medieval Scientific Writings in Latin,* rev. ed. (1963), are in fact a single manuscript: the Salzburg manuscript (St. Peter Inc. Nr. 800) was sold in the inter-war period and is today at Yale (MS 42).

10. A. Raes, *Introductio in Liturgiam orientalem* (Rome, 1947), 26 f.; on pp. 179 ff. it lists information on early printed editions of such "Horologia"; E. Mercenier and E. Paris, *La Prière des Églises de Rite Byzantin,* 2nd ed., part I (no date): reproduction of a Greek "Horologion."

11. For the English language sphere, and based on the linguistic material in the Middle English Dictionary, see A. G. Rigg, "Clocks, Dials and Other Terms."

12. G. G. Meersseman and E. Adda, *Manuale di Computo con Ritmo Mnemotecnico dell'Archdidiacono Pacifico di Verona,* Italia Sacra 6 (Padua, 1966), 145, 169; J. Wiesenbach (chap. 3, note 31), 121.

13. See the references to the frequency of the mention of Gerbert as the inventor of the wheeled clock in D. Enshoff, "Um mittelalterliche Uhren," *Studien und Mitteilungen zur Geschichte des Benediktinerordens und seiner Zweige* 53 (1935): 401–407.

14. "In Magdeburg horalogium fecit, recte illud constituens secundum quandam stellam, nautarem ducem, quam consideravit per fistulam miro modo." Thietmar von Merseburg, *Chronicon* VI.100, ed. R. Holtzmann, MGH Script. rer. Germ. N. S. IX (1935; repr. 1980), 393; see Borst, *Astrolab,* 55.

15. N. Bunow, ed., *Gerberti Opera Mathematica* (Berlin, 1899; repr. 1963), 25–28; H. P. Lattin, "Astronomy: Our View and Theirs," *Medievalia et Humanistica* 9 (1955): 14 ff.; compare Poulle, *Équatoires,* 499.

16. *De gestis regum Anglorum,* ed. by W. Stubbs, Rer. Brit. M. A. script. 90,1 (1887), 196.

17. North, "Monasticism," 397.

18. E. Jünger, *Das Sanduhrbuch,* 110, still mentions the possibility that Wilhelm was the inventor but then decides on Gerbert; J. Wiesenbach (chap. 3 note 31), with illustration p. 120.

19. *Benedicti Regula,* ed. by R. Hanslik (CSEL 75) (Vienna, 1977), 58, 124 f. Engl. transl. by Anthony C. Meisel and M. L. del Mastro, *The Rule of St. Benedict* (New York, 1975), 85, 61.

20. "De perfectione monachorum" 13, chap. 17, ed. by P. Brezzi and B. Nardi, *De divina omnipotentia et altri opusculi* (Florence, 1943), 286 ff.

21. "Totum officium divinum et dispositio domus quoad exercitia," *Caeremoniae Sublacenses,* chap. 4, ed. by J. Biarne, *Temps* (chap. 3, note 12), 117.

22. *Regula Magistri* (beginning of the sixth century), chap. 31.7 and elsewhere, ed. by A. de Vogüé, Sources Chrétiennes 106 (1964); see also J. Biarne, *Temps* (chap. 3, note 12), 117.

23. There are very few exceptions: according to a regulation of the Cistercians, later dropped again, in the summertime the monks should be awakened after their siesta with the help of the "horologium." "Liber usuum," chap. 83, the so-called standard text (1183–1188), ed. by P. Guignard, *Les monuments primitifs de la Règle Cistercienne* (Dijon, 1878); compare the Liber Ordininis Sancti Victoris, chap. 21 (note 61).

24. "Horologium stellare monasticum (saec. XI)," ed. by G. Constable, in CCM VI (1975), 4 f., described as the "clock of the poor," endowed with natural astronomical gifts in Petrus Viret, *Exposition Chrétienne* (Geneva, 1564), dial IX, 179.

25. Cassian, *De institutis coenobiorum* 2.17, ed. by J.-C. Guy (Paris, 1965), 88; S. C. McCluskey, "Gregory of Tours, Monastic Timekeeping and Early Christian Attitudes to Astronomy," *Isis* 81 (1990): 8 ff; "Horologium stellare monasticum," 1–18.

26. "sibi velut horologium metiatur" (note 20).

27. For example, *Regula Magistri* (note 22), chap. 31.7: "horologium conspicere"; compare chap. 56.18–21; *Smaragdkommentar* (ninth century): "frequenter aspicit horologium," ed. by A. Spannagel and P. Engelbert, CCMVIII (1974), 96 f.; E. Zinner, *Alte Sonnenuhren an europäischen Gebäuden* (Wiesbaden, 1964), with a detailed index of the extant pieces; A. A. Mills, "Seasonal-Hour Sundials," *AH* 19 (1990): 147–170.

28. "Bernhard Ordo Cluniacensis" I, chap. 51 (eleventh century), ed. by M. Herrgott, *Vetus Disciplina Monastica* (Paris, 1726); compare the prohibition in St. Albans around 1300 against the keeping of candle clocks ("candelarum et ollarum horologia") in the cells; *Gesta Abbatum Monastarii Sancti Albani,* ed. by H. T. Riley, Rer. Brit. M. A. script. 28.4.2 (1867), 100.

29. E. Wohlhaupter, *Die Kerze im Recht* (Weimar, 1940), 134 ff.

30. *Assers Life of King Alfred,* ed. by W. H. Stevenson (London, 1904), chaps. 103, 104, p. 89; William of Malmesbury, *De gestis regum Anglorum,* lib. II, ed. by W. Stubbs, Rer. Brit. M. A. script. 90.1 (1887), 133; see G. Langenfeldt, *The Historic Origins of the Eight Hours Day* (Stockholm, 1954), 27, who sees in this the legendary beginning of the ideal of the eight-hour workday; in my view his attempt to do that is not convincing.

31. A brief overview in Turner, *Water Clocks,* 117 ff.

32. *Vie de Saint Louis,* ed. by Daunou and Naudet, Recueil des Historiens des Gaulles et de la France, vol. 20 (Paris, 1840), 75 and 79.

33. See below chap. 8, A legendary decree by Charles V of France.

34. Cassiodorus (chap. 2, note 3), 39–42.

35. Codex Carolinus XV, chron 16, Migne PL 98, col. 159.

36. Cassiodorus, *Institutionum,* liber I, chap. 30.5, ed. by R. A. B. Mynors (Oxford, 1963), 77 f.; compare also the formulation later in Fredegar's poem "Desuper cylipsidra," ed. L. Traube, MGH Poetae latini aevi Carolini, part 3 (1896), 323.

37. "Qui haec rationabiliter vult facere, horologium aquae illi necessarium est." *Expositio Regulae ab Hildemaro tradita,* chap. 8, ed. by H. Mittermüller (Regensburg, 1880), 278.

38. *Consuetudines Floriacenses Antiquiores,* chap. 29 (between 1000 and 1026), ed. by K. Hallinger, CCM VII-3 (1984), 42.

39. Outflow clocks continued in use, though, as evidenced by the prohibition from the abbot of St. Albans against the use of "orologia ollarum" in the cells (1302–1308) (note 28).

40. "Ordo Cluniacensis" (note 28) I, chaps. 24, 52: "postquam horologium cecidit"; *Consuetudines Fructuarienses-Sanblasianae* (around 1100): "quando horologium cadit," ed. L. G. Spätling and P. Dinter, CCM XII-2 (1987), 153; the word remained in use for a long time: see, for example, O. Fina, ed., *Klara Steigers Tagebuch. Aufzeichnungen während des Dreißigjährigen Krieges im Kloster Mariastein bei Eichstätt* (Regensburg, 1981), 65: "morgen vor 4 als der wecker gefallen haben wir uns im Chor versamblet"; see also Goethe, *Faust* I, verses 1705–1706: "Die Uhr mag stehn, der Zeiger fallen, es sei die Zeit für mich vorbei" ("The clock may stop, the pointer falling, And time itself be past for me!").

41. An overview of the medieval Vitruvius tradition is given by C. H. Krinksy, "Seventy-eight Vitruvius manuscripts," *Warburg Journal* 30 (1967): 36–70; according to the data in this article, at least twenty-seven manuscripts of Vitruvius are extant from the period up to the end of the thirteenth century. Presumably from the pen of Theoderic, after 1099 abbot of St. Trond near Lüttich, comes a poetic paraphrase of Vitruvius's description of the Tower of the Winds in Athens with its sundials (compare p. 32 f.), mention of a "horoscopus," ed. by M. Manitius, *Neues Archiv der Gesellschaft für deutsche Geschichtskunde* 39 (1914): 157 ff., here 167. In the twelfth century, Petrus Diaconus writes a no-longer-extant excerpt from Vitruvius; see M. Manitius, *Geschichte der lateinischen Literatur des Mittelalters,* vol. 3 (Munich, 1965), 550 ff., 710 ff. H. Koch's *Kleine Untersuchung, Das Nachleben des Vitruv* (Baden-Baden, 1951), does not have more to say; compare note 90.

42. *Consuetudines Fructuarienses-Sanblasianae* (note 40), chap. 4: "Secretarius in nocte surgit, quando horologium cadit, sydera caeli si serenum est aspicit. Et si tempus surgendi est . . . ad horologium pergit, aquam de parva caldaria in maiorem proicit, funem et plumbum sursum trahit, scillam post hec percutit."

43. Nivardus, *Ysengrimus,* ed. by J. Mann, Mittellateinische Studien und Texte XIII (Leiden, 1987), 460.

44. Cistercians: chap. 68, "sonitus orologii"; chap. 74, "audito horologio"; chap. 83, "ad sonitum orologii excitatus"; chap. 96, "inspecta hora in horologio"; chap. 115, "horologium facere sonare." The oldest version (1130–1134),

ed. by B. Griesser in *Analecta S. O. Cist.* 12 (1956): 153 ff.; the so-called Standard Text (note 23); Premonstratensians: "Sacrista debet horologium temperare et ipsum facere sonare . . . ad se excitandum cottidie," ed. by P. F. Lefèvre and W. M. Grauwen, *Les statuts de Prémontré au milieu du XIIe siècle* (Averbode, 1978), 26 f.; *Regularkoniker Arrouaise* (ca. 1135), ed. by L. Milis, CC Cont. Med. 20 (1970), 171.

45. "Consuetudines domus Cisterciensis," chap. 28, ed. by B. Griesser, in *Anal. S. O. Cist.* 3 (1947): 138 ff.

46. *Gesta Abbatum a Thoma Walsingham,* ed. by H. T. Riley, Rer. Brit. M. A. script. 28.4.1 (1867), appendix p. 520.

47. Johannes Belethus, *Summa de ecclesiasticis officiis,* ed. by H. Douteil, CC Cont. Med. 41.41 A (1976), chap. 86, 156; G. Durandus, *Rationale divinorum officiorum* (Nuremberg, 1494), I, 4.11.

48. *Consuetudines Cluniacensium* (Redactio Vallumbrosiana), ed. by K. Hallinger, CCM VII-2 (1983), 136. A review of the other, unedited versions, which are kept in the municipal archive in Florence, has revealed this passage was changed in the thirteenth and fourteenth centuries.

49. *Ordo Cluniacensis* (note 28) I, chap. 51: "ut aliquando (horologium) fallatur ipse notare debet in cereo et cursu stellarum et etiam lune"; compare the list of missing devices for the punctual observance of night offices in Nicole Bozon, *Les contes moralisés* (ca. 1320), 145: "nul oriloge ne fieu ne chaundel," ed. by L. Toulmin and P. Meyer (Paris, 1889), 186.

50. *Opera de vita regulari,* ed. by J. J. Berthier (Rome, 1888), vol. II, chap. 1, p. 69.

51. *Acta Sanctorum,* Apr. 7, vol. I (Antwerp, 1675), 701.

52. The duties of the four "matricularii": "si horologium septimanarius horis debitis non tetenderit 6 d. persolvet"; Sens, Bibl. Mun., MS 6, fol. 16 verso; G. Juliot, "L'Horloge de Sens," *Bull. Soc. arch. de Sens* 9 (1867): 388, describes it as "une horloge à poids et à timbre." It cannot be ruled out that both descriptions might refer to a water clock , though nothing here supports such a supposition. A. Ungerer, *Horloges,* 158, turns this into an "horloge mécanique," which is even more improbable. On the basis of Ungerer's reference, Brusa, *Orologeria,* 19, cautiously places the terminus ante quem non for mechanical clocks at around 1200. This possibility cannot be definitively excluded, but it cannot be substantiated with the clock in Sens, which, judging from the sources, was no different from comparable clocks.

53. "Statuta visitatorum pro monasterio Rothonensi," chap. 3: "Horologium congruum habeant, quo temperato secundum statuta regule signo pulsato surgant fratres . . . ," ed. by B. Griesser, *Anal. S. O. Cist* 8 (1952): 354 f.; Letter of Henry of Newark, York: "Campane, in campanile minus legitime locate, cardas non habent, nec est horologium ecclesie sub debito regimine custoditum." *The Register of William Wickwane, Lordarchbishop of York 1279–1285,* ed. W. Brown, Surtees Soc. 114 (Durham, 1907), 287; the clock was also on a later occasion a topic of the visitations; compare the order (1365) of the visitor Johannes de Lergis—who was sent by Pope Urban V—to procure a clock for

the night offices of the clerics at the church in Beaucaire (Languedoc): "unum horologium bonum et sufficiens per quod possint horas cognoscere tam de die tam de nocte," A. M. Beaucaire, GG 45; compare A. Eysette, *Histoire administrative de Beaucaire,* vol. 2 (Beaucaire, 1889), 216 f.

54. Abelard, ed. T. P. McLaughlin, *Medieval Studies* 18 (1956): 260, 262; other instances where the clock is mentioned among the "ornamenta": Lincoln Cathedral, "Liber Niger" (1236), ed. by H. Bradshaw and C. Wordsworth, *Statutes of Lincoln Cathedral,* part. I (Cambridge, 1892), 285, 386, part 2 (1897), 160; 1295 London, St. Paul's Cathedral, inventory 1295, W. Dugdale, *The History of St. Paul's Cathedral,* 3rd ed. (1818), 330, 331, quoted from O. Lehmann-Brockhaus, *Lateinische Schriftquellen zur Kunst in England, Wales und Schottland* (901–1307), vol. II (Munich, 1956), nos. 2933, 2934; Augsburg, Cathedral of St. Maria, Liber Ordinationum (1305), "orologium bonum et bene instructum," *Mon. Boica* 35 (1847): 147.

55. See the recommendation (in 1238) of Robert Grosseteste, Bishop of Lincoln, to the abbot of Ramsay, Huntingdonshire, to remove the clock from the church before consecrating the latter, if possible; *Epistolae,* ed. by H. R. Luard, Rer. Brit. M. A. script. 25 (1861), 190 ff.

56. On the older discussion about origin and date see Bergmann, *Innovationen,* 69 ff.; a description of the contents of the codex can be found on pp. 243–245.

57. J. Millás Vallicrosa, *Assaig d'història de les idees fisiques i metemàtiques a la Catalunya Medieval,* Estudis Universitaris Catalans (Barcelona, 1931), 316–318; transl. and suggested reconstruction: Maddison, Scott, and Kent, "Medieval Waterclock," reconstruction: Farré-Olivé, "Medieval Catalan Clepsydra."

58. An unpublished text from the Vatican Library in Rome (Vat. lat. 5367) could offer additional clues. E. Zinner, *Frühzeit,* 6, describes it: "A thirteenth-century manuscript originating in France provides instructions for the construction of a wooden (?) water clock with cylinders and cords." According to L. White, Jr., *Medieval Technology,* 120, this was a very simple design: "A cord, with a float at one end and a counterweight at the other, passed around an axle which rotated the dial and operated the alarm." F. Maddison and B. Scott are preparing an edition of the text; see D. Hill, *Arabic Water-Clocks,* 126, note 5.

59. Farré-Olivé, "Medieval Catalan Clepsydra."

60. *Chronica Jocelini de Bracelonda,* ed. by H. E. Butler (London, 1949), 106 f.

61. *Liber ordinis Sancti Victoris Parisienis,* chaps. 21, 54, ed. by L. Joqué and L. Milis, CC Cont. Med. 61 (1984), 94, 224; compare Bilfinger, *Horen,* 148.

62. P. Sheridan, *Les Inscriptions sur Ardoise de l'Abbaye de Villers,* Extrait des Annales de la Société de Bruxelles, vol. 10 (Brussels, 1896); because of the exhaustive commentary and in part variant readings, the reader should also consult A. d'Haenens, "Le Clepsydre de Villers."

63. "Si tardaveris temperare horologim donec sol existat in medietate prime fenestre pones horologium super [l]itteram B," P. Sheridan, *Les Inscriptions,* 16.

64. P. Sheridan, *Les Inscriptions,* 20; one should bear in mind that not all problems of the text have been resolved. For example, the passage "Post festum sancti Martini hyemalis, Secunda stella equorum adherente super linim(ari) fenestre; Thome apostoli, prima rota suppressa, tectum . . ." (17), does not reveal whether the observation of the stars was also used in regulating the clock, or whether the clock might have had in addition also a dial ("rota") with the signs of the Zodiac.

65. *The Itinerary of Benjamin of Tudela,* ed. by M. N. Adler (London, 1907), 30, 32.

66. *Bernardi Abbatis Casinensis in regulam S. Benedicti expositio ad c. 8,* ed. by A. M. Caplet (Monte Cassino, 1894), 172.

67. H. R. Hahnloser, *Villard de Honnecourt,* 30–32 and 349; Hahnloser refers also to the clock case in Beauvais (fig. 82), though in my view that case, with its narrow pedestal construction, bears no resemblance to Villard's sketch; the broad base in the latter seems rather better suited for housing a water tank. On the "earliest clockwork with wheels" that is supposedly also present in Villard's sketchbook, see below chap. 4, The "angelic clockwork" of Villard de Honnecourt.

68. A. J. Turner, *Water Clocks,* 28, gives 1223–1226 as the date; L. White Jr, *Technology,* 120 f., dates it to around 1250.

69. See G. Hentschel, *Kommentar zum Alten Testament, 2. Könige* (Würzburg, 1985), 99; Ahaz, 735–728 B.C.; Hezekiah, 728–700 B.C.

70. C. B. Drover, "A Medieval Monastic Water Clock," *AH* Dec. 1954: 54–58, 63 ff. (repr. in *AH* 12 (1980): 165–170); L. White, Jr., *Technology,* 120 f.; North, "Monasticism," 383 f.; A. W. Sleewyk, "The 13th Century 'King Hezekiah'-Water-Clock," *AH* 11 (1979): 488–494; C. B. Drover, "The 13th Century 'King Hezekiah'-Water-Clock," *AH* 12 (1980): 160–164 (with a reference to an approximately contemporaneous miniature in the "Biblia de Saint Louis" in the cathedral library of Toledo [vol. I, fol. 151], MAS Barcelona 79228, where the drum is divided into twenty-four chambers, and in a copy from around 1300 in London BL Add. MS 18719, fol. 92); J. J. Combridge, "Addenda" (with corrections on the dates), *AH* 12 (1980): 300; idem, "Further Addenda," *AH* 1991: 290 ff. I am limiting myself to presenting the most important arguments. None of the articles mentioned discusses the possible implications of the use of divisions of the day that were equal for technical reasons.

71. See especially Wiedemann and Hauser, *Uhren,* but also E. Wiedemann, *Aufsätze zur arabischen Wissenschaftsgeschichte,* 2 vols. (Hildesheim and New York, 1970); D. R. Hill, *Arabic Water-Clocks,* and the studies by the same author listed in the bibliography. An overview is provided by A. J. Turner, *Water Clocks,* 17–24, where we also find a reference to an "hourly self-sounding bell" which excited the admiration of the pilgrims David, Teilavus, and Paternus in Jerusalem in the fifth or sixth century: ". . . cimbalum, magis famosum quam pulchrum; quia dulci sono videtur excellare omne oranum [organum?]. Perjuros damnat, informos curat, et quod magis videtur mirabile, singulis horis,

nulla movente, sonabat." From a "Book of Llandaff" (twelfth century), in Peter Roberts, *The Chronicle of the Kings of Britain to which are added original dissertations . . .* (London, 1811), Appendix VI, 308a; also quoted by A. J. Turner, *Water Clocks,* 18, note 105; see also the account of Ridwan about the journey of the water clock from the Persians to the Greeks: Wiedemann and Hauser, *Uhren,* 180.

72. In addition to the works of Needham, Price, and Hill, see for an overview L. White, Jr., "Tibet, India, and Malaya as Sources of Western Medieval Technology," *American Historical Review* 65 (1959): 515–526.

73. *Annales regni Francorum,* ed. F. Kurze, MGH Script. rer. Germ. 6 (1895), 123 f.; the similarity holds even if we take into account that when H. Diels was giving a hypothetical reconstruction of the external appearance of the clock of Gaza, his mind's eye was seeing the so-called palace-clock of al-Jazari.

74. Hill, *Arabic Water-Clocks;* idem, in *Journal of Arabic Science* 1 (1977); D. R. Hill, *The Book of Knowledge of Ingenious Mechanical Devices* (Dordrecht, 1974); to which we must add a Greek-Arab construction description, "On the Construction of Water-Clocks," ed. by. D. R. Hill, Turner and Devereux Occasional Paper No. 4 (London, 1976).

75. See A. G. Drachmann, *Ktesibios, Philon and Heron: A Study in Ancient Pneumatics* (Copenhagen, 1948), 36–41. Drachmann does not use the term Pseudo-Archimedes, though D. R. Hill does: *Water-Clocks,* 15 ff.

76. "Tentorium mirifica arte constructum, in quo ymagines solis et lune artificialiter mote cursum suum certis et debitis spaciis peragant et horas diei et noctis infallibiliter indicant." *Chronica Regia Coloniensis Continuatio IV,* ed. by G. Waitz, MGH Script. rer. Germ. 18 (1880), 263; "celum astronomicum aureum gemmis stellatum, habens philosophicum intra se cursum planetarum." *Conradi de Fabaria Casus Galli,* ed. by G. H. Pertz, MGH SS 2 (1829), 178; compare E. Poulle (1980), 500; on the high esteem in which Frederick II held the mechanical arts see Sternagel, *Artes Mechanicae,* 58 ff.

77. Quoted, for example, by Bilfinger, *Horen,* 166; the error was noticed by L. White, Jr., *Technology,* 121, note 4; it was overlooked by Needham, *Science,* vol. 4, sect. 27, p. 532.

78. See R. Hammerstein, *Macht und Klang,* 83; compare M. Sherwood, "Magic and Mechanics in Medieval Fiction," *Studies in Philology* 44 (1947): 567–592.

79. Compare Maurice, *Räderuhr,* I, 42; compare J. Matthews, *DerGral* (Frankfurt, 1981), 24.

80. Compare the poetic description of a scaled water clock by al-Kushagim in Wiedemann and Hauser, *Uhren,* 30.

81. The two accounts are in F. Hirth, *China and the Roman Orient: Researches into Their Ancient and Medieval Relations* (Shanghai, 1885; repr. Chicago, no date), 53, 57, explanations 213 f.

82. Wiedemann and Hauser, *Uhren,* 176 ff.; Turner, *Water Clocks,* 69 ff.

83. D. J. de Solla Price, *Mechanical Waterclocks of the 14th Century in Fez, Morocco* (Ithaca and Paris, 1962).

84. F. Gabrieli, ed., Michele Amari, *Le epigrafi arabiche di Sicilia,* (Palermo, 1971), 29 ff.

85. Among the few smaller divisions of the hour mentioned in this context see the arc degrees (four minutes) mentioned in *The Book of the Scale of Wisdom,* Hill, *Water-Clocks,* 47 ff.; according to Borst, *Computus,* 50, around 980 the circle around Lupitus of Barcelona turned the prevailing usage on its head and called the temporal hours "crooked" or "artificial," and the equinoctial hours "straight" or "natural."

86. See, for example, the outflow clock attributed to Archimedes; Wiedemann and Hauser, *Uhren,* 32 ff.; by the same authors also *Uhr des Archimedes und zwei andere Vorrichtungen,* Nova Acta Leopoldina, Abhandlungen der deutschen Akademie der Naturforscher 103 (Halle, 1918).

87. The following condensed historial sketch, which leaves out many problems of detail and the development and variety of the instruments (equatorial?? Equatorien, quadrants, etc.), is based on Turner, *Astrolabes,* 1–57; and Borst, *Astrolab.* An introduction to the construction and use of the astrolabe is also given by J. D. North, The Astrolabe," *Scientific American* 230 (Jan, 1974): 96–106. The standard work on the history of the astrolabe, now in need of supplementation, is that of R. T. Gunther, *The Astrolabes of the World,* 2 vols. (Oxford, 1932). D. J. de Solla Price provides a provisional list of extant and datable specimens in "An International Checklist of Astrolabes," *Archives Internationales d'Histoire des Sciences* 3, ser. 8 (1955): 243–263, 363–381. The newer literature is given by Turner, *Astrolabes;* the older literature by E. Zinner, *Geschichte und Bibliographie der astronomischen Literatur in Deutschland zur Zeit der Renaissance* (Leipzig, 1941).

88. Wiedemann and Hauser, *Uhren,* 5; Turner, *Astrolabes,* 23.

89. The text of al-Bîruni is in E. Wiedemann, "Ein Instrument, das die Bewegung von Sonne und Mond darstellt nach al-Bîruni," *Der Islam* 4 (1913): 5–13; compare D. J. Price, "Origin of Clockwork," 97 ff.

90. E. J. Millàs Vallicrosa, *Assaig* (note 57), 271 ff.

91. De utilitatibus astrolabii, ed. by N. Bubnow, *Gerberti Opera Mathematica* (note 15), 114 ff.; on the discussion about the attribution of the work see Bergmann, *Innovationen,* 170.

92. "In horologicis et musicis instrumentis et mechanicis nulli par erat componendis," *Annales Bertoldi,* ed. by G. Pertz, MGH SS V (1844), 268; on the "horologium viatorum" see Bergmann, *Innovationen,* 170.

93. In quattuor libros sententiarum IV, 15.3, 4c, in R. Busa, ed., *Opera Omnia,* vol. 1 (1980), 514; Borst, *Astrolab,* 97.

94. After Borst, *Astrolab,* 92 f.

95. Bergmann, but also North and Turner, emphasize, despite all misgivings, the function of the astrolabe as a timekeeper. They are not unaware of the problem of the relatively difficult handling of the astrolabe, the rarity of documented cases, and the question concerning the accuracy of measurement required for astronomical observations, which would be conceivable in large, stationary instruments, but which is unattainable with the extant specimens

with a diameter between ten and forty centimeters. See the skeptical comments about the hour-lines engraved on the backs of astrolabes by J. D. North, "Astrolabes and the Hour-Line Ritual," *Journal for the History of Arabic Science* 5 (1981): 113–114. In contrast, E. Poulle, one of the scholars best acquainted with medieval astronomical literature, has repeatedly emphasized the role of the astrolabe and similar instruments in astronomical teaching and for practical astrology. See E. Poulle, "Le Quadrant Nouveau Medieval," *Journal des Savantes* (1964): 148, 204.

96. C. H. Haskins, "The Reception of Arabic Science in England," *English Historical Review* 30 (1915): 57 f.; a few of such time indications are not clear on the nature of the instruments that were used, for example, Annales Brunwilarenses: "1147 anno, 7. Kal. Novembris, die dominica, accidit eclipsis solis a tercia hora, et perseveravit usque post sextam; qua defectione horam pene integram fixus et immobilis ut in horologio notatum est, stetit," ed. by G. H. Pertz, MGH XVI (1859), 727.

97. Quadrans vetus; on its accuracy see M. Archinard, "The Diagram of the Unequal Hours," *Annals of Science* 47 (1990): 173–190.

98. Edition by E. Poulle, *Les Tables Alphonsines avec les canons de Jean de Saxe,* text, transl., and commentary (Paris, 1984).

99. See table 8, ibid., 123. According to the explanations of Johann of Saxony (around 1320), this table was needed so that one could, after having determined the time of day in equal hours by means of an astrolabe or sundial (?) ("per horologium"), locate the corresponding constellation in the tables that were consistently drawn up according to the minutae diei; ibid. 44 f.

100. Edited by M. Rico y Sinobas, *Libros del Saber de Astronomia del Rey D. Alfonso X. de Castilia,* 5 vols. (Madrid, 1863–1867); *Quecksilberuhr,* vol. IV, 67–76; compare Bilfinger, *Horen,* 150–160; A. J. Cardenas has corrected the text in "The complete Libro del Saber de Astronomia and Cod. Vat. Lat. 8144," *Manuscripta* 15 (1981): 14–22; German transl. by F. M. Feldhaus, "Die Uhren des Königs Alfons X von Spanien," *Deutsche Uhrmacherzeitung* 54 (1930): 608–612; Engl. transl. and description of an attempted reconstruction in A. A. Mills, "The Mercury Clock of the Libros del Saber," *Annals of Science* 45 (1988): 329–344.

101. D. J. Price, "Clockwork," 100–102; Needham et al., *Heavenly Clockwork,* 70 ff.; compare Hahnloser, *Villard,* plate 9 and pp. 24 f., 356 ff.

102. Compare E. Narducci, *Intorno una traduzione fatta nell'anno 1341 di una compilazione di Alfonso X. Re di Castiglia* (Rome, 1865).

103. On the later history of this clock type see S. A. Bedini, "The Compartmented Cylindrical Clepsydra," *Technology and Culture* 3 (1962): 115–141; and Turner, *Water Clocks,* 31–44.

104. "Et si quissieres que se muevan las campaniellas en cada ora, don hy XXIIII estacas, una a cada ora. Et desta guisa farás á dos oras, ó á tres, ó á quantas quissieres, compartiendi et cero á quantas oras quissieres." M. Ricos y Sinobas (note 100), vol. IV, 74.

105. Bilfinger, *Horen,* 156 ff.

106. These cultural contacts have been frequently described: see S. A. Bedini, "Oriental concepts of the measure of time," in Fraser, *Time* II, 452–484; K. Maurice, "Propagatio fidei per scientias. Uhrengeschenke an den chinesischen Hof," in K. Maurice and O. Mayr, eds., *Welt als Uhr,* 30–38; I. Schuster, "Uhren und Automaten: Medien und Spiegel kultureller Kontakte mit Ostasien," *Technikgeschichte* 52 (1985): 1–24; J. Needham et al., *Heavenly Clockwork,* 142 ff.

107. T. Spengler has outlined this image of China all the way to Max Weber and Benjamin Nelson, in Spengler, ed., *Joseph Needham, Wissenschaftlicher Universalismus* (Frankfurt, 1979), 10–33.

108. Needham, *Science and Civilization in China,* vol. 1, 243.

109. Compare J. Needham, "The Missing Link in Horological History: A Chinese Contribution" (1958), in Needham, *Clerks and Craftsmen in China and the West* (Cambridge, 1970), 203–238; Needham, *Science and Civilisation,* vol. 3 (1959), sect. 20 (Astronomy), and vol. 4 (1965), section 27 (Mechanical Engineering, Clockwork). A detailed account with a translation and thorough commentary on all important sources is in J. Needham et al., *Heavenly Clockwork.*

110. This is the phrase used by D. J. Price in many publications.

111. Turner, *Water Clocks,* 62.

112. Needham et al., *Heavenly Clockwork,* 199 ff.

113. To my knowledge there is no account of the history of Japanese clocks and timekeeping in any European language. Here I am relying on some scanty comments in N. H. N. Mody, *Japanese Clocks* (Vermont and Tokyo, 1932; repr. 1977), 23 ff., and on a survey of a number of different Japanese authors which Mrs. Yasuko Toyonaga kindly prepared for me.

114. Needham et al., *Heavenly Clockwork,* 51 ff.; compare, however, the new interpretation of the speed of the rotation (one hundred rotations a day) in Needham, *Science and Civilisation,* vol. 4, 461 f. Speculation about parallels suggests itself: the setting wheel of the clock according to MS Ripoll 225 had a division of the day into one-hundredths. The clock of Villers-le-Ville presumably had 4 × 24 fields on a dial. If the four alphabets were separated by empty fields, we would once again arrive at one hundred fields. Following a suggestion in the sixth century for remedying the irreconcilability of the two systems, in the Liang dynasty the number of "quarters" was reduced to 96; see *Heavenly Clockwork,* 199 note 1.

115. Compare the overview in J. Needham, *Science and Civilization,* vol. 3, 316; and the works of J. H. Combridge cited in Turner, *Water Clocks,* 165.

116. Marco Polo, *Il Milione. Die Wunder der Welt,* ed. by E. Guignard (Zurich, 1984), chap. 85, 135; compare H. Yule and H. Cordier, *The Book of Ser Marco Polo,* 3rd ed. (London, 1903), vol. 1, 378, and vol. 2, 215, where Martin Martinis Atlas Siniensis and other accounts are given.

117. Needham, *Hall of Heavenly Records.*

118. Landes, *Revolution,* 17 ff.

119. Needham, *Heavenly Clockwork,* 197; Needham, *Science and Civilization,* vol. 4, 541.

120. D. R. Hill, "A Treatise on Machines," 39 ff.; compare, for example, also J. H. Combridge, "Chinese Steelyard Clepsydras," *AH* 12 (1981): 530: "The Chinese Steelyard clepsydra per se was in fact a straightforward adaption of an Islamic device to Chinese horary requirements and traditional techniques."

121. See chap. 2, note 31.

122. "Motus horologiorum qui per aquam fiunt et pondera, quae quidem ad breve tempus et modicum fiunt et indigent renovatione frequente et aptatione instrumentorum suorum atque operatione forinseci," De anima, c. I, pars 7a, in *Guillelmus Alvernus, Opera Omnia* (Paris, 1647; repr. 1963), part II, suppl., 72.

123. Needham et al., *Heavenly Clockwork,* 188 ff.; and Needham, *Science and Civilization,* vol. 4, sect. 27, 538 ff.

124. Epistola de Magnete, ed. by G. Hellmann, *Rara Magnetica* (Berlin, 1898), 7 f.; the suggestion may date to before 1248, for Roger Bacon describes such a globe, though without the "drive." Epistola fratri Rogeri Baconis de secretis operibus artis et naturae et de nullitate magiae, chap. 6, ed. by J. S. Brewer, *Opera quedam hactenus inedita,* vol. 1 (London, 1859), Rer. Brit. M. A. script 15, 537; compare Hahnloser, *Villard,* 346–349.

125. Commentarius in Sphaeram, chap. 11: "Nec est hoc possibile, quod aliquod horologium sequatur omnino iudicium astronomie secundum veritatem. Conantur tamen artefices horologiorum facere circulum unum qui omnino moveatur secundum moitum circuli equinoctialis, sed non possunt omnino complere opus eorum, quod, si possent facere, esset horologium verax et valde et valeret plus quam astrolabium quantum ad horas capiendas vel aliud instrumentum astronomie, si quis hoc sciret facere secundum modum antedictum. Modus autem faciendi tale horologium esset, quod homo faceret unum circulum equalis ponderis ex omni parte secundum quod melius possibile esset. Postea quod appendatur pondus plumbeum axi ipsius rote, quod quidem pondus taliter moveat rotam istam quod motus ille compleatur ab ortu solis usque ad ortum preter tantum tempus per quantum oritur unus gradus fere secundum estimationem propinquam veritati." Ed. by L. Thorndike, "Invention;" improved text and transl. in Thorndike, *The Sphere of Sacrobosco and Its Commentators* (Chicago, 1949), 180, 230.

126. A brief explanation of the difference between solar time and sidereal time is given by R.-H. Giese, *Einführung in die Astronomie* (Darmstadt, 1981), 22 ff.

127. Ed. and transl. by J. D. North, "Opus Quorundam Rotarum Mirabilium," *Physis* 8, fasc. 4 (1960): 337–372; on the question of dating, the drive, and a detailed critique see Poulle, *Equatoires,* 500, 502, 641–657; North assumes the period around 1290 as the time of origin, Italy as the place, and thinks the device was intended for a hydraulic drive. Poulle dates it to the beginning of the fourteenth century and believes the drive was manual; in addition, according to Poulle this was presumably the first genuine mechanical simulation of the planetary orbits.

128. J. Labarte, *Inventaire du Mobilier du Charles V., Roy de France,* Coll. Doc. Inéd. sur l'Hist. de France (Paris, 1879), no. 2598; Galvano Fiamma, *Cronica ordinis Praedicatorum,* ed. by B. M. Reichert (Rome and Stuttgart, 1897), 107.

129. J. E. Viard, *Les Journaux du Trésor de Philippe le Bel* (Paris, 1940), nos. 2517–5964, passim, the entries concerning Pierre Pipelart and Gilebertus de Lupara (1299–1308).

130. "Un pieche de fer as ologes" (1300), "pour ouvrages faites as ologes, une roe refaire et i retenail a arrester les plons" (1316), "souffles des orloges" (1322) for a musical mechanism? J.-M. Richard, *Archives Départementales, Archives Civiles, Série A,* parts 1 and 2 (Arras, 1878–1885); J.-M. Richard, *Une petite-nièce de Saint-Louis—Mahaut Comtesse d'Artois et de Bourgogne (1302–1329)* (Paris, 1887), 368; compare M. Sherwood, "Magic and Mechanics" (note 78), 590, with references to later accounts of the park.

131. "Et refait sonner ses orloges / Par ses sales et par ses loges / A röes trop sotivement / De pardurable movement." Verses 21033 ff., ed. by E. Langlois (Paris, 1914).

132. "Indi, come orologio che ne chiami / ne l'ora che la sposa di Dio surge / a mattinar lo sposo perché l'ami / che l'una parte e l'altra tira e urge, / tin tin sonando con si dolce nota . . ." Paradiso X, 139–143; "E come cerchi in tempra d'oriuoli / si giran si, che'l primo a chi pon mente / quieto pare, e l'ultimo che voli," Paradiso XXVI, 13–15, ed. by U. Bosco and E. Reggio (Florence, 1979), 171, 398; compare Petrus de Ceffons Clarevallensis, *Lectura super quattuor libros sententiarum,* ca. 1350; comparison of a work of commentary to a clock: "artificialia contemplabar attendens qualiter ex unius rotae circumductae gyro plurium sequitur per intermedia volubilium rotarum motus," D. Trapp, "Peter Ceffons of Clairvaux," *Recherches de Théologie ancienne et médiévale* 24 (1957): 130; for the sake of brevity I will not discuss X, 145.6, "la gloriosa rota," where Dante could be hinting at an automatically moved procession of figures. XXXIII 144, "come rota ch'igualmente mossa," probably does not deal with a mechanism at all, but with the concept of a ideal circular movement; compare G. Boffito, "Dove e quando poté Dante vedere gli orologi meccanici che descrivi . . . ?" *Giornale Dantesco* 39 (1938): 45–61; and B. Nardi, *Nel Mondo di Dante* (Rome, 1944), 337–350.

133. Compare J. P. Lacaita, ed., *Benvenuti de Rambaldis de Imola Comentum super Dantis Aldighierji Comoedium,* vol. V (Florence, 1887), 47 f., 335 f.; correctly interpreted by E. Morpugo, "L'Orologeria italiana: i primi orologi meccanici (II)," *La Clessidra* 7–8 (1946); repr. in *La Clessidra* No. 3: 24–25.

134. It can be found in the school commentary of M. Porena (Bologna, 1946–1947) as an explanation of an "antico tipo" of the mechanical clock; it also appears in the monumental work *Enciclopediadantesca,* ed. by U. Bosco, vol. IV (Rome, 1973), s. v. "orologio"; also in the edition by U. Bosco and E. Reggio (Florence, 1979), 389; and in the German commentary of H. Gmelin (Stuttgart, 1957), 423.

135. Landes, *Revolution,* 57, where he also speaks erroneously of "these great clocks"; L. White, Jr., *Technology,* 124, note 3.

136. Edmund Colledge, transl. *Bl. Henry Suso. Wisdom's Watch Upon the Hours* (The Catholic University of America Press, 1994), 54., 364 f.; presumably from the second half of the fifteenth century comes the *Andechtig Zitgloegglin* by the Dominican Berthold (first printed Cologne, 1488); from the prologue: "Diss büchlin hat vieruntzwentzig stuck, ussgeteilt nach den xxiiij stunden des natürlichen tags, die der andechtig mensch zuo siner andacht bruchen und betrachten mag," as quoted by P. Künzle, *Heinrich Seuses Horologium Sapientiae* (Freiburg, 1977), 66. On Berthold's controversial biographical dates see *Die Deutsche Literatur des Mittlelalters. Verfasserlexikon,* 2nd. ed. (1978), vol. 1, 801 f.; different dates in Lexikon des Mittelalters I (1980), cols. 203 1 f.

137. K. Maurice, I, 6, considered this a sensible reading. In the meantime, however, the new edition of P. Künzle has appeared. According to Künzle's research, of the 233 manuscripts he examined, only 4 that date from the fifteenth century have the variant "rotis"; compare P. Künzle (note 136), 61. But this does not change the fact that Seuse is hinting at a clockwork with an artful bell work which he must have seen or heard about.

138. *Annales Prioratus de Dunstaplia,* ed. by H. R. Luard, Rer. Brit. M. A. script. 36.3 (London, 1866), 296; addendum to the chronicle of Jean Mathei Cacia, ed. by A. M. Viel and P. M. Girardin, 101.

139. *Annales Colmarienses Maiores,* ed. by H. Pertz, MGH SS 17 (1861), 203; J. Ancelet-Hustache, "Les Vitae Sororum d'Unterlinden," *Arch. d. Hist. Doctrinale et Littéraire du Moyen Age* (1930): 401; *Liber memorandorum ecclesie de Bernewelle,* ed. by J. Wilis Clark (Cambridge, 1907), 220; from Lehmann-Brockhaus (note 54) I (1955), no. 187; J. D. Knowles, *Religious Orders in England* I (Cambridge, 1962), 323; Mâcon: J. Laurent and G. Guigue, *Obituaries de la province de Lyon* (Paris, 1933; repr. 1952), 409; Augsburg: *Mon. Boica* 35 (1847), 147; Chorges: "pro quodam horologio empto de mandato domini nostri et dato monasterio b. Marie ord. Cartus. civit. Caturci," K. H. Schäfer, *Die Ausgaben der apostolischen Kammer unter Johann XXII* (Paderborn, 1911), 772.

140. Thomas de Luda, Thesaurar: "(ecclesiam horologio competenti . . . promisit facere construi) quod ecclesiae alie Cathedrales et Couentaluales vbique fere terrarum regulariter optinere noscuntur," H. Bradshaw and C. Wordsworth, eds. (note 54), pt. 1, 350; pt. 2, CXVI f.

141. 1269–1270 "Orologiarius," S. F. Hockey, *The Account Book of Beaulieu Abbey [Hampshire],* Camden Soc. 4th ser. 16 (London, 1975), 235, 240, 299; "orologiarius/orologus," London 1286, Beeson, *Church Clocks,* 14; "custodibus pro horologio/orologiarius" 1318–1333, J. Houdoy, *Histoire artistique de la Cathédrale de Cambrai* (Lille, 1880; repr. 1972), 158.

142. Zinner, *Räderuhren,* 22, and *Frühzeit,* 6; Bassermann-Jordan, *Uhren,* 358; Lynn White, Jr., *Technology,* 120; Hill, *Water-Clocks,* 126; Turner, *Water Clocks,* 28; Needham, *Science and Civilization,* vol. 4, sect. 27, 544; North, "Monasticism," 382.

143. E. Volckmann, "Köln als ältester Sitz deutscher Uhrmacherei," *Die Rheinlande. Monatsschrift für deutsche Kunst und Dichtung* 28 (1918), 120, 121, and *Alte Gewerbe und Gewerbegassen* (Würzburg, 1921; repr. 1976), 128–131.

144. H. Keussen, *Topographie der Stadt Köln im Mittelaler,* 2 vols. (Bonn, 1910), vol. 1, 442–444.

145. "vvrtafel" (1374), R. Knipping, *Die Kölner Stadtrechnungen des Mittelalters,* 2 vols. (Cologne, 1898), 167; "u(y)rklockenmecher" (1395–1418), K. Militzer, *Die vermögenden Kölner Bürger 1417/1418* (Cologne, 1981), 216, 222.

146. See H. Keussen, 127* on the characteristic features of streets where smiths lived.

147. K. Maurice, *Räderuhr,* I, 43.

148. Leclercq, "Zeiterfahrung," 11 f.

149. M. Lexer, *Mittelhochdeutsches Handwörterbuch,* vol. 2 (Leipzig, 1876), cols. 2007 f.

150. E. Verwijs, J. Verdam, *Mndl. Woordenboek,* vol. 5 (The Hague, 1903), col. 1992.

151. Landes, *Revolution,* 70.

152. "Horologium regis" (note 129).

153. Humbertus de Romanis, *Opera* (note 50), chap. I, p. 69; chap. X, p. 248.

154. Hieronymus de Moravia, *De musica* (before 1304): "ars formandi campanulas sonos musicos exprimentes in horologiis ponendas," ed. by D. Coussmemaker, Scriptorum de musica medii aevi n.s. (Paris, 1864; repr. 1931), 72; Engelbert, Abbot of Admont (Styria), *De musica* tr. II, c. 4: "pondera cymbalorum horologii," ed. by M. Gerbert, Scriptores ecclesiastici de musica II (St. Blasien, 1784; repr. 1963), 301.

155. Exeter: A. M. Erskine, *The Accounts of the Fabric of Exeter Cathedral 1279–1353,* part I (Exeter, 1981), 7 f.; Howgrave-Grahem, "Clocks," 296; Hesdin (note 130); London: F. Madden, *Archeological Journal* 12 (1855): 173 f.; Genoa: compare chap. 5, The first public clocks. Blankenburg: K. Steinacker, *Die Bau- und Kunstdenkmäler des Kreises Blankenburg* (Wolfenbüttel, 1922), 26; Caffa: L. T. Belgrano, "Degli antichi orologi pubblici d'Italia," *Arch. Stor. Italiano* 7 (1868): 53 f.

156. L. Morillot, *Étude sur l'emploi des clochettes chez les anciens et depuis le triomphe du Christianisme* (Dijon, 1883), 139 f.

157. Into this context belongs also a miniature from the Psalter of Belvoir Castle (thirteenth century), which shows a pneumatic organ and a bell-slat; see E. Buhle, *Die musikalischen Instrumente in den Miniaturen des frühen Mittelalters,* vol. 1 (Leipzig, 1903), 70 and fig. XIV; it is also known as the Rutland-Psalter; see Hammerstein, *Tönende Automaten,* 95.

158. *Chronicon Windeshemense* s. a. 1404, ed. by K. Grube (Halle, 1886), 165.

159. See the Latin-German Glossary of 1547 (?): "Cimbalum / cimbel / est instrumentum habens multas campanellas sub proportione magnitudine commensuratas in uno ligno circulari suspensas sicut videtur in quibusdam horologiis monachorum," quoted by W. Theobald, *Technik und Kunsthandwerk im zehnten Jahrhundert* (Berlin, 1933), 441; as a "zimbelstern" the bell-wheel remained a popular side-register in organs right up to the end of the eighteenth century; see Hammerstein, *Tönende Automaten,* 98 ff., where the author also

quotes the severe critique in Arnold Schlick's *Spiegel der Orgelmacher* (1511): "Auch umblauffendt stern mit schnellen klinglen und anders gehört nit in die Kirche / aber wo unser hergot kirchweyhung hält / richtet der teuffel sein schragen darneben uff."

160. The relief is often referred to in the horological literature; most recently Needham described it as a "representation of a clock," *Science and Civilization,* vol. 4.2, 544; a possibly mechanically driven bell-wheel is also depicted in a miniature to Psalm 80 in Oxford Bodl. Libr. MS Douce 211, fol. 265 (fig. 3 in A. W. Sleeswyk, note 70): King David is striking with two hammers a bell driven by a horizontal shaft with fifteen studs.

161. Ed. by A. Chérnel (Caen, 1851), 31.

162. G. Chaucer recommended the astrolabe "for the governance of a clockke" in *A Treatise on the Astrolabe* (1391), ed. by W. W. Skeat (London, 1872; repr. 1872), 3, 17.

163. Compare A. Simoni, "Un nuovo documento per la storia dell'orologeria," *La Clessidra* 24 (1968), No. 4: 18–21, where we also find the comment concerning the architectural similarity with the case of Villard's clock-house; on the miniature compare H. Michel, "L'Horloge de Sapience et l'histoire de l'horlogerie," *Physis* 2 (1960): 291–298; E. P. Spencer, "L'Horloge de Sapience, Bruxelles, Bibliothèque Royale, MS IV, 111," *Scriptorium* 23 (1963): 277–299 and plates 21–30; P. R. Monks, *The Brussels Horloge de Sapience* (Leiden, 1990), 53 ff.

164. Illustrations in K. Maurice, *Räderuhr,* II, figs. 34–46.

165. J. Drumond Robertson, *Evolution,* 15 ff.; North, "Monasticism," 393 ff.; North's view that the Wallingford escapement was historically earlier though mechanically superior and more expensive and for that reason supplanted by the foliot escapement is, in my opinion, not tenable. The cost of two small pinwheels could not have been the cause of the almost complete disappearance of this kind of escapement. We simply do not know whether this represents an older type; see G. Brusa, *L'Arte del Orologeria,* 22 f.

166. Compare Hahnloser, *Villard,* 134 f., 370 f., plate 44, and fig. 92.

167. C. Frémont, "Un échappement d'horloge au treizième siècle," *Comptes-rendus de l'Académie des Sciences* 161 (1915), 69 ff; Usher, *Mechanical Inventions,* 193 f.; A. Lloyd, in Singer, *History of Technology,* vol. 3, 648 f.; Zinner, *Frühzeit,* 7; B. Gille in *Histoire Générale des Techniques,* vol. 1 (Paris, 1962), 593, and *Ingenieurs* (1978 ed.), 26 f.; also J. Gimpel, *Medieval Machine,* 130 f.

168. White, *Medieval Technology,* 173; K. Maurice, *Räderuhr* I, 50; Needham et al., *Heavenly Clockwork,* 195; skeptical also is S. Lilley, *Men, Machines and History* (1948; London, 1965), 56.

169. M. Daumas, "Le faux échappement de Villard de Honnecourt," *Rev. d'Hist. des Sciences* 35 (1982): 43–53; P. Mesnage in *Histoire Générale des Techniques,* ed. by M. Daumas (Paris, 1965), 292.

170. The catchphrase "clock and organ as focal point for automata" has been taken from R. Hammerstein, *Macht und Klang,* 106.

171. A. G. G. Thurlow, "The Bells of Norwich Cathedral," *Norfolk Archeology* 29 (1946): 89 ff.; "Magnum horologium processionibus & spectaculis insigni-

tum & organa mirae magnitudinis in eadem construxit," Chronica sive historia de rebus Glastoniensibus s.a. 1342–1325, ed. by T. Hearne (Oxford, 1726), quoted from R. P. Howgrave-Graham, "Clocks," 288.

172. Paris, Arch. Assistance publique, Fonds St.-Jaques-aux-Pelerins: H. Bordier, L. Brièle, *Les archives hospitalières,* vol. 2 (Paris, 1877), 130 ff.; L. Brièle, *Inv. som. Archives Hospitalières,* vol. 3 (1886); on this see also the handwritten inventory by M. Bordier; London 1344 (note 155); Mondinus: L. De Mas Latrie, *Nouvelles Preuves de l'Histoire de Chypre,* I (Paris, 1837), 56 f.; C. Dehaisnes, *Documents et extraits divers concernant l'histoire de l'art dans la Flandre etc.*, 2 vols. (Lille, 1886), 365 f., H. Houdoy (note 141), 159, 173; Strasbourg: ed. by C. Hegel, Chroniken der deutschen Städte 8 and 9 (Leipzig, 1870–1871), 133, 725, G. Oestmann, "Die zweite Uhr des Straßburger Münsters: Funktion und Bedeutung," Phil. dissertation (Hamburg, 1991); J. Fuchs, "Die Villinger Münsteruhr," *SFAU* 10 (1970–1971): 59 ff.

173. H. Diemar, *Die Chroniken des Wiegand Gerstenberg von Frankenberg* (Marburg, 1909), 433.

174. "Attenta alta fama et auctoritate hujus ecclesie Carnotensis que hactenus pluribus et dulcis extitit jocalibus et notabilibus operibus decorata, capitulum voluit et ordinavit quod de novo zodiacus et quadrens dicte ecclesie reficiatur pulcriori forma . . ." A. Lecocq, "Notice historique et archéologique sur les Horloges de l'église N.-D. de Chartres," *Mém. Soc. Archéol. de Eure-et-Loir* 4 (1867), 329.

175. Compare "Monumentale Astronomische Uhren," in K. Maurice I, 38 ff., from which I have also taken the phrase "monumental astronomical clocks."

176. Galvano Fiamma, *Opusculum de rebus gestis ab Azone Luchino et Johanne vicecomitibus,* ed. by C. Castiglioni, R. I. S. 12.4 (Bologna, 1938), 16.

177. Ibid., 41.

178. *Annales Mediolenses Anonymi,* Muratori 16 (1730), 710.

179. See page 126; quite erroneous in its generalization, and insufficiently informed about England, is the chronicle entry of a monk from Malmesbury under the year 1373: "Hoc anno horologia distinguentia 24 horas primo inventa sunt," Anonyme Continuatio of the Eulogium historiarum (first half of the fifteenth century), ed. by F. Scott Haydon, Rer. Brit. M. A. script. 9, 3 (London, 1863; repr. 1967), 336.

180. This distinction, clearly brought out by G. Bilfinger, *Horen,* 157 and elsewhere, has not always been heeded, and since the notice in the Milanese chronicle (1336) has at the same time become a "locus classicus" for the appearance of mechanical clocks as such, it has led to misunderstandings. For example, H.-U. Grimm ("'Zeit' als 'Beziehungsymbol': Die soziale Genese des bürgerlichen Zeitbewußtseins im Mittelalter," *Geschichte in Wissenschaft und Unterricht* 37 (1986): 199 ff., 219 f.), thinks he must criticize Wendorff (*Zeit und Kultur,* 136) because the latter's conjecture that the appearance of the wheeled clock may have been accidental and should presumably be located in the medieval monasteries "is utterly outside of the research conducted on a social scien-

tific basis." Wendorff's speculation, he claims, can "be neither exactly confirmed nor theoretically substantiated." The first point is true; the second, if we leave aside the somewhat pretentious "theoretically," is wrong. Grimm, by contrast, wants to stick with the "causal chain that is largely undisputed by the experts," namely, that the need to divide the civic day was the cause for the appearance of the new clocks. Grimm overlooks here the difference between "mechanical clock" and its special form the "striking clock." Talk of historically conditioned necessities leads, in my view, inevitably to contradictions of the type "post hoc ergo propter hoc." As one instance of numerous similar misunderstandings this scholarly dispute would not deserve further mention. What makes Grimm's arguments annoying are his confident invocation of the "experts," of "research conducted on a social-scientific basis," and of the "largely undisputed causal chain" it has constructed. Grimm is simply fudging. Nobody has ever tried to prove such causal chains, or even merely a preceding civic need for a division of time.

181. R. E. F. Smith, "Time, Space and Its Use in Early Russia," in *Social Relations and Ideas,* Essays in Honour of R. H. Hilton (Cambridge, 1983), 281; compare E. V. Williams, *The Bells of Russia* (Princeton, 1985).

182. A. G. Keller, "A Renaissance Humanist" (see chap. 1, note 11), 349.

183. Compare the sequence of signals mentioned above chap. 3, The Hours as a time signal.

184. On the different attempts by Richard of Wallingford see North II, 344 ff.

185. Johannes Noviomagus (Jan van Bronckhorst), *Bedae . . . opuscula* (Cologne, 1537), fol. 33v: the Romans did not use equal hours; "nisis ad clepsydram, nos autem iis [equal hours] utimur nunc dato signo per rotarum ferrearum machinas, quae ad temporales horas accomodari non possunt."

186. See N. H. N. Mody, *Japanese Clocks* (1932; Tokyo 1977), 27 f. and plates 14–16.

187. In 1944, Antonio Simoni discovered in a workshop sketchbook of the clockmaking family of the Volpaia a locking-plate with a special stroke sequence (3-2-1-2-3-4) and related it to the ringing of the Hours (A. Simoni, "L'evoluzione del suono nei primitivi orologi," *La Clessidra* 11.1 [1955]); E. L. Edwardes popularized this as "the key to medieval striking," and assumed that it also existed in the clock mentioned in London in 1286 (note 141), *Chamber-Clocks,* 39–46; J. H. Leopold subsequently identified them as auxiliary locking-plates for the 300 daily strokes according to the Italian striking, *Almanus-Manuscript,* 26; A. Simoni thereafter corrected himself, though in an out-of-the-way place, *La Clessidra* 22.12 (1966); compare K. Maurice I, 76 and fig. 20; G. Brusa, *Orologeria,* 18; for the passing on of the misunderstanding see, for example, A. Cordiolani, in C. Samaran, ed., *L'Histoire et ses methodes* (Paris, 1967), 40: "au XVe siècle encore, les balanciers des horloges etaient modifiés soir et matin pour diviser la nuit puis le jour en partie inégales."

188. Bilfinger, *Horen,* 175 ff.

189. What is frequently listed in the horological literature as the earliest ex-

ample of the new reckoning of the hours does not contain a modern hour-indication and probably does not refer to a clock. 1306 in Salisbury (Sarum): according to a regulation concerning the selling of goods issued by the bishop, nobody was to purchase meat, fish, or other foodstuffs before the clock of the cathedral had struck "one." But this is evidently not a clock signal at all, but the ringing of Prime, which we know from many similar regulations; the paraphrase in R. Benson and H. Hatcher, *Old and New Sarum or Salisbury* (London, 1843), 76; for a similar regulation compare, for example, London around 1300, *Liber Albus,* ed. by T. H. Riley, Rer. Brit. M. A. script. 12.1 (London, 1859), 270: "Et que nul regratour de blee, de pessoun, ne de poleterie, achate pur revendre viaund avant houre de pryme sone a Seint Poul."

190. The artistic Byzantine clock in Gaza, which was surely no longer known at the beginning of the fourteenth century, is to my knowledge the only clock for which a striking of the hours according to their sequential counting is attested.

191. Campanus, *Computus maior,* 1518 edition, c. 6, fol. 160 v, written between 1261 and 1264 at the request of Pope Urban IV, discussed the advantages of the equal hours, but mentions only the heavenly phenomena; cited after North II, 339.

192. A more detailed discussion can be found in Bilfinger, *Horen,* part II.

193. Bilfinger, *Horen,* 141 ff.; also Y. Renouard, *Hommes d'Affaires* (chap. 1, note 25), 239.

194. In Orvieto in 1351, a priest was paid for the "scriptura carte horarum mensium pro arlogiis," and a parchment was delivered on which "scripte fuerunt hore mensium et dierum pro arlogiis," according to L. Ricetti, ed., *Il duomo di Orvieto* (Rome, 1988), 213.

195. J. W. Goethe, *Tagebuch der italienischen Reise* (1786), ed. by C. Michel (Frankfurt, 1976), 69 ff.

196. Reduction of the striking work in Sisteron in 1417, E. de Laplane, *Histoire de Sisteron,* vol. 1 (Digne, 1843), 246 ff.; in Chambery in 1514, P. André, "Notice historique sur l'église paroissale de St.-Léger à Chambery," *Soc. Savoisienne d'Hist. et de Arch., Mém. et Doc.* 7 (1863): 104 ff.; "more ultramontanum" in Piacenza in 1534, A. Rapeti, "L'orologeria in Piacenza," *La Clessidra* 7 (1951): 13 ff.; in Savona in 1539, A. Bruno, *Storia di Savona* (1901), 62; installation of a double striking-system in Trento prior to the Council around 1540, A. Gorfer, *Trento, Città del Concilio* (Trento, 1963). Noticeable are also the frequent changes that occurred in the geographic regions where various methods of counting overlapped; they could also indicate changes in political-economic orientation. On the controversy over the introduction of the "orologio oltramontano" in Italy, demanded in the eighteenth century by forces of the European Enlightenment, see Simoni, *Orologi Italiani,* 51 ff.

197. C. T. Gemeiner, *Regensburgische Chronik* (Regensburg, 1803–1821; repr. 1971), II, 181.

198. Such ringing tables and hour-comparison tables are reproduced in K. Maurice, I, 32, 33.

199. L. Guicciardini, *Descrittione di tutti i paesi bassi* (1567), quoted from the Dutch edition (Amsterdam, 1612; repr. 1968), 30.

200. Bilfinger, *Horen,* 253 ff.; Fallet-Scheurer, "Zeitmessung," 297 ff.

201. For example, in the *Annales Matseenses,* ed. by E. H. Pertz, MGH SS 9 (1851), 829, where an earthquake which, according to other sources, took place in the afternoon, was dated to "hora diei naturalis quasi 14" Zinner, *Räderuhren,* 41, speculates here on the use of a wheeled clock, which is entirely unlikely.

202. A. Pinchart in *Archives des Arts, Sciences et Lettres* I.1 (Ghent, 1860), 175.

203. The sandglass is still dated into the Middle Ages by Usher; for example, *Inventions,* 188; Zerubavel, "Hidden Rhythms" (chapt. 3, note 10), 38, sees it in use in monasteries. Though outdated in many details, E. Jünger, *Sanduhrbuch,* is still worth reading; for an up-to-date survey of our current understanding see Turner, "Accomplishment" and *Water Clocks,* 75 ff.; also: R. T. Balmer, "The Operation of Sand-Clocks and Their Medieval Development," *Technology and Culture* 19 (1978): 615–632 (critique in Turner, "Accomplishment"); C. K. Aked, "Kurze Geschichte der Sanduhr," *AU* 1 (1980): 22–37.

204. Barberino: ed. by F. Egidi, vol. 1.3 (Rome, 1913), 124 f.; Sluys: N. H. Nicholas, *A History of the Royal Navy* (London, 1847), 467; a "zegher," a sandglass? appears in a last will in Hamburg in the year 1353, H. D. Loose, *Hamburger Testamente 1351–1400* (Hamburg, 1970), 2; Charles V: "Ung grant orloge de mer, de deux grans fiolles plains de sablon en un grant estuy de boys garny d'archal," J. Labarte (note 128), no. 2120.

205. Beeson, *Perpignan,* 51 f.

206. M. Mauro y Gaudo, ed., *Ordinaciones de la Ciudad Zaragoza* II (Zaragoza, 1908), 465.

207. See on this the works of Poulle and North.

208. "Orologium portabile (or) portatum," K. H. Schäfer, *Die Ausgaben der apostolischen Kammer unter Urban V.* (Paderbron, 1937), 166, 204; "orloge portative," L. Delisle, *Mandements et actes diverses de Charles V. (1364–1380)* (Paris, 1874), no. 1561, p. 778 f.

209. B. Prost, "Liste des artistes mentionnés dans les États de la Maison du Roi et des Maisons des Princes, du XIIIe siècle à l'an 1500," *Arch. Hist. Artist. et Litt.* I (1889–1890), 436; in 1481 the French king paid for a clock "pour porter aveques luy par tous les lieux ou il yra," M. L. Douet-d'Arcq, *Comptes de l'Hôtel des Rois de France au XIVe et XVe siècle* (Paris, 1865), 388; J. Vielliard, "Horloges et Horlogers Catalans à la fin du Moyen Age," *Bull. Hispanique* 63 (1961): 166.

210. This chapter in the history of the clock has been written many times. Detailed accounts: S. G. Atwood, "Die gewundenen Antriebsfedern von Zeitmeßgeräten," *AU* 4 (1981): 83–97; K. Maurice, *Räderuhr,* I, 81 ff.; D. S. Landes, *Revolution,* 85 ff.

211. *The Life of Brunelleschi by Antonio di Tuccio Manetti,* ed. by H. Saalman (University Park and London, 1970), 51.

212. For a depiction of a "snail" with a spring that was pulled out in a spiral shape (ca. 1450–1480), later uncommon, see Maurice, *Räderuhr* II, fig. 79.

213. The Leonardo sketch from the Codex Madrid in Maurice, *Räderuhr,* II, figs. 80a–e; sketches of spring cases with fusees from the same Codex in Atwood, 85; I have passed over another solution to the problem of diminishing force, the "stackfreed," which cannot be securely attested for the fifteenth century and was very rare; see Maurice, I, 84 f.

214. On the "Burgundy clock" see Maurice, *Räderuhr,* I, 85–87, and II, figs. 77a–c; I should also add here the possibly Burgundian table clock from the mid-fifteenth century (London, Victoria and Albert Museum), Maurice, *Räderuhr,* II, figs. 82a–b.

215. V. Gay, *Glossaire archéologique du Moyen-Age et de la Renaissance,* part II (Paris, 1928), 33; compare the inventory of the princes of Viana in 1461, M. De Bofarul y de Sartorio, *Colección de Documentos . . . de la Corona de Aragon,* vol. 26 (Barcelona, 1864), 131; Mantua, 1482: description of a clock with spring drive in A. Bertolotti, "Le arte minori alla corte di Mantua," *Archivio Storico Lombardo* 15 (1888): 19 f.; Medici inventory, 1492, in E. Müntz, *Les collections des Médicis au XVe siècle* (Paris, 1888), 26, 75 f.; in the Almanus manuscript, 1475–1485, ed. by J. H. Leopold, eight of thirty clocks are equipped with spring drive and fusee.

216. See, for example, the depiction of a small Gothic house clock in the portrait of the Burgundian nobleman Jehan Lefevre de St. Remy, painted around 1460 (?), Antwerp, Royal Museum for Fine Arts, Maurice, *Räderuhr,* I, 83.

217. S. Davari, "Notizie storiche intorno al pubblico orologio di Mantova," *Atti e Mem. Accad Virgiliana* (Mantua, 1884): 211–227 (repr. 1974): 18; 1499: construction of "ij petiz ologes" for the Duke of Flanders, who wanted to take them on campaign with him, A. Pinchart in *Mess. Soc. Hist. Belg.* (1859): 301 f.; it should be noted that the term "small clock" was always relative and does not indicate a small, portable clock in the absence of additional qualifications.

218. Gaspare Visconti, *Ritmi* (1493), after R. Renier in *Archivio Storico Lombardo* 13 (1886): 301 f.; see F. Malaguzzi Valeri, *La corte di Lodovico il Moro,* vol. I (Milan, 1913), 424 ff.

219. For example, in *Neue Deutsche Biographie* vol. 8 (1969), 534; later: *Meyers Enzyklopädisches Lexikon,* vol. 11 (1974), 701; the known material is laid out in J. Abeler, *In Sachen Peter Henlein* (Wuppertal, 1980).

220. For example, the discussion initiated by Enrico Morpurego under the title "Richiamo alla Realtà" in the journal *La Clessidra* 1951–1953.

221. From Abeler (note 219), 8, 9.

222. E. von Bassermann-Jordan and H. von Bertele speculate he may have been the inventor of the stackfreed (182) and of a regulating device for balance foliots that used pig bristles (192); there are, however, no good reasons for these assumptions.

223. In addition to the specialized literature on the various kinds of devices, on individual places, or clockmakers, see the overview in Maurice, *Räderuhr,* I, passim and esp. 127 ff.; and Landes, *Revolution,* 85 ff.

Chapter 5

1. J. Renouvier and A. Ricard, *Des maîtres de pierre et des autres artistes gothiques de Montpellier* (Montpellier, 1848), 96 ff.; C. Wilkes, *Quellen zur Rechts- und Wirtschaftsgeschichte des Archdiakonats und des Stifts Xanten,* vol. 1 (Bonn, 1937), 216, 255; A. Diehl, ed., *UB der Stadt Eßlingen,* Würtembergische Geschichtsquellen 4, (1899), 358.

2. Bilfinger, *Horen,* 166, 168; Sombart (chap. 1 note 21) vol. 2, 126 f.; Bloch, "L'heure et l'horloge" (chap. 1 note 24), 217; Renouard, *Hommes d'Affaires,* 2nd ed., 239 ff.; Bec, *Marchands écrivains,* 318.

3. Bilfinger, *Horen,* 160 ff.: "Schlaguhren und damit die moderne Zeiteinteilung sind also von den Vertretern weltlicher Interessen den Kirchen und Klöstern aufgezwungen worden." Sombart (chap. 1 note 21). Elaboration, for example, in E. Morpurgo: "Potente freno agli entusiasmi delle autorità civili era l'opposizione, spesse volte decisiva de clero decisamente contrario a una suddivisione razionale del tempo . . . nei libri comuni bisognava vincere la resistenza delle autorità religiose . . . [The episcopal cities, insofar as they did not belong to the ruling families] chiudevano le porte agli orologi meccanici"; "L'orologeria italiana: di città in città" (1974), in *La Clessidra* 30 (1974, no. 5): 42; compare also W. Hohn, *Die Zerstörung der Zeit. Wie aus einem göttlichen Gut eine Handelsware wurde* (Frankfurt, 1984), 72: "Bankrott der kirchlichen Herrschaft über die Zeit."

4. Le Goff, "Church's time," passim; "Labor Time," 49.

5. Paris: the error appears probably first in J. Fremont, *Orologe à poids* (Paris, 1915); taken over by Ungerer, *Horloges,* 125; by Bassermann-Jordan, *Uhren,* 358; by Leclerq, "Zeiterfahrung," 358; by J. Gimpel, *Machine,* 154, with the nonsensical comment "the first public clock . . . which cost L 6"; compare chap. 4 note 129; Strasbourg: for example, in Sombart (chap. 1, note 21); in Hammerstein, *Tönende Automaten,* 102; Dondi: for example, in G. Gille, *Les Ingénieurs de la Renaissance* (Paris, 1978), 210; a number of such errors also in H. von Bertele, "The Earliest Turret Clock?" *AH* 10 (1977).

6. Letter from Milan to Francesco Nelli, prior of the monastery SS. Apostoli in Florence: (the author is bothered by the loquacity of an uninvited visitor; only the stroke of the clock recues him at the end of the day): ". . . sic totus ille transiisset dies, nisi publicum horologium, quo ultimo invento per omnes fere iam Cisalpinae Galliae civitates metimur temporum, praelium diremisset; admonitus eniem diem ire, surrexit." *Epist. de rebus familiaribus et Variae,* ed. by F. Francassetti (Florence, 1863), Var. 44, p. 419; it was likewise still used by Francesco Maria Grapaldi to describe an innovation of his own day, *De partibus aedium,* book II, chap. 4: "Nostra tempestate horologia & quidem publica rotulis denticulatis . . . ad horas distinguendas & campane aeris . . . fiunt etiam ex aere parvula sed privata"; first printed in Parma, 1494, quoted from Edwardes, *Chamber Clocks,* 24.

7. "Horologium civitatis," "horologium grossum," "h. universitatis," "h. communitatis," "Stadtuhr" and the like are unequivocal and very widely used

expressions. "Horologium magnum [or] grossum" is found above all in France and northwestern Europe, once also in Breslau ("horologium magnum civitatis"). Only in the French-speaking realm did "Grosse Horloge" become the standard term for public clock. This expression should not be confused with the "Große Uhr," "Ganzer Zeiger" in German texts, which always refer to the counting of the hours of the day from one to twenty-four. In addition, "horologium magnum" is a relative term which was in the early period also used for large clocks inside churches.

8. These were very common descriptions to differentiate the striking clocks from other clocks, in particular from alarm clocks.

9. This includes explicit mention of the clock "horae horologii," but also of unequivocal modern hour indications in city chronicles.

10. Public installation site usually means the city tower, the town hall tower, a church tower, or a gate tower. The reported weight indications are difficult to generalize. They vary considerably, depending on the architectural circumstances and the funds the city could spend on the clock. Taking into account that in most instances only the raw weight of the materials is given, we can set the weight of the average heavy clock at between 100 and 500 kilograms. Heavier and correspondingly more expensive than clockworks were the clock-bells. Small bells weighed between 100 and 500 kilograms, but clock-bells weighing in at between 1,000 and 3,000 kilograms were no rarity. Often the weight of the striking hammer will also indicate the weight of the clockwork-bell ensemble. The reach of the time signal depended, of course, on the height and nature of the bell tower and on the weight and plating of the bell. At the time the reach was hardly ever recorded in quantitative terms, but was usually described with the general comment that the clock should be audible throughout the entire city.

11. These texts were discovered by L. Riccetti, "Il cantiere edile negli anni della Peste Nera," in his *Il duomo di Orvieto* (Rome, 1988), 191.

12. *Annales Veteres Mutinensium,* Muratori 11 (1727), col. 78; Johannes de Bazano, *Chronica Mutinense,* ed. by T. Casini, R. I. S. 15/4 (1917), 124.

13. *Chronica Parmensis,* ed. by L. Barbieri, Monumenta ad provincias Parmensis et Placentinam pertinentia III (Parma, 1858), 222 f.: "Alia . . . campana Communis parva, quae erat desuper turrim Communis, sonabat horas diei et noctis," 322.

14. "Johannes de Scilo orologus salariatus communis," Monum. Ragusina I (Zagreb, 1879), 60.

15. G. Espinas, *Documents relatifs à la draperie de Valenciennes* (Paris and Lille, 1931), 240 f., 266, 306 ff., City Archives Valenciennes, CC 701 fol. 10v, 714, fol. 11.

16. See above chap. 4, Hour-striking clocks.

17. "Horas sponte sua designans," P. P. Vergerius, *Vitae Carrariensium Principium,* Muratori 16 (1730), col. 171; "horologium XXIV horarum," *Cortusii Patavini Duo, Historia de novitatibus Paduae et Lombardiae,* Muratori 12 (1728),

cols., 912, 926; A. Gloria, "L'orologio di Jacopo Dondi nella Piazza dei Signori in Padova," *Atti e Mem. della R. Accademia di Scienze, Lettere et Arti in Padova,* n. s. I (1884–1885): 233–293.

18. A. F. Frisi, *Memorie Storiche di Monza e sua corte* (Milan, 1794), vol. 1, 120 f., and vol. III, 112.

19. The passage is not entirely clear in regard to the technology used: "quod campanarii tali hora pulsent ad matutinum quod nostrorum civium reputare debent orelogium et etiam horas alias," G. Mantese, *Memorie storiche della chiesa vincentina,* vol. 3/1 (Vicenza, 1958), 657; Vicenza did not receive a communal "orelogium horarum" until 1378; *Conforto da Costazza,* Framenti di storia vincentina, ed. by S. Steiner, R.I.S. 13.1 (1915), 15 f.

20. "Orarium sive relogium," Statuta Civitatis Terlesti, Addizione 1352, fol. 327. This information was kindly furnished by the Biblioteca Civica Triest (September 15, 1990).

21. "Pulcra et subtilis fabrica ad singulam horam diei noctisque pulsans," G. Stella and J. Stella, *Annales Genuenses,* ed. by G. Petti Balbi, R.I.S. 17,2 (1975), 153.

22. Niccolò Berardi receives 300 gold Florin and is exempted from city taxes "pro constructione orologiorum aptandorum super turri palatii populi florentini pro pulsando horas diei." A. S. Fir., Consigli della Republica, Provvisioni Reg. 40, fol. 175; G. Dati, *Historia di Firenze* (Florence, 1735), 108.

23. W. H. St. John Hope, *Windsor Castle: An Architectural History* (London, 1913), 139 ff., 166 ff.

24. B. Guillemain, *La cour pontificale d'Avignon* (Paris, 1966), 401; R. Michel, "Les premières horloges du palais pontifical d'Avignon," *Melanges d'Archéologie et d'Histoire* 29 (1909): 213 ff.

25. Monumenta Historiae Boemiae III (Prague, 1774), 320; F. Mencik, *Nekolik statutu a narizeni arcibiskupu prazskych* (Prague, 1882), 23; W. Tomek, *Základy mistopisú Prazkého,* vol. 1 (Prague, 1866), 20, 224, 235.

26. "Arlogium pro horis pulsandis," Matheus de Griffonibus, *Memoriale Historicum,* ed. by L. Frati and A. Sorbelli, R.I.S. 18, 2 (1902), 61; Bartolomeus de la Pugliola, *Historia Miscella Bononiensis (1104–1394),* Muratori 18 (1731), col. 444.

27. The sources in Beeson, *Perpignan,* 1356.

28. Rainieri Sardo, *Cronaca di Pisa,* ed. by O. Banti (Rome, 1963), 104, 119, 125, and elsewhere.

29. R. Predelli, *I Libri commemoriali della Repubblica di Venezia, Regesti,* vol. II (Venice, 1878), 292.

30. C. T. Gemeiner (chap. 4 note 197), II, 107.

31. F. de Guilhermy, *Inscriptions de la France,* vol. 3 (Paris, 1877), 24.

32. *Cronaca Senese di Donato di Neri e del suo figlio Neri,* ed. by A. Lisini and F. Jacometti, R. I. S. 15, 6 (1936), 591.

33. There are, in my estimation, several reasons for holding this view: (a) the widespread geographic diffusion of these clocks by the first half of the fourteenth century; (b) in quite a few cases, tower clocks can be shown to

have been installations that were procured in addition to existing, large church clocks; and (c), the testimony of Thomas de Luda (chap. 4 note 140).

34. For example, in London, 1344: "une dyal . . . a moustrer les houres de iour et de nuyt;" Howgrave-Graham, *Clocks,* 268.

35. *Chroniken des Fritsche Closener und Jakob Twinger von Königshofen,* ed. by C. Hegel, Chroniken der deutschen Städte 8 and 9 (Leipzig, 1870–1871), 725.

36. Zinner, *Räderuhren,* 108 f.

37. H. M. Colvin, *History of the King's Works: The Middle Ages* (London, 1963), 262, 509, 802, 997; R. A. Brown, "King Edward's Clocks," *The Antiquaries Journal* 39 (1959): 283–286; J. B. Post and A. J. Turner, "An Account for Repairs to the Westminster Palace Clock," *Archeological Journal* 130 (1973): 217–220.

38. Paris, Vincennes: F. de Guilhermy, *Inscriptions de la France,* vol. 3 (Paris, 1877), 24; Paris, Horloge du Palais: [Charles V] "edificavit . . . et turrem quadratam que est inter palacium et Magnum Pontem . . . ac horologium desuper poni fecit," ed. by P. Moranvillé, Société de l'histoire de France, vol. 2 (Paris, 1893), 396; for the testimony of Jean Golein, see chap. 8, A legendary decree by Charles V of France. Arrêt du Parlement 1452 Jul. 21: "quod defunctus [Charles V] pro decoracione ville notre Parisius, in qua tunc nulla erant grossa horologia et ut nosra parlamenti curia et habitantes ipsius ville melius se regerent et regularent [etc.]," after A. Bossuat, "Doc. inédits sur l'Horloge du Palais," *Bull. Soc. hist. Paris et de l'Île-de-France* 56 (1929): 101; Beauté-sur-Marne: L. Douet-d'Arcq (chap. 4 note 209), 177; Montargis: Ungerer, *Horloges,* 117 f.; Sens: G. Julliot, "L'horloge" (chap. 4 note 52), 386 ff.; Noyon: A. de la Fons de Mélicocq, *Une cité picarde au moyen age* (Noyon, 1841), 109 ff.; Avignon: R. Michel (note 24), 221 f.; Melun: Prost and Reverchon, 46; Angers: P. Marchegay, "Horologe publique à Angers en 1384," in *Notices et documents historiques* (Angers, 1857), 99 ff.; Poitiers: R. Favreau, *La ville de Poitiers à la fin du Moyen Age* (Poitiers, 1978 ff.), 216 ff.; Lyons: M. C. Guigue, *Cartulaire de la ville de Lyon* (Lyons, 1876), 475 f.; Nîmes: A. Bessot de Lamothe, *Arch. com. Inv. som.* vol. 1 (Avignon, 1877); Blois: J. Soyer and G. Trouillard, *Cartulaire de la ville de Blois* (1907), 362.

39. Bourges, La Nonette, Mehun-sur-Yèvre: A. de Champeaux, *Les Travaux d'art exécutés pour Jean de France, duc de Berry* (Paris, 1894), 194 f.; E. de Toulgoêt-Tréanne, *Les comptes de l'Hôtel du duc de Berry 1370–1413, Mém. Soc des Antiquaires du Centre* 17 (1889–1890): 64 ff.; Poitiers (note 38); Riom: E. Clouard, *Les gens d'autrefois. Riom aux XVe et XVIe siècles* (Riom, 1910), 780 ff.; Niort: R. Crozet, "Nouveaux textes et documents relatifs à l'histoire des Arts en Poitou," *Bull. Soc. Antiqu. de l'Ouest,* 4th Series, vol. 3–4 (1956–1957), 583; St. Jean d'Angely: L. Duret, "La tour de la grosse horloge de Saint-Jean-d'Angely," *Recueil Comm. arts et monum. hist. Charente-Inférieure* 12 (= 4th series, 4) (1893–1894): 102–107; Villefranche en Rouerge 1404: Paris BN, Coll. Languedoc (Doat), fol. 213 ff., 97 f., Vierzon: P. Des Chaumes, *Mém. Soc. hist. litt. et scient. Cher* 4e ser. 36 (1927): 6.

40. Rouvres: B. Prost, *Inventaires mobiliers et extratis des comptes des Ducs de*

Bourgogne de la Maison de Valois (1363–1477), 2 vols. (Paris, 1902–1913), II, 59, 72, nos. 366, 427, 428; Montbard: Prost, ibid., II, 84, 99, nos. 497, 588; Villaines, Germolles: M. E. Petit, *Ducs de Bourgogne de la maison de Valois,* vol. 1 (Paris, 1909; repr. 1976), 41; Brussels, Bruges, Ghent: A. Pinchart, *Mess. Sc. Hist. Belg.* (1884): 181 ff.; Arras: J. Finot, *Inv. som arch. dép. Nord,* VII (Lille, 1892), 356; Mâle: L. de Laborde, *Les ducs de Bourgogne,* 3 vols. (Paris, 1849–1852), 2.1, p. LI; La Montoire, Hesdin: C. Dehaisnes, *Inv. som. arch. dép. Nord,* vol. 4 (Lille, 1886), 772; Dijon: J. Froissart, *Chroniques,* ed. by Kervyn de Lettenhove, vol. 10 (1867–1877; repr. 1967), 188; Prost (op. cit), nos. 812, 829, 1054; L. Gouvenain and P. Vallé, *Inv. som. arch. com.,* vol. 2, ser. I (Paris, 1882–), 99; F. Humbert, *Les Finances municipales de Dijon du milieu de XIVe siècle à 1477* (Paris, 1961), 189 f.

41. J. Le Goff, "Merchant's Time," 35; compare also his "Labor Time," 49.

42. Mons: Dehaisnes (note 40); A. Pinchart, *Extraits des comptes relatifs au Hainaut* (Mons, 1884); Lens, Aire-sur-le-Lys: A. Van Nieuwenhuysen, *Les Finances du duc de Bourgogne Philippe le Hardi (1384–1404)* (Brussels, 1984), 420; Sluys, Tamise, Termonde: P. Bonnefant and J. Bartier, *Ordonnances de Philippe le Hardi etc.,* vol. II (Brussels, 1974), nos. 419, 659, 660; C. Rossignol, *Histoire de Beaune* (Beaune, 1854), 249 ff., R. Pretet, *Histoire de Châlon-sur-Saône* (Roanne, 1981), 58 ff.; C. A. Parmentier, *Archives de Nevers,* vol. 1 (Paris, 1842), 224; F. Boutillier, *Inv. som. arch. com.* (Nevers, 1876); Abbé Lebeuf, *Mémoires conc. l'histoire . . . d'Auxerre* (new edition, Paris, 1848–1855), II, 190, VI, 294; Nivelles, Ungerer, *Horloges,* 432.

43. T.-M. Vinyoles i Vidal, *La vida quotidiana a Barcelona vers 1400* (Barcelona, 1985), 31 ff.

44. Verona: Zagata, *Cronaca,* vol. 1, part 2, 221, cited from *Conforto da Costoza, Frammenti,* ed. by C. Steiner, R. I. S. 13.1 (1915), 16, note; 1379: A. Caffaro, *Pinerolensia* (Pinerolo, 1906), 119 ff.; 1376: P. André, "notice historique sur l'église paroissale de St. Léger à Chambéry," *Mém. Soc. Savoisienne d'Hist.* 7 (1863), 104 ff.; Turin: AM Turin, Libri Rationum, A. S. Turin, Comptes des Trésoriers généraux des Princes d'Acayes, Camerale, Inf. 40; F. Rondolino, "Il castello di Torino," *Soc. Piemont. di Archeol. e Belle Arti, Atti* 13 (1932): 13–18; Prato: D. G. M. di Agresti, *Aspetti di vita Pratese del Cinquecento* (Florence, 1976), 35 f.

45. Ordonnance du Roi Henri III . . . sur les plaintes et doléances faites par les Deputes du Royaume, Art. 351: use of "deniers d'octroi et impositions" for "réparations" on instructions from the communal officials, "Entendans toutesfois estre compris en la depense desdictes reparations, celle qui concerne l'entretenement des horloges . . ." In *Recuil de pièces originales . . . ,* vol. 3, Premiers États de Blois en 1576 (Paris, 1789), 623, 624.

46. A. van Nieuwenhuysen (note 42).

47. *Balduini Ninovensis Chronicon,* ad a. 1219, ed. by O. Holder-Egger, MGH SS XXV (1880), 541; on the events see H. Dubrulle, *Cambrai à la fin du Moyen Age* (Lille, 1903), 51 f., 297.

48. Béthune: C. Dehaisnes and J. Finot, *Arch. dép. Nord., Inv. So.* I,2 126 (B 1024); E. de la Quérière, "Notice historique sur l'ancien Hôtel de ville, le

Beffroi et la Grosse Horloge de Rouen," *Bulletin Société d'Émulation, Seine-Inférieure* (Rouen, 1862): 310 ff.; J. B. A. Bostvieux and G. Tholin, *Inv. som. arch. som. d'Agen* (1884), BB 27.

49. Enea Silvio Piccolimini, *Commentarii rerum memorabilium* (Rome, 1584; repr. Frankfurt am Main, 1974), 425 ff.

50. See P. Mesnage, "Construction horlogère" (chap. 4 note 169) 290; C. Cipolla, *Clocks,* 42 f.; Wendorff, *Zeit und Kultur,* 143; Chevalier, *Les Bonnes Villes de France du XIV au XVI siècle* (Paris, 1982), 227 f.

51. "Ut yporiensis civitas . . . que omnium fere est lumine decorata virtutum de virtute in virtutem intellectus et honoris lumine fulciatur . . ."; G. S. Pene-Vidari, *Statuti del Comune di Ivrea* II (Turin, 1969), 400; *Breslauer Urkundenbuch,* ed. by G. Korn (Breslau, 1870), 221; I. Belli, *Guida di Lucca* (1953), 172; F. Marri, "La torre dell'Orologio in S. Gimigniano," *Miscellanea storia della Valldelsa* 33 (1955): 158 ff.; A. Gloria, "L'orologio di Jacopo Dondi" (note 17), 290; M. C. Billanovich, "La Vicenda dell'Orologio di Piazza dei Signori a Padova: commitenti, esecutori, modalità di costruzione," *Arch. Veneto* 5. ser., vol. 133 (1989): 39 ff.

52. Namur: Patart, *Cloches,* 183 f.; H. Q. Janssen, "Het openbare uurwerk te Aardenburg," *Bijdr. tot de Ousheidskunde en Geschiedenis inz. van Zeeuwsch-Flanders* I (1856): 156 ff.; Paris, Lyons: note 38; A. Lecocq, "Notice Historique et Archéologique sur les Horloges de l'Église Notre-Dame de Chartres," *Soc. Archéol. d'Eure-et-Loire, Mém.* 4 (1867): 303.

53. Montpellier (note 1); E. Morpurgo, *Nederlandse klooken- en horlogemakers vanaf 1300* (Amsterdam, 1970), 32, 99; L. J. Meilink-Hoedemaker, Luidklokken en Speelklokken in Delft (Proefschrift) (Utrecht, 1985), 181 f.; Montélimar: L. Fillet, "Les Horloges publiques dans la Sud-Est de la France," *Bull. Archéol. du Comité des Travaux Historiques* (1902): 105; P. Thome de Maisonneuve, "Le Jaquemart de Romans," *Bull. Soc. Archéol. et Stat. Drôme* (1929): 17 ff.; O. Siliprandi, *Notizie sugli orologi pubblici di Reggio* (Reggio Emilia, 1925).

54. L. Decombe, "Notes et Documents concernant la Grosse Horloge de Rennes," *Bull. et Mém. de la Soc. Archéol. du dépt. Ille-et-Villaine* 14 (1880): 175 ff.; J.-P. Leguay, *La Ville de rennes au XVe siècle à travers les comptes des Miseurs* (1968), 45–61.

55. After K. Fischer, "Die Uhrmacher in der Slowakei," *Bohemia-Jahrbuch des Collegium Carolinum* (1969): 406; Coulommiers: *Inv. Arch. dép. Seine-et-Marne,* vol. 4 (1880), 30.

56. Compare M. Warnke, *Bau und Überbau. Soziologie der mittelalterlichen Architektur nach den Schriftquellen* (Frankfurt, 1976), 78–92.

57. P. Wolff, "Pouvoir et investissements urbains," *Revue Historique* 258 (1977): 298. According to R.-H. Bautier and J. Sornay, *Les Sources de l'Histoire Économique et Sociale du Moyen Age* (Provence), vol. II (Paris, 1971), 951, the clock, "véritable symbole de la ville des la second moitié du XIV siècle," was, along with road construction, wells, market halls, and mills, among the important expenses in city budgets in southern France and northwestern Italy.

58. Reims (1380–1397): H. Jadart, *L'Horloge et le carillon de la cathédrale de*

Reims (Reims, 1909); compare Ungerer, *Horloges,* 135 ff.; Rouen (1372): C. de Robillard de Beaurepaire, "Notice sur les Horloges de la ville de Rouen," in *Dernier Recueil de Notes Historiques et archéologiques* (Rouen, 1892), 310 f.; Autun 1377: A. De Charmasse, "L'Horlogerie et une famille d'Horlogeurs," *Mém. Soc. Eduenne n. s. t.* 16 (1888): 174 ff.; Lyons (1379): E. Vial and C. Côte, *Les Horlogers Lyonnais de 1550 à 1650i* (Mâcon, 1928), 1 ff.; Troyes (1380): J.-F. Gadan, *Le Bibliophile Troyen* (Troyes, 1850–1851); Bourges (see note 39); Sens (see chap. 4 note 52); Angers (see note 38); Mainz (1369): *Mitteilungen des Vereins für Geschichte und Altertumskunde in Frankfurt* 5 (1874–1879): 607 f.; Minden (1384): K. Löffler, *Die Bischofschroniken des Mittelalters,* Mindener Geschichtsquellen 1 (Münster, 1917), 212; Xanten (note 1).

59. Ruppert, *Chroniken* (chap. 3 note 36), 203.

60. Compare the explicit remark of the Nuremberg Council when it took over the costs for the "ureglocken" of St. Sebald in 1489: the clock was "nit der kirchen, sunder gemeiner stat notturftig"; A. Gümbel, "Baurechnungen," *Mitteilungen des Vereins für die Geschichte der Stadt Nürnberg* 20 (1913): 17; Rendsburgische Stadt- und Polizeiordnung 1720 Sept. 17: "Wie dann auch eben so wenig die Kirchen mit den gemein Uhren Reparation zu schaffen haben, sondern dieses Onus billig der Stadt-Cassa alleine zufallen muß, nachdem malen der nutzen davon nicht an die Kirche, sondern unstreitig auf die gesamte Bürgerschaft und Einwohner redundierte." *Corp. Const. Regio-Holsaticarum,* ed. by F. D. C. von Cronhelm (Altona, 1749–1757), vol. III, 1753, 841 f. Remarks of a priest: "Ein Uhrwerk stehet zwar mehrenteils, sonderlich in Dörffern, auf dem Kirchthurm, es gehört aber eigentlich nicht zur Kirche . . . , weil es mehr zum weltlichen als zum geistlichen Gebrauch dienet." *Acta Ecclesiastica* . . . (Leipzig, 1727–1752), vol. XIV, 1218 f. Decree of the Neumark Directorate of Church Revenues of Dec. 3 1772: "die Uhren aber mit dem Gottesdienst und der Kirche in keiner nothwendigen Verbindung stehen, sondern einzig und allein zum Besten der Gemeine gereichen"; quoted from G. Arndt, "Die kirchliche Baulast in der Mark Brandenburg," *Jahrbuch für brandenburgische Kirchengeschichte* 15 (1907), 8 f.

61. Paris, Archives Nationales KK 532 n°2ter; the text of the vignette after R. Rodière, "Le clocher de Sainte Saulve de Montreuil," *Mém. de la Comm. dép. des mon. hist. de Pas de Calais* 33 (Arras, 1910): 252:

> Nos maire et eschevins de Monsterrel sur le Mer
> Avons fait cest orloge pour b(ien)n edeffier.-
> Tantost me veissiez ceste tour chi abatre
> Se Saint Sauve ne fust q(ui) n'en vau rien (r)abat(re).-
> Lorloge est b(ien) sea(n)z ychi; Dieu le nos sauve,
> Saint Justin, saint Waloi, sant Macleu et Saint Sauve.

62. Paris (note 38); Magdeburg: ed. by C. Hegel, Chroniken der deutschen Städte 7 (Leipzig, 1869), 378; compare Venice, 1394: "pro honore et consolatione totius civitatis," A. Tessier in *Giornale degli eruditi et curiosi Padova* 4

(1884): 45; Avignon 1461: P. Achard, *Les Horloges publiques et les horlogers à Avignon* (Avignon, 1877), 72.

63. Paris 1419: Bossuat (note 38), 98 f.; Siena: G. Milanesi, *Documenti per la storia dell'arte senese,* vol. 1 (Siena, 1854), 236 f.; Oudenburg 1402: E. Feys and D. van de Casteele, *Histoire de Oudenburg* (Bruges, 1873), vol. 1, 546 ff.; Lüneburg 1445: G. Melbeck, "Lüneburger Uhren und Uhrmacher," *Lüneburger Blätter* 15–16 (1956): 243 ff.

64. Aix-en-Provence 1486: Archives communales, BB 29, fol. 60; this information was kindly supplied by the staff at the archive.

65. Compare V. Carpaccio, "Incontro dei Fidanzati," for the (probably imaginary) dial on the Campanile in Venice, which has been placed higher so as to move it into view; see also the clock tower in "Arrivo degli Ambasciatori Inglesi." Both paintings in Venice, Gallerie dell'Accademia.

66. G. Campori, "Gli oroligieri degli Estensi," *Atti e Mem. delle RR deputazioni di storia patria per la provencie dell Emilia,* new series 2 (1877): 244 ff.; S. Davari (chap. 4 note 217), 9.

67. A. S. Milano, Reg. ducale no. 214; compare also *Archivio Storico Lombardo* 48 (1921): 610 f.

68. M. A. Arnoult, *Boussoit-sur-Haine,* 63 f., 67; C.-L. Diericx, *Mémoires sur la ville de Gand,* vol. 2 (Ghent, 1815), 68.

69. L. Pastor, ed., *Antonio de Beatis, Die Reise des Kardinals Luigi d'Aragona* (Freiburg, 1905); L. Guiccardini, *Descrittione di tutti i paesi bassi* (Amsterdam, 1612; repr. 1968), 30. L. Lautrey, ed., *Michel de Montaigne, Journal de Voyage* (Paris, 1906), 78, 360; it is not entirely clear whether Montaigne means only public clocks when he speaks of the "mancamento d'oriuoli ch'e in questo loco (Luccata) et in la più parte d'Italia." He used the word "orioli" also to describe the clocks (sandglasses?) that were readily available to him in the baths of Lucca; see 334.

70. Compare also the depiction of an unidentified village in western Switzerland: ibid., fol. 93 recto, and C. Pfaff, "Umwelt und Lebensform," in A. A. Schmidt, ed., *Die Schweizer Bilderchronik des Diebold Schilling 1513* (volume of commentary, Lucerne, 1983), 620.

71. Compare the accounts from Torgau, Volkenroda, and Homburg, in D. van Adrichem, "De horologia a. 1525 apud Fratres Minores Torgaviae existentibus," *Arch. Francisc. Hist.* 25 (1932): 105–108; a wealth of material is provided by K. Pallas, ed., *Die Registraturen der Kirchenvisitationen im ehemals sächsischen Kurkreise,* Geschichtsquellen der Provinz Sachsen 41 (Halle, 1906).

72. J. Perrin, "Le Doyenné du Sundgau à la fin de la Guerre de trente ans (1647)," *Archives de l'Église d'Alsace* 23 (= vol. 29, 1976–1979): 117–186, esp. 147 ff.

73. K. Pallas, ed., *Die Registraturen der Kirchenvisitationen im ehemals sächsischen Kurkreise,* Geschichtsquellen der Provinz Sachsen 41 (Halle, 1906), II, 4, 5, 421; from the decree for Bodenteich from a Lüneburg visitation (1568): "Kirche bawfellig vnnd das volck wegerich diesselben zu bessern. Wollen auch di gebottene zulage vom seger, die zu 24 dalern erstiegen, nicht erlegen." In

B. Lange, "Die Generalvisitation im Fürstentum Lüneburg," *Zeitschrift für Nidersächsische Kirchengeschichte* 58 (1960): 53.

74. Codex Augusteus (chap. 10 note 62) I, col. 109; the Ecclesiastical Constitution in E. Sehling, *Die reformatorische Kirchenordnungen des XVI. Jahrhunderts* (Tübingen, 1902 ff.), vol. V, col. 423.

75. Quoted from Cancellieri, *Campane,* 89, who bases himself on S. Sarnelli, *Lettere ecclesiatice,* vol. VII (Venice, 1706), 24.

76. See note 38.

77. See the admonitory letter by the kings of Aragon to Perpignan in 1379, 1387, 1399, Vielliard, "Horloges" (chap. 4 note 209), 163; Beeson, *Perpignan,* 51 ff.; D. Girona Llangostera, "Itinerari del rey Marti (1396–1402)," *Institut d'estudis Catalans, Anuari* 4 (1911–1912): 145; the letter of the Marchese of Mantua to Goito in 1523, Bertolotti (chap. 4 note 215), 289.

78. Lyons 1481: L. Caillet, "Le consulat de Lyon et le clocher de St. Nizier," *Bull. Hist. du Dioc. de Lyon* 12 (1911): 316 ff.; compare the ethical and political tone in the prologue to the first statute of the clockmaker's guild in Paris (1544): "Maintenant que l'invention des orloges a esté trouvée pour vivre et se conduire en reigle et ordre de vertu." Paris Archives Nationales Y 65, fol. 110, printed in A. Franklin, *Vie privé,* 179.

79. J. H. L. Bergius, *Neues Polizey- und Cameralmagazin nach alphabetischer Ordnung,* vol. 6 (Leipzig, 1780), 258; compare G. H. Zincke, *Cameralistikbibliothek,* I. Theil (Leipzig, 1751; repr. 1973), 432 f.; *Grundsätze der Polizeiwissenschaft,* 3rd. ed. (Göttingen, 1782; repr. 1969), § 76, p. 68; compare Marperger, *Horologiographia,* 22 f., 57 f.

80. Ed. by E. Eckertz, "Chronik der Stadt Erkelenz," *Annalen des historischen Vereins für den Niederrhein* 5 (1857): 45.

81. N. van Wervecke, *Kulturgeschichte des Luxemburger Landes* II (Luxembourg, 1924), 86; A. Rigaudière, *Saint-Flour. Ville d'Auvergne au Bas Moyen Age* (Paris, 1982), 754 and 276; J.-P. Leguay, *Un Réseau urbain au Moyen Age* (Paris, 1981), 206 f. and elsewhere.

82. *Breslauer Urkundenbuch* (note 51), 211, 214 f., 221; A. Kerschbaumer, *Geschichte der Stadt Tulln* (Krems, 1874), 374 f.

83. J. Kleyntjes, "Oudste Stadsrekeningen van Culemberg," *De Navorscher* 64 (1915): 81 ff.; J. Jäger, *Urkundenbuch der Stadt Duderstadt* (1885), 453; Pegau: J. Hohlfeld, *Stadtrechnungen als historische Quellen* (Leipzig, 1912), 141 ff.; A. Meerkamp van Embden, *Stadsrekeningen van Leiden 1390–1434,* vol. 1 (Amsterdam, 1913), 39; E. Dupont, "Registre des recettes et dépenses de la ville de Boulogne," *Mém. Soc. Acad. arr. B.* 7 (1882): 186; J. H. W. Unger and W. Bezemer, *Bronnen tot de Geschiedenis van Rotterdam,* III, De oudste stadsrekeningen (Rotterdam, 1869), 70.

84. Zurich 1366; H. Zeller-Wermüller, ed., *Die Stadtbücher des 14. und 15. Jahrhunderts* I (Leipzig, 1899), 412; Letter of safe-conduct for clockmakers from Delft, 1368, "De horologiorum artificium exercendo," in T. Rymer, *Conventiones, Literae . . . ,* 3rd. ed. (Farnborough, 1967), vol. III, part II, 145; E.

Lemaire, *Archives anciennes de la ville de St. Quentin,* vol. II, 1910, vol. III (MS), nos. 701, 744 ff. (information kindly furnished by the archive); in Lille Pierre Demileville was employed at the city clock from 1381, D. Clauzel, *Finances et politique à Lille pendant la période bourgignonne* (Dunkirk, 1982), 143.

85. Lucca (note 51); Montpellier (note 1); F. de Potter and J. Broeckhardt, *Geschiedenis der Stad Aalst,* part 2 (Ghent, 1874), 77 f.; Brunswick 1385–1386: Städtisches Archiv Braunschweig, Copialbuch fol. 54 v., 65 v; Lucerne 1385: P. X. Weber, *Das älteste Luzerner Bürgerbuch (1357–1479)* (Stans, 1921), 176 f.; Santiago de Compostela: F. Landeira de Compostela, *Theatro Chronométrico del Noroeste Espanol* (Madrid, 1957), 7; Moscow (chap. 4 note 181); Caffa 1374: M. Balard, *La Romanie Génoise,* vol. 1 (Genoa and Rome, 1978), 208 f.

86. Mondinus 1334: L. de Mas Latrie, *Nouvelle preuves de l'Histoire de Chypre,* vol. 1 (Paris, 1837), 65 f.; journey of the company of Giovanni Loredan 1338: R. S. Lopez, "Venezia e le grandi linee dell'espansione commerciale nel secolo XIII," in *La Civiltà Veneziana des secolo di Marco Polo* (Venice, 1955), 53, 77; C. Potvin, ed., *Ouvres de Ghillebert de Lannoy* (Löwen, 1878), 67; Kurz, *European Clocks,* 20 f.; Kurz downplays the event of 1338 for no good reason by interpreting the "relogium" as a sandglass, which would hardly have been mentioned in a property liquidation.

87. On the clocks as part of the tributes paid to the Turks see the thorough work by G. Mraz, in *Welt als Uhr,* 39–54.

88. "Neque enim horae sunt Turcis, quibus temporum, neque millaria, quibus locorum spatia distinguant. Talismanes habent hominum genus templorum ministerio dicatum; hic mensuris utuntur ex aqua. Quibus postquam adventari auroram cognoverunt, clamorem tollunt e celsa turri in eum usum constructa; [division of the day into four unequal segments according to the seasons] Noctis omne tempus incertum es." . . . "Non enim facile gentem aliam minus piguit aliorum bene inventa ad se transferre. Testes majoresminoresque bombardae, multaque alia, quae a nostris excogitata ipsi ad se avertunt. Ut libros tamen typis excuderent, horologia in publico haberent, nondum adduci potuerunt: quod scripturam, hoc est suas literas sacras, non amplius scripturam fore, si excuderetur; & si horologia publice haberentur, aliquid de aedituorum suorum & prisci ritus authoritate diminutum iri arbitrentur." Ogier Ghiselin des Busbecq, *Opera quae extant omnia* (Basel, 1740; repr. 1968), 28 f. The lack of interest on the part of the Muslims is confirmed in the introduction of Taq ad-Din, in Wiedemann and Hauser, 12.

89. The report about Murad III comes from a MS in the monastery of Koutloumousi, R. M. Dawkins, *The Monks of Athos* (London, 1936), 311; R. Lubenau, *Beschreibung der Reisen,* ed. by R. Sahm (Königsberg, 1914), 98 (Jagodina, 1587); see Kurz, *European Clocks,* 100; H. W. Duda, "Balkantürkische Studien," *Österreichische Akademie der Wissenschaften, Philosophisch-Historische Klasse* vol. 226,1 (Vienna, 1949): 18, 28, 60 f.

90. Following are the dates on the geography of the spread of public clocks: (a) abbey/monastery (c) residence/palace:

—by 1360: Orvieto, Parma, Valenciennes, Milan, Modena, Padua, Monza, Windsor (c), Avignon (c), Florence, Genoa, Bologna, Perpignan (c), Prague, Vincennes (c), Regensburg, Siena, Durham.

—by 1370: Nuremberg, Frankfurt, Munich, Ferrara, Freiburg im Breisgau, Brussels, Augsburg, Beucaire, St. Quentin, Zurich, Fano, Queensborough (c), Breslau, Sheen (c), King's Langley (c), London, Ivrea, Troppau, Udine, Utrecht, Mainz, Schweidnitz, Brieg, Deventer, Gloucester, York, Paris.

—by 1380: Verona, Forli, Ghent, Middleburg, Ely (a), Mons, Mechelen, Golzinne (c), Rouen, Cologne, Colmar, Strasbourg, Tulln, Stralsund, Aachen, Exeter, Caffa, Feodosija, Culembourg, Bourges, Bridgewater, Perugia, Xanten, Avignon, Toledo, Chambéry, Douai, Dunes (c), Assisi, Vercelli, St.-Jean-de-Maurienne, Stade, Görlitz, Vienna, Montreuil-s.-M., Tournai, Termonde, Lier, Autun, Auxerre, Castres, Sens, Beaute-s.-M (c), Vicenza, Ieper, Mehun (c), Nonette (c), Argentan, Valencia, Pinerolo, Rostock, Lüneburg, Klosterneuburg (a), Lyons, Noyon, Riez, Ripon, Pisa, Bamberg, Basel, Bern, Neuveville, Neufchâtel, Zwolle, Nieppe (c), Cambrai, Macon, Tours, Troyes, Reims, Montargis (c), Rouvres (c), Montbard (c), Cornillon (c), Montserrat (a).

—by 1390: Blankenburg, Vorau (a), Löwen, Mâle (c), Lille, St. Mihiel, Senlis, Hamburg, Bruges, Nijmegen, Courtrai, Zutphen, Elbing, Fritzlar, Oudenburg, Dijon, Laon, Venice, Friedberg, Minden, Lübeck, Amiens, Angers, Greifswald, Lucerne, Tonnere, Burgos, Reggio, Emilia, Würzburg, Brunswick, Neusohl, St. Omer, Hesdin (c), Evreux, Villaines (c), Palma de Mallorca, Salisbury, Muggia, Albenga, Cracow, Poitiers, Beauvais, Germolles (c), Barcelona, Oxford, Reinbek (a), Béthune, Lerida, Göttingen, St. Flour, Rottenmann, Châlons-sur-Marne, Schlettstadt, Wittingau (a), Ragusa (Dubrovnik), Gubbio, Preetz (a), Luxembourg, Hal, Leiden, Grenoble, Ottery (a), Sagan (a).

—by 1400: Lucca, Chieri, Metz, Melun (c), Riom, Damme, Chartres, Wells, Olmütz, Hannover, Ochsenfurt, Beaumont, Namur, Rodez, Olite (c), Aire-sur-la-Lys, Doberan (a), Esslingen, Speyer, Lausanne, Enghien, Aalst, Beaune, Santiago-de-Compostela, Mantua, Sluys, Niort, Magdeburg, Seville, Città di Castello, Ulm, Duderstadt, Asniere-sur-Seine (c), L'Isle-sur-Sorgue, Caen, Fougeres, Châteauneuf-sur-Loire, Aardenburg, Saaz (a), Koscian, Warrington (a), Marienburg, Leitomischl, Geraardsbergen, Le Quesnoy (c), Locquignol (c), Peronne, Montpellier, Ingolstadt, Kaster (c), Pegau, Überlingen, Yverdon, s'Hertogenbosch, Lens, Apt, Nîmes, Winchelsea, Auxonne, Piacenza, Brescia, Ancona, Rovato, Winterthur, Chinon, Compiègne, Abbeville, Moulins, Coutances, Jugny, St. Jean-d'Angely.

For reasons of space it is not possible to list all localities. They have been collected in a source inventory on the history of public clocks in late medieval Europe; the inventory is being prepared for duplication. Interested readers are invited to address question and inquiries to the author.

91. J. Le Goff, "Ordres mendiants et urbanisation de la France médiévale," *Annales: Economies. Sociétés. Civilisations* 25/2 (1970): 937.

92. P. Contamine, "Contribution à l'Histoire d'un mythe: Les 1,700,000

clochers du royaume de France (XV–XVI siècles)," in *Économies et Sociétés au Moyen Age (Melanges E. Perroy)* (Paris, 1973), 414–421; P. Meyer, "Les dix-sept-cents mille clochers de la France," *Romania* 7 (1878): 104–106.

93. N. Machievalli, "Ritracto de Cose di Francia (1512/1513)," in *Opere* II (Verona, 1979), 338, where the figure is given as 1,000,700 (incl. seven abbeys); the German edition, ed. by H. Floerke, vol. 2 (Munich, 1925), 196, takes over this figure with the comment that the French edition of Guiraudet (Paris, 1798) corrected it to 17,000.

94. L. Génicot, "Les grandes villes de l'occident en 1300," in *Économies et Sociétés* (note 92), 199–219.

95. H. Aubin and W. Zorn, eds., *Handbuch der deutschen Writschafts- und Sozialgeschichte,* vol. 1 (Stuttgart, 1971).

96. An "arlogio all commune" is mentioned in Fano after 1366; this information was kindly furnished by the town archives; Troppau (note 82); Brieg 1370: I. Grünhagen, *Urkundenbuch Brieg,* Codex diplomaticus Silesiae 9 (Breslau, 1870), no. 295; Scheidnitz (note 51); Ivrea (note 49); Tulln (note 82); Görlitz 1377: R. Hecht, *Görlitzer Ratsrechnungen,* Cod. dipl. Lusatiae superiores, part II (Görlitz, 1896), 42; Stade ca. 1377: J. Bohmbach, *Urkundenbuch Stade* (Hildesheim, 1981), 101; Stadthagen 1388: R. Brosius, *Stadthagener Stadtrechnungen 1378–1401* (Bückeberg, 1968), 44 ff.; Culembourg (note 83); Castres 1377: L. Barbaza, *Annales de la ville de Castres* (Castres, 1886), 133; Donaustauf before 1400: C. T. Gemeiner (chap. 4 note 197), III, 107.

97. On the demography see W. Preuvenier, "La démographie des villes du comte de Flandre aux XIIIe siècles," *Revue du Nord* 65 (1983): 255–275; A. Derville, "Le nombre d'habitants de l'Artoise et de la Flandre Wallone (1300–1400)," ibid. 277–299. A comparison with the cities invited to the general meeting of the Estates General further reinforces this impression.

98. Chevalier, *Bonnes Villes* (note 50), 37 ff.; C. H. Taylor, "Assemblies of French Towns in 1316," *Speculum* 14 (1930): 275–299.

99. J. Le Goff, "Ordres" (note 91).

100. Recently the usefulness of the "critère mendiant" for comparative purposes in the research on urbanization has been questioned. A. Derville has pointed out that while there is little doubt that four or more convents presuppose a certain urban size, it is not true that every large city, at least in northwestern Europe, was home to a corresponding number of convents. He also points to the appearance of demographically and economically insignificant localities with convents of mendicant orders.

101. N. Bulst, "Vers les états modernes: Le tiers état aux États généraux de Tours en 1484," in R. Chartier and D. Richet, eds., *Représentation et vouloir politiques* (Paris, 1982), 11 ff; N. Bulst, *Die französischen Generalstände von 1484. Prosopographische Untersuchungen zu den Deligierten,* Beihefte zu Francia 26 (Sigmaringen, 1992).

102. Favreau (note 38).

103. Leguay (note 81), 206 f.

104. L. Fillet, "Horloges publiques dans le Sud-Est de la France," *Bull.*

Archéol. du Comité des travaux Historiques (1902): 101–119; R. Bailly, "Horloges et Beffrois Vauclusiens," *Mém. de l'Académie de Vaucluse* 5. ser. 8 (1961–1962): 65–129.

105. A. Bovelli (note 27); Juan Alemany in Valencia (1378): L. Montañés Fontenla, *Capitulos de la Relojeria en Espana* (Madrid, 1954), 61 f.; Antonio Core from Bologna and Matheu Alemani in Lérida (1390, 1406), Roberto de Malina (Mecheln) in Valldigna (1435): L. Montañés, 68 f.

106. Palermo: H. Bresc, *Un monde méditerranén. Économie et société en Sicile 1300–1450,* vol. 2 (Paris, 1986), 721; Diaria Napoletana, in Muratori 21 (1732), cols. 1043 ff.; Ragusa 1389: D. Petrovic (the first public clock and the first clockmakers in Dubrovnik), Museum für die dekorativen Künste 13 (Belgrade, 1969): 59–66; My earlier conjecture that no dates existed for this region must be corrected; see G. Dohrn-van Rossum, "The Diffusion of the Public Clocks in the Cities of Late Medieval Europe 1300–1500," in B. Lepetit and J. Hoock, eds., *La ville e l'innovation* (Paris, 1987), 38.

107. "Veruntamen confitendum erit, ob imperitam antiquitatem et simplicitatem vetusti seculi, tardissime ad gentes Aquilonares eam dierum, horarumque distictionem, qua nunc alie nationes utuntur (immo et ad usum campanarum) pervenisse. Nunc autem in Aquilonari plaga fere tota adeo perfecta, facilia et iusta horologia, tam domestica, quam externa arte fabricata sunt, ut nil amplius, quam sagax directio eisdem competere videatur . . . Quod autem tardius modernorum usus pervenerit ad Septentrionales regiones, mirum non est, cum Roma omnium rerum curiosissima, longissimo aevo indiscretam lucem habuisse perhibeatur, donec Scipio Nasica . . . [there follows an account based on Pliny, *Hist. nat.* VII, 215]." Olaf Magnus, *Historia de gentibus septentrionalibus* . . . (Rome, 1555; repr. 1971), 52 f.

Chapter 6

1. Maurice, *Räderuhr* I, 26.

2. A. Hauber, *Planetenkinder und Sternbilder* (Strasbourg, 1916). Since edited textual sources are relatively rare, I should like to give one example of the description of the children of Mercury: "Sin kind ist gar wol gesprech vnd maisterlich red vnd ru(e)mat sich selber geren vnd fraget gern nach grossen kunsten vnd wurt gewunlich ain maister der redkunst vnd predigens der astronomi der geometri der mensur des ertrichs vnd der figur des gestirnes, der speren des hymels vnd der elementen beschayden kan ain disputirer vnd ain maister naturlichen sachen vnd zaubry vnd ander ku(e)nsten mit den man kunftigew ding bevindet. vnd ain maister hu(e)bscher spru(e)ch vnd ain mayster dichtens vers machens vnd grosser rechnung, malens snitzens vnd ergrabens vnd entwerffens vnd hat wunderlich betrachtung vff die kunst vnd ainen vnmu(e)ssigen sin vnd begrift was er wil vnd behebt es lang vnd is vnstat vnd beweglich vnd kumpt gern in fremdw lant." *Astronomisch-astrologisches Lehrbüchlein,* after a text (dated 1404) in the F. Öttingen-Wallersteinschen Bibliothek zu Maihingen, ed. III (German) 14° I, fol. 25, quoted from V. Stegemann, *Aus einem mittelalterlichen deutschen astronomisch-astrologischen Lehrbüchlein*

(Reichenberg, 1944; repr. 1973), 53; compare R. Dieckhoff, "Antiqui-moderni. Zeitbewußtsein und Naturerfahrung im 14. Jahrhundert," in A. Legner, ed., *Die Parler und der schöne Stil* (Exhibition Catalog), vol. 3 (Cologne, 1978), 83.

3. From "De Sphaera," Biblioteca Estense Modena, MS lat. 209, quoted from "Arti, Mestieri, Tecnichi" (Exhibition Catalog) (Modena, 1983), 65.

4. An account of the theoretical literature in Sternagel, *Artes Mechanicae.*

5. The comparison of cosmos-creator with clockwork-clockmaker in Nicolas Oresme, *Tractatus de commensurabilitate* . . . (ca. 1360), ed. by E. Grant, *Nicole Oresme and the Kinematics of Circular Motion* (Madison, 1971), 294; Nicolas Oresme, *Le livre du ciel et du monde* (1377), ed. by A. D. Menuit and A. J. Denomy (Madison, 1968), I.29 and II.2, p. 288; Heinrich von Langenstein, *Lectura super Genesim* (1385–1393), comparison of creator-god with clockmaker, creatures with clockwork, unpublished; compare N. H. Steneck, *Science and Creation in the Middle Ages: Henry of Langenstein* (died 1397) (Notre Dame, Ind., and London, 1976), 92, 112, and elswhere.

6. J. Abeler, *Meister der Uhrmacherkunst* (Wuppertal, 1977); G. H. Baillie, *Watchmakers and Clockmakers of the World,* vol. 1 (London, 1972); J. L. Bassanta Campos, *Relojeros de España. Diccionario Bio-Bibliografico* (Museo de Pontevedra, Pontevedra, 1972); B. Loomes, *The Early Clockmakers of Great Britain* (London, 1982); E. Morpurgo, *Dizionario degli orologiai italiani,* 2nd ed. (Milan, 1972); idem, *Nederlandse klokken—en horlogemakers vanaaf 1300* (Amsterdam, 1970); Tardy (H. Lengellè), *Dictionnaire des Horlogers Français,* 2 vols. (Paris, 1972).

7. L. White, Jr., "The Expansion of Technology 500–1500," in C. Cipolla, ed., *The Fontana Economic History of Europe,* vol. 1 (New York, 1972), 160.

8. On the role of clockmakers as machine builders during the early phase of industrialization see A. E. Musson and E. Robinson, *Science and Technology in the Industrial Revolution* (Manchester, 1969).

9. See, for example, Cipolla, *Clocks,* 47–49.

10. Goldsmiths as clockmakers according to K. Maurice, *Die französische Pendule des 18. Jahrhunderts. Ein Beitrag zu ihrer Ikonographie.* Neue Münchner Beiträge zur Kunstgeschichte, vol. 9 (Berlin, 1967), 93; clockmakers as a small elite from the fifteenth century: D. Sella, in C. C. Cipolla and K. Borchardt, eds., *Europäische Wirtschaftsgeschichte,* vol. 1 (Stuttgart, 1978), 238.

11. Clockmakers and bell-founders as traveling craftsmen: K. Maurice, *Räderuhr,* I, 43; on the wanderings of specialized craftsmen see Sella (note 10), 255.

12. Hermann Josef (chap. 4, note 51); Beaulieu, St. Paul's in London (chap. 4 note 141); Robertus Anglicus (chap. 4 note 125).

13. Landes, *Revolution,* 56 f.; see also North, "Monasticism," 384 f.

14. See chap. 4 note 122.

15. H. Géraud (chap. 3 note 25), 162; he is listed in Tardy, *Dictionnaire* (note 6) as the earliest name in the French-speaking regions.

16. F. Godefroy, *Lexique de l'Ancien Français* (Paris, 1901), 18; *Le livre de taille de Paris l'an 1296/1297,* ed. by K. Michaelsson, Romanica Gothoburgensia 7, 9 (Göteborg, 1958), 1962.

17. On Pipelart and Gilebert see chap. 4 note 129.

18. Sens: Julliot (chap. 4 note 52), 389; he is described as an English clockmaker in Paris in Tardy, *Dictionnaire;* the frequency of the last name "l'anglais" is apparent from the tax rolls of 1296 and 1297 (note 16) as well as from the roll of 1313, ed. by K. Michaelsson (Göteborg, 1951), passim.

19. Noyon 1333–1334: "cura de pulsatio orologiorum . . . magister orlogiorum . . . ad videndum si esset aliquid corrigendum dictis orologiis," Archives de département Oise, G. 1380, compare *Revue des Soc. savantes* 3 (1862): 587; Hesdin (chap. 4 note 119); J. Houdoy, *Histoire artistique de la Cathédrale de Cambrai* (Lille, 1880), 158.

20. Avignon 1329: K. H. Schäfer, *Die Ausgaben der apostolischen Kammer unter Johann XXII* (Paderborn, 1911), 772; B. Guillemain, *La cour pontificale d'Avignon 1309/1376* (Paris, 1966), 401; Mondinus of Cremona (chap. 4 note 172), Johannes de Scilo (note 14); R. Ebeling, ed., *Das älteste Stralsunder Bürgerbuch (1319–1348)* (Stettin, 1926), 71.

21. I. H. Jeayes, ed., *Court Rolls of the Borough of Colchester,* vol. 1 (1310–1352) (Colchester, 1921), 35 f.; W. Hudson, ed., *The Three Earliest Subsidies for the County of Sussex,* Sussex Record Society 10 (Lewes, 1910), 229; see G. Fransson, *Middle English Surnames of Occupation 1100–1350* (Lund, 1935), 136.

22. Compare *Middle English Dictionary,* vol. 7 (Ann Arbor, 1980), 296.

23. C. Wordsworth, *Statutes of Lincoln Cathedral* (chap. 4 note 140), pt. 2, p. CXVI f.; Norwich 1322–1325 (chap. 4 note 171); St. Albans: *Gesta Abbatum,* ed. by H. T. Riley, Rer. Brit. M. A. script. 28 (1867), 335, 383, 385.

24. For the early period see, in addition to G. Fransson (note 21), also T. Erb, "Die Berufsbezeichnung der Handwerker im Mittellatein," dissertation (Berlin, 1973); idem, "Probleme der Terminologie der Handwerksberufe . . . ," Ungarische Akademie der Wissenschaften, ed., *Internationales Handwerksgeschichtliches Symposion* (Veszprém, 1978), 395.

25. On the construction of astronomical clocks see esp. the works by E. Poulle and J. D. North; L. White, Jr., "Medical Astrologers and Late Medieval Technology," *Viator* 6 (1975): 295–308.

26. The sources in J. D. North, *Richard of Wallingford.*

27. Chap. 5, note 17.

28. Whether bestowed by someone, attributed, or assumed within the family, such honoring epithets were nor unusual in this period; compare in Venice (chap. 5, note 29), Johannes ab horelegio, Mantua 1402, Davari (chap. 4 note 217), 7. Such epithets also appear in the German-speaking regions; however, the boundary between this and an occupational designation that is not inherited is indistinct: Peter van der uyrclocken (Aachen 1373), J. G. Laurent, *Aachener Zustände im 14. Jahrhundert auf Grund von Stadtrechnungen* (Aachen, 1876), 237, 249; Magister Johannes de orologio (Klosterneuburg 1429–1430), F. Röhrig, "Das kunstgeschichtliche Material aus den Klosterneuburger Rechnungsbüchern des 14. u. 15. Jahrhunderts," *Jahrbuch des Stifts Klosterneuburg* N. F. 6 (1966), 157; the Ulm patrician Otto Roth (died 1422) was given the epithet Horglogg presumably because he donated a clock, from E.

Zinner, "Zur Ulmer Sternkunde und Uhrmacherei im Mittelalter," *Berichte der Naturforschenden Gesellschaft Bamberg* 39 (1947): 6 f.

29. On the biography of Giovanni Dondi and on the history of the astrarium see Bedini and Maddison, *Mechanical Universe;* the text has been edited by E. Poulle, *Opera omnia Jacobi et Johannis de Dondis. Fac-simile du manuscrit, edizione critica della versione A, trad. franc.* (Padua and Paris, 1987–.); older edition: A. Barzon, E. Morpurgo, A. Petrucci, and G. Francescato, *Giovanni Dondi dall'Orologio, Tractatus Astrarii* (Bibl. Capitolare di Padova, Cod. D. 39) (Rome, Vatican City, 1960); compare Poulle, *Équatoires;* bibliography: G. Bozzolato, "Le opere edite e inedite, le fonti e la bibliografia su Jacopo e Giovanni Dondi dall'Orologio," *Boll. del Centro Internazionale A. Beltrame di Storia dello Spazio e del Tempo* 2 (1984): 75–102; rev. ed. in *Padua Sidus Preclarum. I Dondi dall'Orologio e la Padova dei Carraresi* (Padua, Edizioni 1+1, 1989). Dondi's astrarium was a clockwork-driven planetarium, as was presumably also Richard of Wallingford's Horologium Astronomicum. Unlike clocks with simple astronomical indications and simulations (lunar phase, signs of the Zodiac) and astrolabe dials, genuine planetaria were rare! In the Middle Ages the descriptive designation astrarium was given only to Dondi's clock and to the copy of this clock by Regiomontan at the end of the fifteenth century; on these distinctions see Poulle, *Équatoires.* Because of the prominent place in which they have been published, it might be useful to point out the considerable accumulation of mistakes in the entry "Astrarium" in *Lexikon des Mittelalters,* vol. 1 (1980), cols. 1134 f. The opening part of the discussion emphasizes as the basic "intellectual idea" behind the construction of the astrarium in 1344 (Dondi's calculations were in fact for 1365; the author has presumably confused his clock with the tower clock that Jacopo Dondi built in Padua) N. Oresme's conception of the universe as a clockwork (which he formulated around 1377!). Subsequently all of the more or less well documented astronomical simulations are included among the astraria. Vitriuvius (IX.9.1) supposedly mentioned a planetary clock (in fact Vitrivius is talking about a water-clock-driven astrolabe IX.8.8–15). As an example of royal patronage we are told about the astronomical clock that was sent to Frederick II in 1332 (correct date 1232!). The event is cited following the account of Trithemius, which is later and sounds more mechanical. Next the automata clock in Damascus is grouped among the astraria, something the cited author E. Wiedemann never asserted. Even the "orolei" in the poetic work *Jüngerer Titurel* is included as a diminished form. Next the clock of Richard of Wallingford, abbot of St. Alban in Hastingshire (correct: St. Albans in Hertfordshire), is singled out for its economic importance owing to its indication of the tides, a claim that is open to debate given that the device was so far from the sea. For the clock on (correct: in) the Strasbourg Minster in 1352 (better: 1352–1354) the article names as the constructor the legendary Jean Bournave, unknown in the Middle Ages, who "is said to have learned his craft among the Arabs" (on this see Maurice, *Räderuhr* I, 36 ff.). Finally, some public clocks are included in the list of astraria.

30. E. Zinner, *Leben und Wirken des Joh. Müller von Königsberg, genannt Regi-*

omontanus, 2nd ed. (Osnabrück, 1968), 180, 215 ff.; E. Poulle, "L'Horloge planétaire de Regiomontanus," in G. Hammann, ed., *Regiomontanus-Studien* (Vienna, 1980), 335–341.

31. L. Thorndike, *A History of Magic and Experimental Science,* vol. IV (New York, 1934), 150 ff.; F. D. Prager, "Fontana on Fountains," *Physis* 13 (1971): 341–360; E. Battisti and G. S. Battisti, *Le macchine cifrate di Giovanni Fontana* (Milan, 1984); a depiction of a rather plain wooden water clock can be found in a Turin manuscript of Robert Valturio's *De re militari,* ed. by S. Ricossa, *Le macchine di Valturio* (Turin, 1988), facs.—fol. 49 v.

32. Cambrai, "Compotus orologiorum" 1348–1349: "pro quodam nuntio misso Ambianis ad magistrum Johannem de Linieres," Dehaisnes, *Documents* (chap. 4 note 172), 365 f. Amiens was the birthplace of Johannes de L., who was active in Paris between 1320 and 1335 and died before 1355. Of course this does not confirm the identity of the person, since this surname was not rare in Amiens at that time.

33. E. Poulle, *Un Constructeur;* idem, "Un atelier parisien de construction d'instruments scientifiques au XVe siècle," in P. Benoit and D. Cailleaux, eds., *Hommes et Travail du Métal dans les villes médiévales* (Paris, 1988), 61–67.

34. A. Pinchart, "Archives des Arts, des Sciences et des Lettres," in *Mess. Sc. Belg.* (1884): 188 f.; Poulle, *Constructeur,* 27; idem, *Équatoires,* 503.

35. Symon de Phares, *Receuil des plus célèbres astrologues et quelques hommes doctes,* ed. by E. Wickerheimer (Paris, 1929), 261; cited from G. Le Cerf and E. R. Labande, *Instruments de Musique du XVe siècle. Les Traités d'Henri-Arnaut de Zwolle* (Paris, 1932), p. XIII.

36. A. L. Isaacs, *The Jews of Majorca* (London, 1936); guides to the extensive specialized literature are D. Romano, "Les Juifs de la Couronne d'Aragon, avant 1391," *Revue des Études Juives* 141 (1982): 169–182; and A. Alvaro Santamaria Arandes, "Mallorca en el Siglo XIV," *Anuario de estudios medievales* 7 (1970–1971): 165–238.

37. Jewish clockmakers and instrument builders at the court of Aragon: A. Rubió y Lluch, *Documents per l'historia de la Cultura Catalana Mig-eval,* 2 vols. (Barcelona, 1908–1921), I, nos. 201, 202, 284; II, nos. 79, 85, 88, 106, 118, 129, 141, 143, 144, 175, 181, 197, 253, 258, 268, 290, 292.

38. M. Millás Vallicrosa, in *Serafad* 6 (1946): 165.

39. Bartolomeo Manfredi (died 1478): Davari (chap. 4 note 217); E. Poulle, "L'Équatoire de Guillaume Gilliszoon de Wissekerke," *Physis* 3 (1961): 223–251; Johannes Stöffler: Zinner, *Räderuhren,* 34, 57; C. Maccagni, "The Florentine Clock and Instrumentmakers of the della Volpaia Family," *Actes du Congrès Intern. d'Histoire des Sciences* 10a (Paris, 1968): 65–73; Poulle, *Équatoires,* 653 ff.

40. See Turner, *Astrolabes,* 37 ff.

41. See the critical and apt remarks by G. Bayerl, "Technische Intelligenz im Zeitalter der Renaissance," *Technikgeschichte* 45 (1978): 336–353.

42. Galvano Fiamma, *Opusculum* (chap. 4 note 176), 41.

43. L. Osio, *Documenti diplomatici tratti dagli archivi Milanesi,* vol. 1 (Milan, 1864), 117.

44. Compare chap. 4, The first small clocks. H. Saalman, ed. (chap. 4 note 211), 51 ff., 103.

45. E. Cantù, ed., *Annali della Fabbrica del Duomo di Milano* (Milan, 1877 ff.); Francisco Pessono: I, 225, 256 f.; A. Angelucci, *Documenti inediti per la storia delle armi da fuoco italiane* (Turin, 1869), 107 ff.; Giovanni di Zellini: *Annali* I, 262, 264; Filippo da Modena: *Annali,* Appendici I, 306.

46. G. B. Gallicciolli, *Delle memorie venete . . .* (Venice, 1795), 358, after D. Calabi and P. Morachiello, *Rialto: Le fabriche e il ponte* (Turin, 1987), 95; F. Scaramella, "Il primo orologio pubblico a Verona," *La Clessidra* 8/3 (1952): 22 f.

47. Bedini and Maddison, *Mechanical Universe,* doc. 13, p. 51 f.; M. Savanarola, "Commentariolus de laudibus Patavii," Muratori 24 (1731), col. 1164.

48. Bedini and Maddison, *Mechanical Universe,* 25.

49. Antonio de Tritio: *Dizionario biografico degli Italiani,* vol. 1 (Rome, 1960), s. v. "Antonio da Trezzo," Ragusa (chap. 5 note 108).

50. For the most part these experts undoubtedly stood outside of the social structure of feudal society. However, G. Bayerl's inquiring conjecture that they may also have contributed to the transformation of feudal society (note 41, 340 f.) does not strike me as plausible. By compensating for structural obstacles to modernization, for example, they may also have contributed to stabilizing it.

51. A. Bossuat (chap. 5 note 38), 101.

52. The view of Rolf Sprandel: "Die Ausbreitung des deutschen Handwerks im mittelalterlichen Frankreich," *Vierteljahresschrift für Sozial- und Wirtschaftsgeschichte* 151 (1964): 66–100, 93.

53. Avignon (chap. 5 note 38).

54. On the significance of lead and the profession of plumbers for medieval construction see P. Benoit, "Le plomb dans le bâtiment en France à la fin du Moyen Age," in P. Benoit and O. Chapelot, eds., *Pierre et Métal dans le bâtiment au Moyen Age* (Paris, 1985), 339–355.

55. Delft: Rymer, "Foedera" (chap. 5 note 84); Ragusa (chap. 5 note 108); Valencia (Lerida: chap. 5 note 107).

56. Pierre Merlin (chap. 4, note 52, chap. 5 note 38); Girardin Petit: P. Pansier, *Histoire de la Langue Provençale à Avignon,* 5 vols. (Avignon, 1924–1932; repr. 1974), II, 109; Renouvier and Ricard (chap. 5, note 1), E. de Laplane, *Histoire de Sisteron,* vol. 2 (Digne, 1843), 246 ff.; "Henricus Halder horelogifex di Basilea," H. Witte, ed., *Urkundenbuch Strassburg,* vol. 7 (Strasbourg, 1900), 459 f.; as citizen and locksmith in Basel 1385–1410, Fallet-Scheurer, *Zeitmessung,* 294 ff.; Weber, *Luzerner Bürgerbuch* (chap. 5 note 85); Claus Gutsch: F. X. Kraus, *Kunst und Altertum im Unter-Elsaß* (Strasbourg, 1876), 385; Villingen (chap. 4 note 172).

57. Mons 1372–1379: A. Wins, *L'Horloge à travers les âges* (Paris, 1924), 111, Dehaisnes, *Documents* (chap. 4 note 172), 556; Cologne and Frankfurt 1372: Zinner, *Räderuhren,* 108; Orléans 1452–1454: P. Vreyer et al., *Inv. som arch. com. Orléans* (1907); G. Lavergne, *Moulins inédit. L'enfance de Jaquemart* (Meaux, 1908).

58. Meister Marquardt: Municipal Archive Braunschweig B 1, no. 2 (cartu-

lary), fol. 54 v., 65 v.; Werner Hert: K. Zülch, *Frankfurter Künstler 1223–1700* (Frankfurt, 1935); Lazarus Kreger: I. A. Jenzen, "Die Monumentaluhr im Frankfurter Dom," in idem, ed., *Uhrzeiten,* 43 ff.; Andernach: LHSTA Koblenz, Bestand 612, according to information kindly furnished by the archivist.

59. C. Milanesi, *Documenti per la storia dell'arte senese* (Siena, 1854), 326 f.; *Memorie originali italiane risguardanti le Belle Arti,* ser. IV a (Bologna, 1843), 145 ff.

60. Hinrik von dem Hagen: K. F. Leonhardt, *Das älteste Bürgerbuch der Stadt Hannover* (Leipzig, 1933), 96; Leonard Wunderlich 1456: *Bürgerbuch,* ed. by K. Kaczmarczyk, Wydawnictwa Archivum Aktów Dawnych Miasta Krakowa (Cracow, 1904–1915), V, 211, no. 6340; Hans Graff: H. Dopsch, ed., *Geschichte Salzburgs,* vol. 1/2 (Salzburg, 1983), 1417.

61. Giovanni de Fabriano 1375 in Orvieto: L. Fumi, *Il Duomo di Orvieto e suoi ristauri* (Rome, 1891), 459; Onofrio de Fabriano 1433 in Iesi: G. Annibaldi, *Il Lucagnolo. Ovvero sull'oreficeria di Jesi* (1879), 14; Nicolo di Bonandrea de Fabriano 1443 in La Spezia: Belgrano, *Antichi orologi* (chap. 4 note 155), 52 f.; fr. Jacobus de Fabriano 1460–1480 in Rome: E. Müntz, *Les Arts á la Cour des Papes* I (Paris, 1878), 305; *Almanus-MS,* ed. Leopold, no. 19; Tebaldus Persiani from Fabriano 1523 in Todi: P. Alvi, *Todi, città illustre nell'Umbria, cenni storici* (Todi, 1910), 305.

62. Bourges (chap. 5 note 39); Leguay (chap. 5 note 81), 120, 206 f., 272; compare R. Sprandel, "Ausbreitung" (note 52), 79.

63. For Rome see C. L. Maas, *The German Community in Renaissance Rome 1378–1523* (Rome, 1981).

64. Florence 1407 ff.: A. S. Florenz, *Arte del chiavaiuoli, ferraiuoli e calderai,* no. 3; Cracow 1410: J. Ptasnik, *Cracovia Artificium 1300–1500* (Cracow, 1917), nos. 177, 400; Basel: from 1413 Hensli Halde in the manual of the guild of smiths, Fallet-Scheurer, *Zeitmessung,* 291; Andernach: "Mitgliederliste der Bruderschaft der Schmiedezunft (15 Jh.)," ed. by E. Schulte, *Vierteljahresschrift für Wappen-, Familien- und Siegelkunde* 40 (1912): 141, 143; Nuremberg: Maurice I, 299 f.; E. Groiss, "Das Augsburger Uhrmacher-Handwerk," in Maurice, *Welt als Uhr,* 63 ff.

65. G. Hertel, ed., *Urkundenbuch der Stadt Magdeburg,* vol. 2 (Halle, 1894), 146; H. v. Voltelini, "Urkunden und Regesten aus dem K. U. K. Hof- und Staatsarchiv," *Wiener Jahrbücher* 19 (1898): I, II.

66. Chap. 5 note 44.

67. See G. Dohrn-van Rossum, "Uhrmacher," in H. Reith, ed., *Lexikon des alten Handwerks,* 2nd ed. (Munich, 1991), 246–252 with additional references to the literature.

68. See P. Sasson, "Quelques aspects sociaux de l'artisanat Bruxellois du métal," *Les Cahiers Bruxellois* 6 (1961): 98–111. As one example for the "valets du chambre" and their activities for the court see the material on Pierre Lombard, Pinchart (note 34), 189 ff.

69. W. Reininghaus, ed., *Quellen zur Geschichte der Handwerksgesellen im*

spätmittelalterlichen Basel (Basel, 1982), 45; Abbé Requin, "Documents inédits sur les origines de la typographie," *Bull. Hist. et Philol. du Comité des travaux historiques* (1890): 32–50; A. Swierk, "Was bedeutet 'ars artificialiter scribendi'?" in H. Widmann, ed., *Der gegenwärtige Stand der Gutenberg-Forschung* (Stuttgart, 1972), 241–250; A. Bardon, *Histoire de la ville de Alais* (Nîmes, 1896), 207, 282.

70. In a charter (dated July 31, 1593) signed by the citizens of Ems, we find this among the members of the guild of smiths: "derwyl jan Uhrwerker nicht schryven kann so hefft syn marck [a sign] hyr onder schrivet," *Zeitschrift für Geschichte und Altertumskunde Westfalens* 10 (1872): 244, 258 with plate.

Chapter 7

1. Examples: in Tournoi the city council ordered that the clock in the belfry should strike the "bancloque" (1377), A. Hocquet, "Chroniques de Franche, d'Engleterre . . ." *Soc. bibliophiles belges* 38 (Mons, 1938): 204 f. In Enghien in 1430 an order went out for "ung orloige sour le grand cloque," E. Mathieu, "Histoire de la ville d'Enghien," *Soc. des Sciences, des Arts et des Lettres du Hainaut* I (1876): 294 ff.

2. See Battard, *Beffrois;* Michel, *Hôtels de Ville;* Patard, *Cloches,* passim.

3. Quoted from Michel, *Hôtels de Ville,* 22; see Patard, *Cloches,* 81; compare the city charter given to St. Valery (county of Artois) in 1376 by Jean of Artois: "Item nous avons donné et accordé Echevinage, Ban-cloque grande et petite, pilori, scel, et banlieue . . ." F. Ragueau, *Glossaire du Droit François . . .* (rev. ed. by E. de Laurière and L. Favre; repr. Geneva, 1969), s. v. "Beffroy," 80.

4. H. Keller, *Die Entstehung der italienischen Kommunalpaläste als Problem der Sozialgeschichte* (1976), 206 f.

5. On *Herrschafts-, Gerichts-* and *Pfarrbezirk* see *Deutsches Rechtswörterbuch* IV, cols. 951, 952 (s.v. "Glocke" B II–IV) with many references; factory novel by Walter Bloem, *Das jüngste Gericht* (Leipzig, n.d.), 118.

6. It would seem that punishments became increasingly harsh: Keure von Middleburg 1217: "Quicunque campanam sine communi consilio pulsaverit, emedabit comiti III lib., oppido libram," W. S. Unger, *Bronnen tot de Geschiedenis van Middelburg,* part 1 (The Hague, 1923), 6; Statutes of Savona (1345) I, chap. 178: misuse of the "campana Brandalis" to be punished by amputation of the right arm, unless it could be shown that it was done for the good of the city, compare II, chap. 86, L. Balletto, ed., *Statuta antiquissima Saone* (Bordighera, 1971), vol. 1, 212; compare vol. II, 98; in Commines in 1361 we find the threat of capital punishment if the bell is rung as a call to revolt; ORF vol. IV, 209, cited from J. Le Goff, "Labour Time," 47; on the punishments see also E. Lippert, *Glockenläuten,* 47 ff.

7. Galvano Fiamma, *Chronicon extravagans,* ed. by A. Ceruti, in Miscellanea di Storia Italiana 7 (1869), 453.

8. *Chroniken deutscher Städte* 4 (Augsburg, vol. 1), ed. by K. Hegel (Leipzig, 1865; repr. 1965), 21.

9. Such punitive measures against the city tower and bells were common.

In addition to the above-mentioned incidents in Cambrai, Lyons, Béthune, and Agen, comparable occurrences were reported for Hesdin 1197 (see Le Goff, "Labor Time," 48), Fosses 1302, Corbie 1310, Laon 1331, Tournai 1332, Peronne 1360, Roye 1373, Montpellier 1379, Ghent 1539; for the sake of brevity I have dispensed with giving references to each case.

10. Numerous examples can be found in visitation reports and in damage reports from the Peasants' War, for example, W. P. Fuchs, *Akten zur Geschichte des Bauernkriegs in Mitteldeutschland,* vol. II (Jena, 1942; repr. 1964), 30, 76, 512, 524, 526, 598, 708, 929; see also D. Stockmann, "Der Kampf um die Glocken im deutschen Bauernkrieg," in *Der arm man 1525. Volkskundliche Studien,* ed. by H. Strobach (Berlin, 1975), 309–340.

11. Jean Bodin, *Les six livres de la République* IV.7 (Paris, 1583; repr. 1961), 657.

12. L. Fèbvre, *The Problem of Unbelief in the Sixteenth Century,* transl. by Beatrice Gottlieb (Cambridge, Harvard University Press, 1982), 351.

13. The felicitous expression "place of the norm" ("Ort der Norm") has been taken from C. Meckseper, *Kleine Kunstgeschichte der deutschen Stadt im Mittelalter* (Darmstadt, 1982), 195 ff.

14. See L. Röhrich, *Lexikon der sprichwörtlichen Redensarten* (Freiburg, 1973), s.v. "Glocke."

15. H. Keller (note 4); J. Paul, "Italienische Kommunalpaläste des Mittelalters in Italien" (dissertation, Freiburg, 1963; Cologne, 1963); P. Racine, "Les palais publics dans les communes italiennes (XIIe–XIIe siècles)," in *Le paysage urbain au Moyen-Age* (Lyons, 1981), 133–153.

16. In Metz, for example, the tower of the cathedral was until 1907 the property of the city; R. Bour, *Histoire de Metz* (Metz, 1983), 105; A. Erler, *Das Sraßburger Münster im Rechtsleben des Mittelalters* (Frankfurt, 1954), 34 ff.

17. The expression "acoustic environment" ("akustische Umwelt") has been taken from D. Stockmann (note 10); the same author deals with the problems involved in research in "Die Glocke im Profangebrauch des Spätmittelalters," in *Studia instrumentorum musicae popularis* III (Festschrift E. Emsheimer), ed. by G. Hilleström (Stockholm, 1974), 224–232.

18. T. Esser, "Das Ave-Maria-Läuten," *Historisches Jahrbuch der Görres-Gesellschaft* 23 (1907): 27.

19. See Bilfinger, *Horen,* 165; and Stockmann, "Profangebrauch," 225 f.

20. Esser, "Ave-Maria-Läuten" (note 18); for the development in Italy see A. Lattes, "La campana serale nei secoli XIII e XIV secondo gli statuti delle città italiane," in F. Novati, *Indagini e Postille Dantesche, serie prima* (Bologna, 1899), 163–176, and F. Novati, "La 'squilla di lontano' (Dante, Purg. VIII.5) è quella dell'Ave Maria?" ibid., 139–150.

21. See P. Rossiaud in *Histoire de la France urbaine,* vol. 2, La ville médiévale, ed. by J. Le Goff (Paris, 1980), 562.

22. R. Davidsohn, *Geschichte von Florenz,* vol. 1 (1896; repr. Osnabrück, 1969), 731 f.

23. Bonvesin de la Riva, "De magnalibus urbis Mediolani," ed. by A. Paredi,

Bonvesin de la Riva, Grandezze di Milano, Latin-Italian (Milan, 1967), 43; compare A. Murray, *Reason and Society in the Middle Ages* (Oxford, 1978), 183.

24. Ed. by F. Gianani, *Opicino de Canistris L'Anonimo Ticinese" (Cod. vaticano Palatino latino 1893)* (Pavia, 1927), 110 f.

25. In the vast literature on campanology there are unfortunately few attempts to document the municipal civic use of bells; helpful in this regard is H. Otte's *Glockenkunde,* 2nd ed. (Leipzig, 1884); E. Lippert, *Glockenläuten,* compiles the material of the German legal dictionary; Patart, *Cloches,* has assembled the sources for the civic use of bells in three Belgian cities.

26. Based on Patart, *Cloches,* 75–141.

27. Based on A. Marchesan, *Treviso Medievale* I (Treviso, 1923; repr. 1971), 111–120.

28. After R. Cessi, *Deliberazioni del Maggior Consiglio di Venezia* II, III (Venice, 1931–1934), passim; G. Monticolo, ed., *I capitolari delle Arte Veneziane . . . dalle origini al MCCCXXX,* Fonti per la Storia d'Italia, vol. 27, 2, (Rome, 1905), 664–670; and F. Sansovino, *Venetia. Città nobilissima* (Venice, 1663; repr. 1968), 294 ff.

29. See, for example, the persistent conflict between the knights of the Teutonic Order and the Augustinian friars in Münnerstadt about ringing, services, and preaching between 1280 and 1284; A. Zumkeller, ed., *Urkunden und Regesten zur Geschichte der Augustinerklöster Würzburg and Münnerstadt,* Regesta Herbipolensia V, part 2 (Würzburg, 1967), nos. 879, 880, 888. In Piacenza, for example, it was decreed in 1297 that all city churches and monasteries, including those of the predicant monks, should give precedence to the "ecclesia matrix" in ringing the Hours; the regulation was renewed in 1337, apparently because a Carmelite monastery had been set up in the city; *Statuta varia Civitatis Placentiae,* ed. by G. Bonora (Parma, 1860), 532, 541; on the episcopal cities see, for example, C. Trexler, *Synodal Law in Florence and Fiesole 1306–1518* (Rome, Vatican City, 1971), 31, 209, 237.

30. *Paradiso* (chap. 4 note 132), XV, verses 97–99.

31. A public clock in Florence around 1325 is cautiously suggested by Y. Renouard; see *Les hommes d'affaires* (chap. 1, note 25), 240, and "Affaires et Culture" in his *Études d'histoire médiévale* (Paris, 1968), 491. This detail is not insignificant for Renouard's argument about the pioneering role of the mercantile metropolises in the introduction of the new method of reckoning the hours. A. Sapori, in his review of J. Le Goff's "Temps de l'Église" turned this into a fact and elaborated upon it with some fanciful details: the first clock at the Palazzo dei Signori, he says, struck the equal hours since 1325 and indicated them on a dial with 12 (!) sections. V. Stamm, *Ursprünge der Wirtschaftsgesellschaft* (Frankfurt, 1982), 109, also believes that this phantom clock justifies some critical objections to G. Bilfinger's work. Bilfinger did encourage the misunderstanding when (*Horen,* 181–183) he interpreted a time indication concerning an earthquake in the chronicle of Giovanni Villani (1325, May 21: "dopo il suono delle tre ore") as an indication of a night hour, though he did

not interpret it as a reference to a clock. Bilfinger used an old edition of the chronicle: *Giovanni Villani, Florentini Historia Universalis,* Muratori 13 (1728), col. 571. A newer edition, *Cronica di Giovanni Villani,* ed. by F. Gherardi Dragomanni, vol. 2 (Florence, 1845; reprint 1969), 324, only has the wording "dopo il suono delle tre," and this could mean the third hour of the night but also a three-fold striking of the bell.

32. *Benvenuto da Imola,* ed. by J. P. Lacaita, vol. 5 (Florence, 1877), 144 f.; *Iacopo della Lana* (Milan, 1865), 403; on the Florentine work bell see chap. 9, Work bells (*Werkglocken*)

33. R. Davidsohn, *Forschungen zur Geschichte von Florenz* IV (Berlin, 1908), 386.

34. L. Frati, ed., *Statuti di Bologna dall'anno 1245 all'anno 1267,* vol. 3 (Bologna, 1877), 230: 1260–1262: "(custodes) sonent campanam (ad sogam) tantum quod possit iri per unum miliare." 1267: "(debeant sonare) ad martellum dando . . . XV. percussiones raras una post aliam et v. percussiones spissas." Compare the regulations concerning work bells in Amiens and York, 160 f.

35. G. Bonora, ed. (note 29), chap. I, 35, p. 231; statutes drawn up between 1313 and 1322, confirmed in 1391.

36. A. Giffard, ed., *Ordonnances de J. d'Ableiges pour les métiers d'Evreux (1385–1387)* (Caen, 1913), 14.

37. See the ecclesiastical statutes of Lüneburg (1564), in Sehling (chap. 5 note 74), vol. VI, 541.

38. Sehling, *Kirchenordnungen* (chap. 5 note 74), vol. I, 177 f.; vol. VI.1, 180.

39. Hermann, *Ernestinische Lande,* 106 f.

40. J. B. Götz, *Die erste Einführung des Kalvinismus in der Oberpfalz 1559–1576* (Münster, 1933), 119.

41. Municipal Archive Nuremberg.

42. See Borst, *Computus,* 73; Wendorff, *Zeit und Kultur,* 145.

43. Drawing: R. Mielke for the exhibit "Stadt im Wandel," (Brunswick, 1985), Cat. no. 939.

44. L. Devillers and E. Matthieu, *Chartres du Chapitre de Sainte-Waudru de Mons,* vol. 4 (Brussels, 1913), nos. 1895 and 1931.

45. J. Spöring (chap. 4 note 5), 29.

46. Marperger, *Horolographia,* 33.

47. After K. Pfaff, *Geschichte der Reichstadt Eßlingen* (Eßlingen, 1840; repr. 1979), 369.

48. P. Thome de Maisonneuve, *Le Jaquemart de Romans* (chap. 5 note 53), 169.

49. H. Plath, *Räderwerke alter Turmuhren* (Hannover, Hist. Museum, 1973), 10–12.

Chapter 8

1. Cipolla, *Clocks,* 41; P. Usher, *History,* 208 f.; J. Gimpel, *Medieval Machine,* 168, with the infelicitous phrase "authoritarian equinoctial hours;" Landes, *Revolution,* 75; J. Le Goff, "Labor Time," 50; K. Pomian, *Ordre du Temps* (Paris, 1984), 262.

2. C. de Pisan, *Livre des Fais et Bonnes Meurs . . .*, ed. by S. Solente (Paris, 1936–1940), vol. 1, 42 f.

3. Franklin, *Vie privée,* 61 f.

4. "Et ce a ordone le roy Charles premier a Paris les cloches qui a chascune heure sonnent par poins a maniere d'arloges (variant: orloges). Si comme il apiert en son palais et au boys et a sint pol. Et a fait venir ouvriers d'estrange pays a grans fres pour ce faire afin que religieux et autres gens sachent les heures et aient propres manieres et devocion di iour et de nuit (variant: pour Dieu servir). Comment que par devant on sonnast une fois a prime et deux (variant: trois) fois a tierce si n'avoit une ?mie? si certaine cognaissance des heures comme on a et puet on dire d'icelui Charles le VIe (variant: Ve) roy de france que sapiens dominabitur astris. Car luise le soleil ou non l'on scet touz jours les jours les heures sanz defaillir par ycelles cloches bien attrampes (variant: atrempées) . . ." I have quoted BN MS fr. 176, fol. 13 verso, dated 1372, in parentheses the variants in BN MS fr. 931, fol. 17 verso (MS of the fourteenth century).

5. Strangely enough the sources rarely provide any information about the acoustic indication of hour segments. We can confidently say that the half-hour stroke appeared in the fifteenth century; reports about quarter-hour strokes in the fourteenth century are usually somewhat dubious. In the fourteenth and fifteenth centuries, hour segments were determined mainly with sandglasses.

6. The literature on bells and laws relating to them known to me does not give a single comparable case.

7. G. Bilfinger, *Horen,* 224–226; references to the volumes of the Chronicles are to J. M. B. C. Kervyn de Lettenhove, ed., *Oeuvres de Froissart* (Brussels, 1867–1877; repr. Osnabrück, 1967).

8. A. Diehl, ed., *Urkundenbuch der Stadt Eßlingen,* vol. 2 (Stuttgart, 1899; repr. Osnabrück, 1967), no. 1740, p. 358, comp. no. 1743, p. 360.

9. "Hora nona de mane vel quasi secundum motum horologii ecclesiae Coloniesis," I. Joester, *Urkundenbuch der Abtei Steinfeld* (Cologne and Bonn, 1976), no. 430, pp. 358 ff.

10. This very brief sketch is based primarily on the clear discussion by P.-J. Schuler, *Geschichte des südwestdeutschen Notariats von seinen Anfängen bis zur Reichsnotariatsordnung von 1512* (Bühl and Baden, 1976).

11. P.-J. Schuller emphasizes the role of ecclesiastical courts in the spread of the notaryship, against F. Wieacker, who stresses the growing need for documents primarily in the southern German cities that engaged in trade and commerce with northern Italy; see Schuler (note 10), 29 and elsewhere; F. Wieacker, *Privatrechtsgeschichte der Neuzeit* (Göttingen, 1967), 102.

12. Their value for social history has long been recognized. So far only a few collections have been analyzed. Apart from single documents quoted above, editions exist above all for the Italian notarial protocols from the early period. However, a systematic analysis of these sources for our question is unfortunately often hampered by modern editorial practices, which drop or modern-

ize in the interest of the readers the formal elements which are of sole interest to us. The observations that follow are thus based on scattered series of notarial datings which happen to have been edited.

13. M. Chiaudano, *Contratti commerciali Genovesi del secolo XII* (Turin, 1925), 23 ff.; M. Moresco and G. P. Bognetti, *Per l'edizione dei notai liguri del sec. XII* (Turin, 1938), 50; M. Ferrando Bongionni and G. Cattaneo Cardona, "Contributo allo studi degli usi notarili medioevali: I Cartolari di Nicolo de Porta," in *Studi di Storia Medioevale e di Diplomatica* 5 (Milan, 1980), 163, 189; M. Balard, *Gênes et l'Outre-Mer,* vol. 1: Les actes de Caffa du notaire Lamberto di Sambuceto 1289–1290, pp. 17, 30 f., and vol. 2: Actes de Vilia du notarie Antonio di Ponzo 1360, p. 15 f. (Paris and The Hague, 1973, 1980); R. Mosti, ed., *I Protocolli di Johannes Pauli, un notaio del '300 (1348–1379)* (Rome, 1982); S. A. Epstein, "Business Cycles and the Sense of Time in Medieval Genoa," *Business History Review* 62 (1988): 238–260.

14. For example, "hora qua domini auditores sacri palatii causarum surrexerant de audiendis relationibus publicis" and the like, often referred to in short as "hora causarum"; these examples are from Avignon in 1340; see R. Salomon and J. Reetz, *Rat und Domkapitel von Hamburg um die Mitte des 14. Jahrhunderts,* vol. 2 (Hamburg, 1975), 121 ff.; there are many more similar time indications in this volume.

15. "Ob res scribendas bene noscit festa calendas / Mensis habet nonas, ydus per tempus et horas / Mensis et hora dies locus annis ac homo princeps." Orfinus Laudensis, "Poema de Regimine et Sapientia Potestatis," ed. by A. Ceruti, *Miscellanea di Storia Italiana* 7 (1896): 86.

16. See, for example, C. R. Cheney, *Notaries Public in England in the 13th and 14th centuries* (Oxford, 1972), 117, 137 f.

17. F. von Voltelini, *Die Südtiroler Notariatsimbreviaturen des 13. Jahrhunderts,* part 1 (Innsbruck), p. XXXI.

18. G. Orlandelli, ed., *Il Salatiele Ars Notarie,* 2 vols. (Milan, 1961); Wilhelm Durantis, *Speculum iudiciale illustratum et repurgatum a Giovanni Andrea et Baldo degli Ubaldi* (Basel, 1574; repr. 1975), 633.

19. Quoted from Schuler (note 10), 266.

20. On the requirements see *Notariatbuch / Wes einem Notarien oder Schreiber / aller seiner Practic / . . . / zu wissen / zu beraten . . . sei* (Frankfurt, Chr. Egenolf, 1543), fol. 2 verso; compare, on the other hand, *Artis notariatus, hoc est de officio exercitioque tabellionum* (Frankfurt, Chr. Egenolf, 1539), part 2, fol. 3 verso ("An autem hora sit de substantia").

21. One example from the acts of the parliament in Paris from the fifteenth century in F. Autrand, *Les Dates,* 164.

22. F. Tadra, ed., *Soudni akta konsistore Prazke (1307–1408),* vols. 1–6 (Prague, 1893–1900); vol. 2, 228; vol. V, between pp. 99 and 407.

23. K. Fischer, "Die Uhrmacher in Böhmen und Mähren zur Zeit der Gotik und der Renaissance," *Bohemia—Jahrbuch des Collegium Carolinum* 7 (1966): 33 ff.

24. R. Doehard and C. Kessemans, *Les Relations commerciales entre Gênes, la*

Belgique et l'Outremont d'après les archives notariales génoises 1400–1440 (1952); A. Roccatagliata, *Notai Genovesi in Oltremare. Atti rogati a Pera e Mitelene,* vol. 1 (Genoa, 1982), no. 37, a sandglass? "hora secunda intrantis noctis, quod est quartum horalogium noctis ineuntis"; modern hour indications also in nos. 38, 44, 93, and elsewhere.

25. F. Luschek, *Notariatsurkunde und Notariat in Schlesien von den Anfängen (1482) bis zum Ende des 16. Jahrhunderts* (Weimar, 1940).

26. C. Cenci, *Documentazione di vita assisiana 1300–1500,* 3 vols. (Grotoferrata, 1974–1976), 1125.

27. J. Le Goff, "Church's Time," 41.

28. V. Branca, ed., *Mercanti Scrittori. Ricordi nella Firenze tra Medioevo e Rinascimento* (Milan, 1986), XVI.

29. Ulman Stromer, *Püchel von meim geslechet und von abentewr,* ed. K. Hegel, Chroniken der deutschen Städte 1, Nuremberg I (Leipzig, 1862; repr. 1961), 67–70.

30. In: V. Branca (note 28), 153, 159, 278 and more; Branca's comment that Morelli, in the passage quoted last, was one of the first authors to use the "ora laica," though still in conjunction with the "ora canonica," is not accurate; such passages are exceedingly numerous.

31. C. Bec, ed., *Il Libro degli affari proprii di casa de Lapo di Giovanni Niccolini de' Sirigatti* (Paris, 1969), 60–140.

32. Leon Battista Alberti, *Il Libri della Famiglia,* in *Opere Volgari,* vol. 1, ed. by C. Grayson, Scrittori d'Italia no. 218 (Bari, 1960), 119; English transl. by Renée Neu Watkins, *The Family in Renaissance Florence* (Columbia, South Carolina, 1969), 123.

33. Compare I. Origo, *"Im Namen Gottes und des Geschäfts." Lebensbild eines toskanischen Kaufmanns der Frührenaissance,* 2nd ed. (Munich, 1986), 106, 127 and fig. 8; the last quote is missing in the German edition and is from the Italian translation of I. Origo's book (Milan, 1958), quoted from G. Brusa, *Orologeria,* 30.

34. In V. Branca (note 28), 372, 346, 347.

35. *Tagebuch des Lucas Rem aus den Jahren 1494–1541,* ed. by B. Greiff (Augsburg, 1861), 64–70; on several occasions the author also notes precisely how many hours he spent in the medicinal baths.

36. F. C. Lane, "Ritmo e rapdità di giro d'affari nel commercio veneziano del Quattrocento," in *Studi in onore di Gino Luzzato,* vol. 1 (Milan, 1949), 254–273; U. Tucci, "Alle origini dello spirito capitalistico a Venezia," in *Studi in onore di Amintore Fanfani,* vol. 3 (Milan, 1962), 547–557.

37. Of terminological interest is, for example, the expression "conto di tempi" (attested since 1459) for forward merchandise dealings that usually ran for one year; see F. Edler, *Glossary of Medieval Terms of Business, Italian Series 1200–1600* (Cambridge, Mass., 1934; repr. New York, 1970), 88 and 377 f.

38. G. Bilfinger, *Horen,* 160–162.

39. J. Martène, in Migne PL 66, cols. 411 ff.

40. *Statua Capitulorum Generalium Ord. Cist.,* ed. by J. Canivez, vol. IV (Löwen, 1936), 337.

41. Archives communales Beaucaire, GG 45; compare A. Eyssette, *Histoire administrative de Beaucaire,* vol. 2 (1889), 216 f.; M. Fournier, *Les Status et privilèges des Universités Françaises,* vol. 2 (Paris, 1891), 305 ff.; Magistretti, ed., *Beroldus sive Ecclesiae Ambrosianae . . . Kalendarium et Ordines* (Milan, 1888), 156; E. Zinner, "Die alten Domuhren in Bamberg," *32. Bericht der Naturforsch. Gesellschaft* (Bamberg, 1959): 13 ff.; paraphrase of the Colmar statues in F. Goehlinger, *Histoire du Chapitre. . .* (Colmar, 1951), 72–84; York: for example, "ad septem de orologio," "circa octo de le cloke," L. MacLachlan and J. B. L. Tolhurst, *The Ordinal and Customary of the Abbey of St. Mary York* (London, 1936), 49, 158.

42. See, for example, Vicenza (chap. 5 note 19).

43. See Introduction, The mechanical clock and the transformation of time-consciousness.

44. "Biau filz, quelle heure est-il maintenant? / Vel sic: Qu'est ce qu'a sonnee de l'oriloge?—Vel sic: Quant bien a il sonne de l'oriloge? . . . / Mon sr., si Die m'ait, je vous sai dire, mais je panse bien qu'il a sonnee dis, car il y a bien un heure passée depuis qu'il sonna neuf." "La maniere de Langage qui enseigne a parler et a écrir le français (1396)," ed. by P. Meyer, *Revue Critique d'Histoire et de Littérature* 5 (1870): 395.

45. J. Grimm and W. Grimm, *Deutsches Wörterbuch,* vol. 15 (Leipzig, 1956), col. 568 f.

46. Unfortunately the voluminous surviving statutes in Cologne can be supplemented only by a relatively meager documentation on the history of the clock in this city. A public clock probably existed in Cologne prior to 1372. The Frankfurt city accounts attest that a clockmaker went to take a look at a "werg der orglocken" in Cologne; Zinner, *Räderuhren,* 108 f.; ibid. also the hitherto ignored references to public clocks in St. Martin, St. Loren, S. Maria im Capitol, and St. Jakob. This reference and the following references are taken from two editions which, for reasons of space, are listed in summory form: W. Stein, *Akten zur Geschichte der Verfassung und Verwaltung der Stadt Köln im 14. u. 15. Jahrhundert,* 2 vols. (Bonn, 1893); therein: Tagelöhnerordnung [statute on day laborers] (1374) II, 41; Schöffenschrein- und Schöffengerichtsordnung [statute on the register and the court of the city assessors] (1335–1395) I, 555 f., 573; Salmenhandel [salmon trade] (1385) II, 69 f.; Polizeistundenregelung [regulations of curfew hours] (1398–) I, 206 f., 233, 249, 257, 388 ff., 681; II, 91 f., 98–102, 147 f., 282, 295. All other references, in particular the regulation of work time in the *Amtsbriefe* (1397), can be found in H. von Loesch, *Die Kölner Zunfturkunden nebst anderen Kölner Gewerbeurkunden bis zum Jahre 1500,* 2 vols. (Bonn, 1907).

47. On the use of "ure" to mean "hour" see Stein (note 46) I, 610 and Loesch (note 46) I, 17, II, 215.

48. Stein (note 46) I, 556; still the case in 1435 in a statute on the court of assessors: "dat soll syne, ee de sunne an den yseren payll komen sij," Stein I,

766. I have not found similar regulations in municipal statutes from the fourteenth and fifteenth centuries.

49. In Nuremberg, for example, the master artisans had themselves exempted in 1502 from attending the sessions of the council because they felt that the sessions reduced their earning potential; see A. Jegel, *Alt-Nürnberger Handwerksrecht und seine Beziehung zu anderen* (Neustadt, 1965), 3.

50. According to L. Gilliodts van Severen, ed., *Coutumes des pays e comté de Flandre* (Coutumes de la Prévoté de Bruges), vol. 1 (Brussels, 1887), 32.

51. ORF, vol. I, 728.

52. G. S. Pene-Vidari, ed., *Statuti* (chap. 5 note 51), 15 ff.

53. G. Bonora, ed., *Statuta* (chap. 7 note 35), 229; compare the time regulation for the election of the podestà, 217.

54. H. Zedler-Werdmüller, ed., *Stadtbücher* I (Leipzig, 1901), nos. 88 and 139; the editor mentions the ringing of the council bell in the nineteenth century at eight o'clock and assigns the medieval eating time to eleven o'clock, which is probably correct.

55. "Fragment d'un Répertoire de Jurisprudence Parisienne au XVe Siècle," *Mém. Soc. Hist. de Paris et de l'Ile-de-France* 17 (1891): 63.

56. J. François and N. Tabouillot, *Preuves de l'Histoire de Metz,* vol. VI (Metz, 1781), 140.

57. ORF vol. II (Paris, 1723), 1 ff.

58. *Corpus Constitut. Prussico-Brandenb.* (note 92), II.I, no. 5, col. 32; 1621 "Edict wegen Besuchung der Audienten," ibid., 20, col. 116; 1558 "Gerichtsordnung," ibid., no. 2, col. 150.

59. *Mém. Soc. hist. et litt. de Tournai* 7 (1861): 23 ff.

60. W. Schultheiß, ed., *Satzungsbücher und Satzungen der Stadt Nürnberg aus dem 14. Jahrhundert* (Nuremberg, 1965), 331; compare the similar regulation for the Commission of Three from 1390, ibid., 302.

61. *Weberchronik des Clemens Jäger* s.a. 1519, ed. by F. Roth, Chroniken der deutschen Städte 34, Augsburg 9 (Stuttgart, 1929), 244 f. note 2.

62. F. Geier, ed., *Oberrheinische Stadtrechte,* 2. Section 2. H. (Überlingen) (Heidelberg, 1908), 386; compare 400 ff.

63. J. Gény, *Schlettstädter Stadtrechte* (Heidelberg, 1906), 839.

64. F. Hauß, ed., *Zuchtordnung der Stadt Konstanz 1531* (Lahr, 1931), 132 f.

65. O. Feger, *Vom Richtebrief zum Roten Buch. Die ältere Konstanzer Ratsgesetzgebung,* Konstanzer Geschichts- und Rechtsquellen VII (Constance, 1955), nos. 11, 12, 45, 60, 109, 119, 137, 208, 220, 224, 287, 348, 352, 380; compare also O. Feger, ed., *Die Statutensammlung des Stadtschreibers Jörg Vögeli,* Konstanzer Stadtrechtsquellen IV (Constance, 1951), no. 7.

66. O. Feger, *Richtebrief* (note 65), no. 380, p. 116.

67. Stein (note 46) I, 757.

68. K. T. Eheberg, *Verfassungs-, Verwaltungs- und Wirtschaftsgeschichte der Stadt Straßburg bis 1681* (Strasbourg, 1899), no. 201, p. 437 f.

69. *Urkundenbuch Braunschweig,* vol. 1, ed. by L. Haenselmann and H. Mack (Brunswick, 1872; repr. 1975), 154.

70. Stein (note 46) I, 345, compare 418, 434, 437, 459, 523, 526, 539.

71. K. O. Müller, *Nördlinger Stadtrechte* (Munich, 1933), 44, 72, 152 et al.; addendum to the Stadtbuch of Wolfach 1479, in *Fürstenbergisches Urkundenbuch* 7, ed. by S. Riezler (Tübingen, 1891), 12; A. Voigt, ed., "Thorner Denkwürdigkeiten," *Mitteilungen des Copernicus-Vereins* 13 (1904), 122, 124; Riga 1502: F. G. v. Bunge, ed., *Liv-, Est- u. Kurländisches Urkundenbuch* (1853–1914; repr. Aalen, 1967–1981), Section 2, vol. 2, nos. 412, 19, 1534; Section I, vol. 2, nos. 950, 55, and 77; R. Rau and J. Sydow, *Die Tübinger Stadtrechte von 1388 und 1493* (Tübingen, 1964), 8; Asperg 1510: F. C. J. Fischer, *Versuch über die Geschichte der deutschen Erbfolge,* vol. 2 (Leipzig, 1778), 140; Kaufbeuren (sixteenth century?): R. Zech, *Das Stadtrecht von Kaufbeuren* (1951), 49.

72. K. O. Müller, *Die ältesten Stadtbücher von Leutkirch und Isny* (Stuttgart, 1914), 279.

73. Feger, *Statutensammlung* (note 80), no. 265, § 7; compare no. 227 § 5, no. 359 § 12; P. Meisel, *Die Verfassung und Verwaltung der Stadt Konstanz im 16. Jahrhundert,* Konstanzer Geschichts- und Rechtsquellen VII (Constance, 1957), 35–37.

74. F. Geier (note 62), 35–37.

75. F. Gény (note 63).

76. C. Koehne, ed., *Oberrheinische Stadtrechte,* vol. 7: Bruchsal (Heidelberg, 1906), 927.

77. G. Schrepfer, *Dorfordnungen im Hochstift Bamberg* (Erlangen, 1941), 49; compare also the diary of Thomas Wirsing, pastor of Sinnbronn, from A. Gabler, *Altfränkisches Dorf- und Pfarrhausleben* (Nuremberg, 1952), 58.

78. G. Schrepfer (note 77), 50; compare the Gemeindeordnungen Unterhaid 1677 §34, from K. S. Kramer, *Volksleben im Hochstift Bamberg* (Würzburg, 1967), 27.

79. "The Abbey of the Holy Ghost," in G. G. Perry, *Religious Pieces in Prose and Verse* (Early English Text Society o.s. 26), expanded edition 1913 (repr. New York, 1969), 60.

80. Ed. Veesenmeyer (Tübingen, 1889), 82.

81. Lippert, *Glockenläuten,* 58–61.

82. The "Statuti della Podesteria" in Ponteassieve (Tuscany) from the year 1523 contain a vivid justification for the spread of such regulations: chap. 43: "Et perchè ell'è cosa inhumana che uno mercato di quella qualità non habbia modo et ordine come gli altri mercati simili"; market selling was prohibited prior to the bell signal; the bell should be rung "ore 15" in the summer, "ore 18" in the winter (Italian striking), from P. Benigni and F. Berti, *Statuti del Ponte a Sieve* (Ponteassieve, 1982), 144 f.

83. See, for example, H. Schubert, "Unterkauf und Unterkäufer in Frankfurt am Main im Mittelalter. Ein Beitrag zur Geschichte des Maklerrechts" (dissertation, Frankfurt, 1962).

84. H. Bruder, *Die Lebensmittelpolitik der Stadt Basel im Mittelalter* (Ackern i. Br., 1909); H. Kimmig, *Das Konstanzer Kaufhaus im Mittelalter,* Konstanzer Geschichts- und Rechtsquellen VI (Constance, 1954); E. Nübling, *Ulms Kaufhaus im Mittelalter* (Ulm, 1895).

85. O. Richter, *Verfassungs- und Verwaltungsgeschichte der Stadt Dresden,* vol. 2 (Dresden, 1891), 366; compare also the relatively complicated regulations in the new organization of the market for the products of the "draperie" in Douai in 1403; G. Espinas and J. Pirenne, *Recueil des documents relatifs à l'histoire de l'industrie drapière en Flandre,* vol. II (Brussels, 1961), 317 ff.

86. B. Schmidt, *Frankfurter Zunfturkunden bis zum Jahre 1612,* vol. 1 (Frankfurt, 1914), 391.

87. One example from Regensburg in 1391 can be found in C. T. Gemeiner (chap. 4 note 197), vol. I/II, 283 f.

88. In Constance, for example, fishermen could not offer the fish they brought to market for sale for more than three hours: "die gankfish und stuben, die sy an markt pringent, nit lenger feil haben dann 3 stund"; after O. Feger and P. Rüster, *Das Konstanzer Wirtschafts- und Gewerberecht zur Zeit der Reformation* (Constance, 1961), no. 14.

89. See "Recess der Tagfahrt der Stände von Pomerellen in Stargrad, Mai 1401," in M. Toeppen, *Akten der Ständetage,* vol. 2 (Leipzig, 1880; repr. 1974), 329: "Sal men in den steten keynen fanen uffstecken adir glocken luthen, ymand zcu hyndern an seyme kouff."

90. "Die Kölner Marktordnungen," from vols. 3 and 4 of the Cologne council statutes collected as broadsheets, Cologne Municipal Archive.

91. C. O. Mylius, ed., *Corpus Const. Marchicarum* (Berlin, 1737–1755), 2nd. Con., Section I, Part I, no. 5, col. 49.

92. C. O. Mylius, ed., *Novum Corpus Const. Borussico-Brandenburgensium praecipue Marchicarum . . .* (Berlin and Halle, 1751–1806), V. 3, no. 25, col. 133; compare also ibid. IX, no. 10, cols. 1897–1898 (market regulations for Brunswick 1794).

93. To my knowledge a relevant study on this topic exists only for Vienna: A. Gigl, *Geschichte der Wiener Marktordnungen vom 16. Jahrhundert an bis zum Ende des 18. Jahrhunderts* (Vienna, 1865).

94. J. C. Troll, *Geschichte der Stadt Winterthur,* vol. 7 (Winterthur, 1848), 23 ff.; K. Pallas, *Geschichte der Stadt Herzberg* (1901), 133.

95. One exception is Josef Dolch, *Lehrplan des Abendlandes. Zweieinhalb Jahrtausende seiner Geschichte* (repr. of the third edition, Darmstadt, 1982).

96. R. H. Rouse and M. A. Rouse, "Statim invenire: Schools, Preachers, and New Attitudes to the Page," in R. L. Benson and G. Constable, eds., *Renaissance and Renewal in the Twelfth Century* (Oxford, 1982), 201–225; compare Schreiner MS, 21 f.

97. A selection of texts in W. H. Woodward, *Vittorino da Feltre and other Humanist Educators* (1897; repr. New York, 1947); Leon Battista Alberti (note 32).

98. The connection between "variety and scheduling" that becomes clear here was pointed out by Ricardo Quinones, *Discovery,* chap. 5: "Children, Education and Time," here p. 192.

99. Petrus Paulus Vergerius, *De ingenuis moribus,* in *Atti e memorie della R. Accademia di scienze, lettere ed arti in Padova,* New Series, vol. 34 (1918); English transl. in Woodward (note 97), 112.

100. Reference to this miniature has been taken from Turner, "Accomplishments," 163; a reproduction appears also in T. Metzger and M. Metzger, *Jüdisches Leben nach illuminierten hebräischen Handschriften* (Würzburg, 1983), plate 301; I would like to thank Dr. Neubauer for help in identifying the line of text (which does not relate to the picture).

101. M. Güdemann, *Geschichte des Erziehungswesens und der Cultur der Juden in Frankreich und Deutschland* (Vienna, 1880); B. Straßburger, *Geschichte der Erziehung und des Unterrichts bei den Israeliten* (Breslau, 1885), 58 f.; A. Berliner, *Aus dem Leben der deutschen Juden im Mittelalter* (Berlin, 1900), 8 f.

102. J. Levy, no. 228, J. Weil, no. 130, I. Bruna, no. 116, from A. Berliner (note 101), 9, 34; the biographical notes from J. Fürst, *Bibliotheca Judaica* (Leipzig, 1849), 1851.

103. M. Güdemann, *Quellenschriften zur Geschichte des Unterrichts und der Erziehung bei den deutschen Juden* (Berlin, 1981), 264; compare the discussion of Moses ben Ahrons, "Wie ist die Lehrordnung?" (Lublin, 1653), ibid., 93.

104. J. Dolch (note 95) §§ 19–26; G. Mertz, *Das Schulwesen der deutschen Reformation im 16. Jahrhundert* (Heidelberg, 1902); G. Strauss, *Luther's House of Learning: Indoctrination of the Young in the German Reformation* (Baltimore and London, 1978).

105. On this see esp. J. Dolch (note 95), §21.

106. From Johannes Müller, *Vor- und frühreformatorische Schulordnungen und Schulverträge in deutscher und niederländischer Sprache* (Zschopau, 1885–1886), 125, 126.

107. Ibid., 145 ff.

108. Luther (chap. 1 note 14), vol. 15, 47; on the duration of class time in the Protestant school statutes see G. Mertz (note 104), 361 ff.

109. From J. Dolch (note 95), 200.

110. Ibid., 307.

111. From G. Mertz (note 104), 622.

112. J. Dolch (note 95), 189.

113. For example: Herzogliche Württembergische Schulordnung 1729, chap. 5: for a "good order with the time and the hours," a "fourfold sandglass" shall be maintained, in R. Vormbaum, *Evangelische Schulordnungen,* vol. 3 (Gütersloh, 1864), 331; Nordhausener Schulordnung 1583: detailing of student watchers from the third grade who are to listen to the "clock" and to announce the hour stroke in the classrooms, in G. Mertz (note 104), 625; "In the villages where no striking clock exists," the church clerk should give a bell signal for the beginning of school, Fürstlich-Oelßnische Kirchenordnung 1664, in H. Jessen and W. Schwarz, *Schlesische Kirchen- und Schulordnungen von der Reformation bis ins 18. Jahrhundert* (Görlitz, 1938), 425 ff.

114. Schulordnung für Militsch 1709, in Jessen and Schwartz (note 113), 497.

115. For example, in Bologna 1317–1347, in H. Denifle and F. Ehrle, "Die Statuten der Juristenuniversität Bologna vom Jahre 1317–1347," in *Archiv für Literatur- und Kirchengeschichte* III (Berlin, 1887), 313 f.

116. Compare H. Rashdall, *The Universities of Europe in the Middle Ages,* vol. 1, new ed. (Oxford, 1936; repr. 1969), 216 f.

117. See the years of quarreling over the right to give the "lectio matutina," in H. Denifle, *Chart. Univ. Parisiensis,* vol. III (Paris, 1894; repr. 1964), 425–439, 468–477, here 425 f.

118. See, for example, Montpellier 1339, in M. Fournier (note 41), vol. 2, 50 f., and Perpignan 1380–1390, 655 ff.

119. K. Rückbrod, *Universität und Kollegium. Baugeschichte und Bautyp* (Darmstadt, 1977).

120. See, for example, the Paris statutes 1367, in Du Boulay, *Hist. Univ. Paris,* vol. IV (Paris, 1668; repr. 1966), 412 f. and 428 (1370).

121. Howgrave-Graham, "Clocks," 268.

122. M. Marianai, *Vita universitaria pavese nel secolo XV* (Pavia, 1889), 105 f.

123. Fournier (note 41), 425, 532 f.

124. J. H. Beckmann and R. Feger, *Statuta Collegi Sapientiae* (Lindau, 1957), 85.

125. Martin Luther, "Deudsche Messe und ordnung des Gottisdiensts" (1526), (chap. 1, note 14), vol. 19, 72.

126. Ecclesiastical constitution Augsburg 1537; extract in Sehling (chap. 5, note 74), vol. 12, 65.

127. J. Calvin, *Institutio Christianae Religionis,* German edition ed. by O. Weber (Neukirchen, 1963), III, 20, 30, IV, 5, 18, V, 10, 29; Gottesdienstordnungen 1541–1561: *Corpus Reformatorum* 38 (1872; repr. New York, 1964), 21, 99.

128. Compare, for example, F. X. Himmelstein, *Synodicon Herbipolense* (Würzburg, 1855) (Eccl. Const. for the episcopacy of Würzbrug 1589, 1693); H. Hollerweger, *Die Reform des Gottesdienstes zur Zeit des Josephinismus in Österreich* (Regensburg, 1970), 124 ff., 231 ff.

129. Luther, *Tischreden* (chap. 1 note 14), II, no. 2643; V, nos. 5200, 6400; III, no. 3419; "Von der Ordnung des Gottesdienstes und der Gemeinde zu Wittenberg" (1523); "Predigt zum 3. Adventssonntag" (Dec. 11, 1530) (chap. 1 note 14), vol. 32, 247.

130. In Sehling (chap. 5 note 74), IV, 74.

131. Ibid., IV, 232.

132. Ibid., XIII, 510; consideration for the councilors is mentioned in the Ecclesiastical Constitution of Zwickau 1533, ibid., I, 723.

133. In B. von Mehr, *Das Predigtwesen in der Kölnischen und Rheinischen Kapuzinerprovinz im 17. und 18. Jahrhundert* (Rome, 1945), 104.

134. From K. Pallas (chap. 5 note 71), introduction.

135. In *Corpus Const. March* (note 91), I.i, no. 96, cols. 513, 514.

136. Ibid., no. 96, cols. 527, 528.

137. B. von Mehr (note 133), 105.

138. See, for example, W. Diehl, *Zur Geschichte des Gottesdienstes und des gottesdienstlichen Lebens in Hessen* (Gießen, 1899), 126, 132, 177, 197.

139. Johannes Jacob Schudt, *Jüdische Merkwürdigkeiten* (Leipzig, 1714), lib. 4, § 40.

140. Compare the report about the wild invective of a preacher brought to Regensburg from Nabburg in 1543. It closes with the words: "In this tone he preached for three hours, from 10 to 1 o'clock. And the following Sunday he

said that this invective was prompted by an inspiration from the Holy Spirit." In J. B. Götz, *Die religiöse Bewegung in der Oberpfalz von 1520–1560* (Freiburg, 1914), 84.

141. Instructions for the local visitations 1577–1586, in Pallas (chap. 5 note 71), I, 144 f.; compare Sehling (chap. 5 note 74), I, 393; on the control of preaching times by the authorities see G. Heinrich, "Amtsträgerschaft und Geistlichkeit," in G. Franz, ed., by *Beamtentum und Pfarrerstand* (Limburg, 1972), 213.

142. Doubts about the admissibility of preaching limitations appear already in Luther: "Predigt zum 3. Adventssonntag 1530": "Ante 10 annos cogitabam, si mihi praedicandum: est umb 1 stund zu thun. Sed iam cogitandum: hoc officium quod hic fure, non est meum, sed in meo nomine, tum bin ich des teuffels"; in Luther (chap. 1 note 14), vol. 32, 247.

143. M. Schian, *Orthodoxie und Pietismus im Kampf um die Predigt* (Giessen, 1912), 8.

144. In P. von Ludewig, *Gelehrte Anzeigen* 102, vols. 1–3 (Halle, 1743–1749), 310–315 (explanations of the preaching limitations in Brandenburg-Prussia 1714–1717); in the eighteenth century the historical circumstances of the preaching limitations were not always understood. Mooser, for example, jokes about the court-preacher who "may not exceed his hour-glass," and considers this an indication of pedantic regimentation, in *Deutsches Hof-Recht,* vol. 2 (1776), 370; on external appearance: H. Grötzsch, "Kanzelsanduhren aus der Sammlung des staatlichen Mathematisch-Physikalischen Salon," in *SEAU* 5 (1965): 16–20.

145. Compare on this Schian (note 143), 6 ff.

146. "Quanta autem clepsydrae mensura fuit, mihi non constat." G. Budaeus, *Annotationes in Pandectas,* vol. III (Basel, 1557), 373.

147. Bernard Ferrarius, *De ritu sacrarum Ecclesiae veteris Concionum* (Utrecht, 1692), lib. I, c. 34, 156–160.

148. R. Cruel, *Geschichte der Predigt im Mittelalter* (Detmold, 1879), 635 ff. (excessively long sermons in the fourteenth century); compare also the account of Johannes von Capistran in Magdeburg (1453) in the *Magdeburger Schöppenchronik,* ed. by C. Hegel, Chroniken der deutschen Städte 7, Magdeburg 1 (Leipzig, 1869), 391 f.

149. T.-M. Charland, *Artes Praedicandi: Contribution à l'histoire de la rhétorique au moyen âge* (Paris and Ottawa, 1936), 223, 238 f.

150. Johannes Ulrich Surgant, *Manuale Curatorum* (Basel, 1506), lib. 1, cons. 23.

151. "Decreta Synodalis Provincialis Camaracensis (1586)," in K. von Hardouin, *Conciliorum collectio regia maxima,* vol. 9 (Paris, 1714), col. 2156.

152. For example, in the Ecclesiastical Constitution of Würzburg 1589, in Himmelstein (note 128).

153. C. Ripa, *Iconologia: Overo Descrittione di diverse Imagini cavate dall'antichità, 6 di propria inventione* (Rome, 1603; repr. 1970), 127, s.v. "Eloquenza": "Il libro, e l'orologio [da polvere] . . . è indicio che le parole sono l'istromento

dell'eloquente: le quali però devono essere adoperate in ordine e misura del tempo, essendo dal tempo solo misurata l'oratione," quoted from L. Cheles, *The Studiolo of Urbino: An Iconographic Investigation* (Wiesbaden, 1986), 60.

154. Francesco Petrarca, *Prose,* ed. by G. Martellotti (Milan, 1955), 568; compare Quinones, *Discovery,* 106 ff.; W. Liebenwein, *Studiolo. Die Entstehung eines Raumtyps und seine Entwicklung bis um 1600* (Berlin, 1977).

155. Sundial: *Ser Lap Mazzei, Lettere di un notaro a un mercante del secolo XIV con altre letttere e documenti,* ed. by C. Guasti, 2 vols. (Florence, 1880), I, 169, CXXVI: ". . . e voi attendiamo a Firenze per mostrarvi la nuova casa ove sono tornato, e la mia biblioteca o studiolo, ove questo scrivo a una spera o razzo di sole che mi conforta."

156. The picture "Jerome in his Study" (around 1440) that was begun by Jan van Eyck and finished by Petrus Christus seems to have become prototypical; it used to be in the Medici palace in Florence; see Liebenwein (note 154), 134. Examples after the middle of the fifteenth century can probably be counted in the hundreds.

157. F. Petrarca (note 154), 310 f.

158. See, in contrast, M. Schneider, "Zeit und Sinnlichkeit. Zur Soziogenese der Vanitas-Motivik und des Illusionismus," *Kritische Berichte* 4/5 (1980): 8–34.

159. Ignatius of Loyola, *Das geistliche Tagebuch (1544),* ed. by A. Haas and P. Knauer (Freiburg, 1961), 147, 157, 180, 184, 200; idem, *Exercitia Spiritualia,* transl. by F. Weinhandl (Munich, 1978), passim; on the rendering of accounts see esp. 76 ff.

160. A good account for Hamburg in Nahrstedt, *Freizeit,* 87 ff., 156 f.; also: "Nürnberger Ordnung 1700," Municipal Archive Nuremberg, Mandate N, 121; "Frankfurter Torsperrordnungen ("gate-closing regulations") 1724, 1788, 1832"; "Aufhebung des Sperrgeldes" ("abolition of the closing fee") 1836, Municipal Archive Frankfurt, Ratsverlässe; "Torsperr-Ordnung Hannover 1765," in *Hannoveranische Geschichtsblätter* 10 (1907), 97–100; "Hildesheimer Torsperrordnung 1792," Municipal Archive Hildesheim 257.

161. T. G. Glick (chap. 2 note 26); idem, *Irrigation and Society in Medieval Valencia* (Cambridge, Mass., 1970), 188–222.

162. See, for example, H. Boos, *Geschichte der rheinischen Städtekultur unter bes. Berücks. der Stadt Worms,* part 3 (Berlin, 1899), 168.

163. I shall refer only to the oldest begging statutes with such time regulations: Nuremberg 1478, in J. Bader, *Nürnberger Polizeigesetzgebung aus dem 13. bis 15. Jahrhundert* (Stuttgart, 1861), 316 ff.; a brief introduction to the problem of permitted begging by T. Fischer, "Armut, Bettler, Almosen. Die Anfänge städtischer Sozialfürsorge im ausgehenden Mittelalter," in C. Meckseper, *Stadt im Wandel,* vol. 4 (exhibition catalogue) (Stuttgart, 1985), 271–286.

164. For example, the bride shall be in church "wan de klocke negen sleyt" (1717), H. Sievers, *Kieler Burspraken* (Kiel, 1953), 179 ff.; compare *Hamburgische Burspraken,* part 2 (Hamburg, 1960), 43 ff.; more detailed discussion by W. Starke, "Zu den Motiven, Kontrollinhalten und Kriterien sozialer Klassifizierung in den Hochzeitordnungen des Spätmittelalters und der frühen Neu-

zeit" (MS, Bielefeld, 1983); to the source lists there can be added: Hochzeitsordnung Dresden 1595 ("wedding statute"), in Richter II (note 85), 137 f.; Würzburg 1617, in E. Specker, "Die Reformtätigkeit der Würzburger Fürstbischöfe," *Würzburger Diözesangeschichtsblätter* 27 (1965): 29–125, 96 ff.; Eßlingen 1604, 1659, in K. Pfaff (note 8); Ravensbergische Landespolizeiordnung, *Jahrbuch des Vereins für die historiche Verfassung der Grafschaft Ravensburg* 13 (1899).

165. Ecclesiastical Constitution Danzig 1590–1595, in Sehling (chap. 5 note 74), IV, 192 ff.; Stendaler Hochzeitsordnung 1622 ("wedding statute"), in L. Goetze, *Urkundliche Geschichte der Stadt Stendal* (Stendal, 1873), 421 ff.

166. E. Schmidt, *Inquisitionsprozeß und Rezeption* (Leipzig, 1940), 60 ff.; P. Fiorelli, *La tortura guidiziaria nel diritto commune* (Varese, 1953–1954); J. Gilissen, "La preuve en Europe du Xive au début du XIXe siècle - Rapport de Synthèse," *Recueil de la Soc. Jean Bodin* XVII.2 (1965), 788; J. H. Langbein, *Torture and the Law of Proof: Europe and England in the Ancien Régime* (Chicago, 1977).

167. Paulo Grillando, *De quaestionibus et torturae* (Lyons, 1536), and in *Tractatus universi iuris,* vol. XI.1 (Venice, 1584), 295 verso ff.

168. *Magnum Bullarium Romanum* (Rome, 1745; repr. 1965), vol. IV.1, no. 59, p. 237; Pietro Follerio, *Practica Criminalis* (Venice, 1557); Hortensi Cavalcani, *Tractatus de brachio regio* (Marburg, 1605), 232; Tranquilo Ambrosini, *Processus informativus* lib. IV, c. III (Rome, 1647), here quoted according to the Venice edition of 1722, 185; J. V. Kirchgeßner, *Tribunal Nemesis, Juste Judicantis oder Richter-Stuhl der recht richtenden Gerechtigkeit* (Nuremberg, 1706), part 4, chap. 7; C. Wildfogel, *Diss. jur. de arbitrio judicis circa torturam* (Jena, 1736) (regarding the Saxon rescript of July 26, 1705), 34 ff.; *Constitutio Criminalis Theresiana* (Vienna, 1769), art. 38, § 12, 13.

169. F. Merzbacher, *Die Hexenprozesse in Franken,* 2nd. expanded ed. (Munich, 1970), 141 ff.

170. E. Hubert, *La Torture aux Pays-Bas Autrichiens pendant le XVIIIe siècle* (Brussels, 1897), 49; R. C. van Caenegem, "La preuve dans l'ancient droit belge, des origines à la fin du XVIIIe siècle," *Recueil de la Soc. Jean Bodin* XVII.2 (1965): 421.

171. H. Beaune and J. D'Arbaumont, *Mémoires d'Olivier de la Marche,* vol. 3 (Paris, 1885), 130 ff.; "Ein spanischer Bericht über ein Turnier in Schaffhausen im Jahr 1436," *Basler Zeitschrift für Geschichte und Altertumskunde* 14 (1915): 153 ff.; M. Kloeren, *Sport und Rekord. Kultursoziologische Untersuchungen zum England des 16. bis 18. Jahrhunderts* (Leipzig, 1935); H. Eichberg, *Leistung, Spannung, Geschwindigkeit. Sport und Tanz im gesellschaftlichen Wandel des 18./19. Jahrhunderts* (Stuttgart, 1978).

172. An early example in *The Household of Edward IV: The Black Book and the Ordinance of 1478,* ed. by A. R. Meyers (Manchester, 1959), esp. 211 ff., §§ 8, 40, 50; rich material in A. Kern, ed., *Deutsche Hofordnungen des 16. und 17. Jahrhunderts,* 2 vols. (Berlin, 1905–1907); G. Liebe, "Die Kanzleiordnung Kurfürst Albrechts von Magdeburg, des Hohenzollern (1538)," A. Kern, ed.,

Forschungen zur Brandenburgischen und Preußischen Geschichte 10 (1898): 31–54; H.-J. von der Ohe, *Die Zentrale- und Hofverwaltung des Fürstentums Lüneburg (Celle) und ihre Beamten* (Celle, 1955), 24 ff.; K. Plodeck, *Hofstruktur und Hofzeremoniell in Brandenburg-Ansbach vom 16. bis zum 18. Jahrhundert* (Ansbach, 1972), 116 ff.

173. Compare the complicated compromise between the city, the city church, and the abbey of Saint-Waudrau concerning the ringing for services in Mons in 1531, where the communal clock provides the temporal framework that is immune to challenge; in L. Devillers and E. Matthieu, *Chartres du Chapitre de Saint-Waudrau de Mons,* vol. 4 (Brussels, 1913), nos. 1895 and 1931.

174. Compare, for example, the altar endowment for the Church of St. Catherine in Brunswick in 1295: the mass was to be held during matutin "tum pro serventibus, tum pro itinerantibus, tum pro victualia manuali opere querentibus vel labori necessario incumbentibus, ut audita missa redeant ad labores"; similarly in the confirmation of the endowment of a mass in 1300: "ut negotiatores et mechanici, viatores et mendici hac missa maturius audita efficatius consequantur prosperitatem et salutem vitae presentis et eterne," in *Urkundenbuch Braunschweig* (note 69), vol. 2, 199, 242; examples can be multiplied almost at will from the sources of many European cities; compare also Schreiner MS, 35; a common formula for the time of early mass is "in aurora ante operariorum exitum."

175. Compare, for example, the conflict in Niederwenigern in 1654, in F. Darpe, "Die Anfänge der Reformation und der Streit um das Kirchenvermögen in den Gemeinden der Grafschaft Mark II," *Zeitschrift für vaterländische Geschichte und Altertumskunde* 51 (1893), 12 and 63 ff.

176. Pallas (chap. 5 note 71) gives numerous examples: 1528 Prettin (II, 3.3), 1531 Bitterfeld (II, 2, 7), 1555 Zahna (II, I, 385), 1671 Seyda (II, I, 187), and more; complaints, for example, in F. Kumm, *Die Befreiung des Lehrers vom niederen Küsterdienste. Ein Beitrag zur Hebung des Gedeihens der Volksschule und der sozialen Stellung der Lehrer* (Bielefeld, 1890), 14: "Gerade diese Seite der Küsterfunktionen . . . macht dem Lehrer den meisten Ärger" (the arduousness of climbing up the tower in winter, the endless Sisyphus task of regulating the clock; if the clock was running poorly the people thought it was the sacristan's fault); "Nun ist die Uhr zehn Minuten zurückgeblieben. Wann soll der Lehrer sie richtig stellen? Während der Arbeitszeit, dann ist's dem gnädigen Herrn nicht recht, der 40 Arbeiter auf dem Felde hat und durch das Stellen 40 mal 10 Minuten gleich 6 2/3 Stunden Arbeitszeit verliert. Während der Freistunde, so wird diese den Arbeitern verkürzt."

177. C. Gerber, *Historie der Kirchenceremonien in Sachsen, Dresden und Leipzig* (1772), 237 f.; according to Gerber, the "Ambassadeurs und Gesandten fremder Potentaten" also oriented themselves by the clock for their private services.

178. Rothwell pointed this out in 1959 based on his studies of late medieval French literature, idem., "Hours of the Day," 249; compare Autrand, "Dates," 165 f.

179. A. C. Crombie, *Augustine to Galileo: The History of Science,* A.D. 400–

1650 (London, 1952), 187 f.; Mumford, *Technics,* 16; compare Thorndike, *History* III, 290, 344 f.

180. Crombie (note 179), 150–151.

181. Daniel C. Boorstin, *The Discoverers* (New York, 1983), 39.

182. See I. Peri, "Omnia mensura et numero et pondere disposuisti: Die Auslegung von Weish. 11, 20 in der lateinischen Patristik," in *Mensura, Maß, Zahl, Zahlensymbolik im Mittelalter,* Miscellanea Medievalia 16/1 (Berlin and New York, 1983), 1–21.

183. A. C. Crombie, "Quantification in Medieval Physics," in H. Woolf, ed., *Quantification: A History of the Measurement in the Natural and Social Sciences* (Indianapolis, 1961), 13–30.

184. See Maurice, *Räderuhr* I, 5–15; O. Mayr, "Die Uhr als Symbol für Ordnung, Autorität und Determinismus," in Maurice, *Welt als Uhr;* F. C. Haber, "Zeit, Geschichte und Uhren," ibid., 10–20; O. Mayr, *Uhrwerk und Waage. Autorität, Freiheit und technische Systeme in der frühen Neuzeit* (Munich, 1987); S. Macey, *Clocks and the Cosmos: Time in Western Life and Thought* (Hamden, Conn., 1980); on the notion of the primum mobile as an absolute clockwork see Pierre Duhem, *Le Système du Monde: Histoire des doctrines cosmologiques de Platon à Copernic,* vol. 7 (Paris, 1956); compare chap. 6 note 5.

185. See, for example, A. Murray, *Reason and Society in the Middle Ages* (Oxford, 1978), part II.

186. Crombie (note 183), 25.

187. "De omnibus mundi balneis," quoted from L. Thorndike, *A History of Magic and Experimental Science,* vol. 4 (New York, 1934), 203.

188. D. Beaver, "Bernhard Walther: Innovator in Astronomical Observation," *Journal for the History of Astronomy* 1 (1970), 39 ff.

189. D. W. Waters, *The Art of Navigation in Elizabethan and Stuart Times* (New Haven, Conn., 1958), 58 ff., 164, 538 ff.

190. Michaele Savonarola, *De febribus, de pulsibus, de urinis . . .* (Venice, 1498), fol. 80 recto ff., quoted from W. F. Kümmel, "Zum Tempo der italienischen Mensuralmusik des 15. Jahrhunderts," *Acta Musicologica* 42 (1970): 150–163, 153.

191. Nikolaus von Kues, *Philosophisch-theologische Schriften,* Latin-German, ed. by L. Gabriel, transl. by D. Dupre and W. Dupre, vol. III (Vienna, 1967), 616 f.

192. Ibid., 630 ff.

193. E. Battisti and G. S. Battisti, *Le macchine cifrate di Giovanni Fontana* (Milan, 1984).

194. References in Kümmel (note 190), 160.

195. Hieronymus Cardanus, *Opus novum de proportionibus numerorum, motuum, ponderum, sonorum, aliarumque rerum mensurandarum . . .* (Basel, 1570), 50; compare 249, cited after W. F. Kümmel, "Der Puls und das Problem der Zeitmessung in der Medizin," *Medizinhistorisches Journal* 9 (1974): 1–22, here p. 5; this essay gives the older literature on the history of pulse measurement;

we can add R. H. Shryock, "The History of Quantification in Medical Science," in Woolf, *Quantification* (note 183), 85–107.

196. From the account of Viviani in *Le Opere di Galileo Galilei* (Edizione Nazionale, vol. 19; repr. Florence, 1968), 647–659.

197. Johannes Kepler, "Astronomia Pars. Optica," in *Opera Omnia,* ed. by C. Frisch, vol. 2 (1859), 334; idem, "Epitome Astronomiae Copernicanae," ibid., vol. 6 (1866), 248; cited from Kümmel, "Puls," 6.

198. Elias, *Zeit,* 92 f.; he also discusses (ibid., 83 f.) another experiment where, instead of time being measured, water was weighed.

199. For the horological aspects of the "Scientific Revolution" see S. A. Bedini, in Maurice, *Welt als Uhr,* 21–29.

Chapter 9

1. Compare P. Dubuis, "Les paysans médiévaux" (chap. 1 note 19); T. C. Smith, "Peasant Time and Factory Time in Japan," *Past and Present* 111 (1986): 165–197.

2. N. McKendrick, "Josiah Wedgwood and Factory Discipline," *Historical Journal* 4 (1961): 30–55; S. Pollard, "Factory Discipline in the Industrial Revolution," *Economic History Review,* 2nd series, 16 (1963–1964): 254–271; K. Thomas, "Work and Leisure in Preindustrial Society," *Past and Present* 29 (1964): 50–66; E. P. Thompson, "Time"; A. Lüdtke, "Arbeitsbeginn, Arbeitspausen, Arbeitsende. Skizzen zu Bedürfnisbefriedigung und Industriearbeit im 19. und frühen 20. Jahrhundert," in G. Huck, ed., *Sozialgeschichte der Freizeit* (Wuppertal, 1980), 95–122; E. Hopkins, "Working Hours and Conditions during the Industrial Revolution: A Re-Appraisal," *Economic History Review* 35 (1982): 52–66.

3. Le Goff, "Merchant's Time," 36.

4. Le Goff, "Labor Time," passim.

5. This is, in my view, how we should understand an introductory remark in the sermon "De temporis venditione" by Saint Bernhard of Siena (1380–1444). Invoking Giraldus ("De contractibus" q. 15), he makes the following distinction: "tempus dupliciter considerari potest. — Primo modo, ut est quaedam duratio quae consequitur primum motum, seu est mensura motus mutabilium rerum; et hoc modo tempus est quid commune omnium et nullo modo vendi potest . . . Quod commune non est particulariter singulorum. — Secundo modo, ut est quaedam duratio applicabilis alicui rei; quae duratio atque usus est alicui concessus ad eius opera exercenda; et hoc modo tempus est proprium alicuius, quemadmodum annus equi mihi accomodati dicitur esse meus. Et huius modi tempus licite vendi potest," ed. by A. Sépinsky, *Opera Omnia,* vol. IV (Florence, 1956), 165 f.; compare Schreiner MS, 45.

6. Johannes Heinrich Zedler, *Großes vollständiges Universallexicon aller Wissenschaften und Künste,* vol. 41 (1744), col. 1479, s.v. "Tagelöhner."

7. B. Geremek, *Le Salariat dans l'artisanat parisien aux XIIIe–XVe siècles* (Paris, 1969), 82–84; F. P. Caselli, *La Costruzione del Palazzo dei Papi di Avignone*

(1316–1367) (Milan, 1981), 171; S. Beissel, *Die Bauführung des Mittelalters. Studie über die Kirche des Hl. Viktor zu Xanten, II, Geldwert und Arbeitslohn,* 2nd ed. (Freiburg, 1889; repr. 1966), 158; M. Baulant, "Le salaire des ouvriers du bâtiment à Paris de 1400 à 1726," *Annales* 26 (1971), 470 ff.; G. Pinto, "Il personale, le balie e i salariati dell'ospedale di San Gallo di Firenze negli anni 1395–1406," *Ricerche Storiche* 4 (1974): 143.

8. See the sketches on the history of the concept of working time (German: Arbeitszeit) in Nahrstedt, *Freizeit,* 38 ff. According to Nahrstedt, the first references to "Arbeitszeit" in the meaning of yearly working time are found after 1548. Nahrstedt is here relying on Grimm's German Dictionary. We know that non-literary sources prior to 1500 are hardly included in Grimm, and this finding by itself therefore does not mean much. "Arbeitszeit," with the meaning of day work time, is then found in the large dictionaries from about 1800 on. The number of word forms using "Arbeit-" doubles between 1700 and 1807, quadruples between 1807 and 1854, and increases thirty-fold in the period 1807–1961. As early as 1876 the number of word forms was called "inexhaustible."

9. E. Perroy, "Wage Labour in France in the Later Middle Ages," *Economic History Review,* 2nd series 7 (1955): 232–239.

10. See W. Abel, *Agrarkrisen und Agrarkonjunktur,* 3rd. ed. (Hamburg, 1978), 61 ff.; N. Bulst, "Der Schwarze Tod. Demographische, wirtschafts- und kulturgeschichtliche Aspekte der Pestkatastrophen 1347–1352. Bilanz der neueren Forschung," *Saeculum* 30 (1979): 55; R. Greci, "Forme di organizzazione di lavoro . . . ," in R. Elze and G. Fasoli, eds., *La città in Italia e in Germania nel Medioevo* (Bologna, 1981), 110; G. Pinto, "L'organizzazione del lavoro nei cantieri edili (Italia Centro-Settentrionale)," in *Artigiani e Salariati. Il mondo del lavoro nell'Italia dei secoli XII–XI* (Pistoia, 1987), 69–101, esp. 85 ff.; in Châlons-sur-Marne in 1369, demands for shorter work times and wage increases were rejected with explicit reference to the plague, in ORF V, 193 f.

11. See, for example, "Statute of Labourers" 1349–1351, 23 Edward III, in *The Statutes of the Realm* I (1810; repr. 1963), 307–316; "Ordonnance concernant la Police du Royaume" of the French King Jean I, in ORF I, 350 ff., with the work time regulations for the "Vignerons et autres Manovriers," 367 f.; Duke Albrecht II of Austria, "Satzung des Lohns für den Weingartenbau der Stadt Wien," in K. Weiss, ed., *Geschichtsquellen der Stadt Wien* vol. I (Vienna, 1879), no. 47.

12. Compare the ordonnance of the prévôt of Paris concerning the length of day work working time in 1395, in R. de Lespinasse, *Les métiers et corporations de la Ville de Paris,* vol. 1 (Paris, 1886), 52.

13. G. Fagniez, *Études sur l'industrie et la Classe industrielle à Paris au XIIIe et XIVe siècle* (Paris, 1877), 336; compare 381; this regulation uses an ancient formula for differentiating dawn, compare "ea hora, qua incipit homo hominem posse cognoscere, id est prope luce ante tamen quam lux fiat," in "S. Silvae peregrinatio" (4th century), ed. by P. Geyer, *Itinera Hierosolemitana saeculae III–III,* CSEL 39 (Vienna, 1898), 75.

14. R. de Lespinasse (note 12), vol. III (Paris, 1897), 87; compare G.

Fagniez, *Documents relatifs à l'histoire de l'industrie et du commerce en France,* vol. I–II (Paris 1898–1900), vol. I, 56 f.

15. O. Rüdiger, *Die ältesten Hamburgischen Zunftrollen und Brüderschaftsstatuten* (Hamburg, 1874), no. 48.

16. R. de Lespinasse and F. Bonnardot, eds., *Etienne Boileau, Le Livre des Métiers, XIIIe siècle* (Paris, 1897; repr. 1980); passim; as late as 1371, time signals from three different churches appear in the working time regulations for the cloth-weavers of Paris; see Fagniez, *Études* (note 13), 340 f.

17. D. Oschinsky, *Walter of Henley and Other Treatises on Estate Management and Accounting* (Oxford, 1971), 314 ff.

18. See Schreiner MS, 27 with detailed note 128; Geremek, *Salariat,* 79 f.; Verona 1319: explicit permission for weavers "de quolibet tempore laborare de nocte ad suam voluntatem," L. Simoni, *Gli antichi statuti delle arti Veronesi* (Venice, 1914), 69, 357.

19. This account and the quotes follow E. Maugis, "La Journée de Huit Heures et le vignerons de Sens et d'Auxerre devant le Parlement en 1383–1393," *Revue Historique* 145 (1924): 202–218 (reproduction of the ordonnance and the acts of the parliament); the local parallel examples in M. Delafosse, "Les Vignerons d'Auxerrois (XIVe–XVIe siècle)," *Annales de Bourgogne* 20 (1948): 7–41; Delafosse used primarily a draft-like note from the burghers in Auxerre, which was later reworked by jurists for presenting their position to the parliament in Paris (AD Yonne E 523); see also C. Demay, "La Sonnerie pour les Vignerons et les Laboureurs à Auxere," *Soc. Sci. Hist. et Nat., Bull.* 3rd series II (1887): 129–147.

20. "(De) faire et esmouvir une maniere de jacquerie et maillecterie," Delafosse (note 19), 28; on the choice of words see Autrand, *Dates,* 175 ff.

21. "Ad evocacionem none que vulgariter cliquetus nuncupatur"; compare Du Cange II, 374, s.f. "cliquetum."

22. Compare the story, difficult to verify, of a work time–reducing act of grace by Jean de Berry, in E. Pasquier, "De ce mot 'Tintamarre,'" in his *Recherches de la France* VII, 52, in *Oeuvres completes* vol. I (1723; repr. 1972), cols. 853 f.

23. Orfinus Laudensis (chap. 8 note 15), 57, 60:

> Semper, ut est mori, resonet campana laboris
> Artibus impletis paveat campa quietis.
> / . . . /
> In solitis horis resonet campana laboris,
> Tinniat expletis gestis campana quietis.

24. Ferrara, *Statuta Ferrarie,* ed. by W. Montorsi (Ferrara, 1955), 194; Novara: F. A. Bianchini, *Le cose rimarchevoli della città di Novara* (Novara, 1828; repr. Bologna, 1974), 157; Padua: R. Cessi, *Le Corporazioni dei Mercanti di Panni e della Lana in Padova* (Venice, 1908), 43 f.; Verona: L. Simoni (note 18), p. LXXII; Parma: *Chronica Parmensis* (chap. 5 note 13), 202 f.; Treviso: A. Marchesan, *Treviso Medievale* I (Treviso, 1923; repr. 1971), 111 f.

25. Douai: G. Espinas and J. Pirenna (chap. 8 note 85), vol. II.2, 68 ff., and elsewhere.

26. Therouanne: T. Duchet and A. Giry, *Cartulaires de l'Église de Thérouanne* (St. Omer, 1881), nos. 214, 264; 242, 266, 267; compare Espinas and Pirenne (chap. 8 note 85), vol. I.3, 394 f.

27. Bruges: Espinasse and Pirenne (chap. 8 note 85), 561; Provins: F. Bourquelot, *Histoire de Provins,* vol. I (Paris, 1839), 61, 239 ff., 428 f.

28. "Clocke des oevriers/weerclocke," H. Michelant, ed., *Le Livre des Mestiers. Dialogues français-flamands, composés au XIV siècle* (Paris, 1875); "belle of werkemen," W. Caxton, *Dialogues in French and English,* ed. by H. Bradley (London, 1900), 30.

29. Municipal Archive Florence, Arte della Lana, Statuti 6 (1362–1427), IV.4, work time statute 1361.

30. In the texts giving permission for their use for Montreuil-sur-Mer 1371, ORF V, 529, and for Beauvais, see note 49.

31. See, for example, the work bell regulations for Comines 1359–1361, in ORF IV, 208 f.; compare Espinas and Pirenne (note 25), vol. I, 620 f.

32. "Elc wevere es sculdich werc te latene binnen der clocke daer hie onder behort"; Espinas and Pirenne (chapt. 8 note 85), 397 ff.

33. (1295), ibid., 561; in Courtria the other guilds have to follow the "weversclocke" (1340), ibid., 653.

34. "Dat si nemmermeer ander weercklocke hebben en zullen dan de clocke daer mede ghemeenlike mede gaat te weercke," ibid., 592, 597.

35. Text in Espinas and Pirenne (chap. 8 note 85), 56.

36. I believe J. Le Goff allowed himself to be misled by the broadly displayed material in Espinas and Pirenne; see his "Labor Time," 46. Le Goff also cites Fagniez, *Études* (note 13), 84, but Fagniez only wrote that in cities lacking dominating trades, such as Paris, work bells were not to be found.

37. F. G. von Bunge, *Die Stadt Riga im 13. und 14. Jahrhundert* (Leipzig, 1878), 129; "Heckerglocke" in Würzburg from the year 1333: H. Hoffmann, *Würzburger Polizeisätze, Gebote und Ordnungen des Mittelalters* (Würzburg, 1955), 57, 77, and elsewhere; Sisteron 1366: the "fossores et alii opera ruraliter facientes quibus hodie salaria magna dantur tarde ad opera incedentes et hora de vespere nimis brevi redeuntes minime faciunt quod deberent" shall henceforth no longer feign "ignorancia," and for that reason the council can stipulate "certas horas" for work; E. de Laplane, *Histoire de Sisteron,* vol. 1 (Digne, 1843), doc. XXIII, 518 f.

38. London: L. F. Salzmann, *Building in England down to 1540,* 2nd ed. (Oxford, 1967); Windsor: W. H. St. John, *Windsor Castle* (London, 1913); Moor End: Salzmann; York: J. J. Raine, ed., *The Fabric Roll of York Minster,* Surtees Society 35 (Durham, 1859), 181; compare D. Knoop and G. P. Jones, "Overtime in the Age of Henry VIII," *Economic History, Supplement to the Economic Journal* (1938): 13 ff.; Florence: S. Guasti, *S. Maria del Fiore* (Florence, 1887), 162; Mi-

lan: *Annali* (chap. 6 note 45), Appendix I, 132, 145, and *Annali* I, 112; Pavia, Certosa: L. Beltrami, *Storia documentata della Certosa di Pavia,* vol. 1 (Milan, 1896), 129, 183.

39. A. Thiery, *Recueil des Monuments inédits de l'Histoire du Tiers État,* 1st ser., vol. 1 (Paris, 1850), 456 ff.

40. The regulations, which Thiery believed to have been lost, can be found in E. Maugis, *Recherches sur les Transformations du Régime politique et social de la ville de Amiens* (Paris, 1906), 346.

41. G. Chaucer, *A Treatise on the Astrolabe (1391),* ed. by W. W. Skeat (Oxford, 1872; repr. 1968), 8; on the measure indications see Oschinsky (note 17), 354 f. with further literature.

42. J. J. Raine (note 38), 171–173.

43. H. Caffiaux, "Le Beffroi et la Cloche des Ouvriers en 1358," *Mém. Soc. Hist. Valenciennes* 3 (1873): 17.

44. In ORF IV, 588 f.

45. L. Riccetti, *Duomo* (chap. 6 note 45).

46. *Annali* I (note 38), 79, *Annali Appendici* I, 318.

47. Hamburg: see note 15; Frankfurt: B. Schmidt, *Frankfurter Zunfturkunden bis zum Jahre 1612,* vol. 1 (Frankfurt, 1914), 29, 355 f.; Paris: Lespinasse (note 12), vol. III (1897), 107; Cologne: see chap. 8: The change of the temporal order in the cities.

48. Genoa, first half of the fifteenth century: L. Balletto, "I Lavoratori nei Cantieri Navali (Liguria, sec. XII–XV)," in *Artigiani e Salariati* (note 10), 147.

49. Text: Paris AN X[iA] 36, fol. 179–183 verso; slightly shortened in Fagniez, *Documents* (note 14), 135–144.

50. Ibid., fol. 182 verso, 183; Fagniez, ibid., 141 ff.

51. Paris: Lespinasse, *Métiers* (note 12), 108; Orléans: Fagniez, *Documents* (note 14), vol. 2, 187 f.; T. Boutiot, *Histoire de la ville de Troyes,* vol. 2 (Paris, 1872), 339 f.; C. T. Gemeiner (chap. 4 note 197), III, 108.

52. C. Mollwo, *Das Rote Buch der Stadt Ulm* (Stuttgart, 1905), 204 f.; C. de Robillard de Beaurepaire (chap. 5 note 58), 310 f.

53. The statute on the working time of the journeymen in the iron wire mills in Nuremberg calls for sandglasses, since the mills were often too far away from striking clocks and their signals were not audible because of the work noise; R. Stahlschmidt, *Die Geschichte des eisenverarbeitenden Gewerbes in Nürnberg* (Nuremberg, 1970), 188 f.

54. J. Beneyto Perez, "Regulacion del trabajo en la Valencia del 500," *Annuario de Historia del Derecho Espanol* 7 (1930), 268 ff. and 297.

55. *The Complete Works of St. Thomas More,* ed. by E. S. Surtz, S. J., and J. H. Hexter, vol. 4 (New Haven, 1965), 126; in Campanella's "City of the Sun," only four hours of a day spent in an eventful and pleasant manner were given to work: "Ast in Civitate Solis dum cunctis distribuuntur ministeria, et artes, et labores, et opera: vix quatuor in die hora singulis laborare contingit: reliquum tempus consumatur in addiscendo iucunde, disputando, legendo, nar-

rando, scribendo, deambulando, exercendo ingenium et corpus, et cum gaudio"; *F. Thomae Campanellae Appendix Politicae Civitas Solis* (Frankfurt, 1623), 435.

56. Quoted from A. Zycha, *Das böhmische Bergrecht des Mittelalters auf der Grundlage des Bergrechts von Iglau,* 2 vols. (Berlin 1900), here vol. 2, 12–15.

57. Ibid., 187–189.

58. (GLA Karlsruhe, A. Münstertaler Bergwerke, copy from the fifteenth century), quoted from Gothein in *Zeitschrift für Geschichte des Oberrheins* Neue Folge 2 (1887), 447.

59. Compare on this R. Müller, *Das Bergrecht Preussens und des weiteren Deutschlands* (Stuttgart, 1917), 376 ff.; R. Dietrich, "Untersuchungen zum Frühkapitalismus im mitteldeutschen Erzbergbau und Metallhandel," *Jahrbuch für die Geschichte Mittel- und Ostdeutschlands* 9/10 (1961): 172 ff., 191; A. Koch, *Arbeitsrechtliche Bestimmungen am steirischen Erzberg im 16. Jahrhundert* (Graz, 1942), 54 f.; "frie zeit" 1469, H. Ermisch, ed., *Freiberger Urkundenbuch* II (Leipzig, 1883–1891), 197; J. Köhler, *Die Keime des Kapitalismus im sächsischen Silberbergbau (1168–1500),* Freiburger Forschungshefte Kultur und Technik D 13 (Berlin, 1955), 66, 102 f.

60. *Zwölf Bücher vom Berg- und Hüttenwesen,* ed. by H. Schiffner (Munich, 1977), 78; compare Koch, 54; Köhler, 105; C. Neuburg, *Goslars Bergbau bis 1552* (Hannover, 1892), 230.

61. In Schneeberg in 1478: O. Hoppe, *Der Silberbergbau zu Schneeberg bis zum Jahre 1500* (Heidelberg, 1918), 121 f.; in Schreckenberg in 1499: H. Helbig, ed., *Quellen zur älteren Wirtschaftsgeschichte Mitteldeutschlands* IV (Weimar, 1953), 117; in Tarnowitz, however, his "lordship" declared himself willing in 1532 to pay for a clock for the magistrate's office (*Amtshaus*), K. Wutke, ed., *Schlesisches Berg- und Hüttenwesen. Urkunden und Akten (1529–1740),* Codex diplomaticus Silesiae XXI (Breslau, 1901), no. 469 (14).

62. Altenberg 1544–1545: H. Löscher, ed., *Das erzgebirgische Bergrecht des 15. und 16. Jahrhunderts,* part 1 (Berlin, 1957), 67.

63. Ed. by C. R. Dodwell (Oxford, 1961), lib. I, c. 25; lib. II, c. 4, 7, 8; III, c. 36, 54; the numbering of the chapters is different in the incomplete edition of W. Theobald (Berlin, 1933).

64. S. J. von Romocki, *Geschichte der Explosivstoffe* (Berlin, 1895), 127 ff.; W. Hassenstein, *Das Feuerwerkbuch von 1420* (Munich, 1941), 65; fuses: Romocki, 186 f.

65. Romocki (note 64), 278, 290 ff.

66. *The Three Books of the Potters Art,* ed. B. Rackham (London, 1934), 41; *Biringuccios Pirotechnia,* ed. by O. Johannsen (Brunswick, 1925); compare the account of the Prussian factory commissioner F. A. A. Eversmann about a Dutch linen bleachery, in which the women worked in long rows under the command of the manufacturing mistress, who was sitting on the platform with a sandglass, in *Technologische Bemerkungen auf einer Reise durch Holland* (Freiberg and Annaberg, 1792).

67. Michael de Massa, a predicant monk in Paris, noted in his commentary

on Matthew (around 1330) that it would be better to pay workers a piece wage, "ad mensuram operis," instead of time wage, "ad mensuram temporis." Since time wage workers were paid at the end of the day as though they worked diligently, it was difficult to prevent cheating. Vienna National Library MS lat. 1512, fol. 180 vb, quoted from W. J. Courtenay, *Schools and Scholars in 14th-Century England* (Princeton, New Jersey, 1987), 4.

68. R. A. Goldthwaite, "The Building of the Strozzi Palace: The Construction Industry in Renaissance Florence," *Studies in Medieval and Renaissance History* X (1973): 99–194, here 173: work day, "opera," with divisions into fourths, sixths, and twelfths.

69. B. Dini, "I Lavoratori dell'Arte della Lana nel XIVe à XVa secolo," in *Artigani e Salariati* (note 10), 27–68, here 45 ff.

70. Casalini (chap. 3 note 4), 3; compare L. Ricetti, *Duomo* (chap. 5 note 11), 214.

71. B. Dini takes them to be modern hours and in his recapitulation he thus expresses his agreement with Le Goff: "I lavoratori erano completamente determinati dal nuovo modo di concepire il tempo." For this view Dini also refers to the legendary public clock in Florence in the year 1325 (see chap. 7 note 31).

72. See L. Riccetti (chap. 5 note 11).

73. D. Balestracci, "'Li lavoranti non congnosciuti.' Il salariato in una città medievale (Siena 1340–1344)," *Bolletino senese di storia patria* 82/83 (1975–1976): 67–157, here 119.

74. See S. Battaglia, *Grande Dizionario della Lingua Italiana,* vol. IV (Turin, 1971), 977, s.v. "Dotta."

75. See P. Braunstein, "Les débuts d'un chantier. Le Dôme de Milan sort de terre 1387," in P. Benoît and O. Chapelot, eds., *Pierre et Métal dans le bâtiment au Moyen Age* (Paris, 1985), 81–102; idem, "Les salaires sur les chantiers monumentaux du Milanais à la fin du XIVe siècle," in X. Barral y Altet, ed., *Artistes, Artisans et production artistique au moyen âge,* vol. 1 (Paris, 1986), 123 ff.

76. *Annali* I (note 38), 79, Annali Appendici I, 318, II, 29.

77. ". . . [Q]uia quando magistri recedunt a suis laboreriis, recedunt medii ipsorum, et vacant ab operibus suis per tres horas, et facto computo quod ipsi habent salarium den. 8 pro qualibet hora, dicitur quod fabrica portat damnum dicta occasione saltem gross. 12 1/2 singulo die." *Annali* I, 78, 309; the published source selections are themselves a large quarry; I am grateful to P. Braunstein for references.

78. Hourly calculation of transportation services at the construction site of the Certosa in Pavia (1396), Beltrami (note 38), 160.

79. See above note 61, fol. 183 verso, Fagniez, *Documents* (note 14), 143; the text mentions another possibility of piece(?) or time(?) wage whose meaning is not clear to me: "ad taelnam," from "tallia" = cut, cut piece? Orléans: Fagniez, ibid., 187 f.

80. L. Fumi, *Statuti e Regesti dell'Opera di S. Maria d'Orvieto* (Rome, 1891), statutes 1421, § 36, p. 37.

81. "Lutz Steinlingers Baumeisterbuch vom Jahre 1452," *Mitteilungen des Vereins für die Geschichte der Stadt Nürnberg* 2 (1880): 60 ff.

82. M. Lexer and F. von Weech, *Endres Tuchers Baumeisterbuch der Stadt Nürnberg (1464–1475)* (Stuttgart, 1862; repr. 1968), 60–69, "stundgelt" p. 68, compare 272–277.

83. Ed. by A. Gümbel, *Repertorium für Kunstwissenschaft* 32–34 (1909–1911).

84. Ed. by A. Gümbel (chap. 5 note 60), 19–94, pp. 35, 41, 57 and elsewhere.

85. F. Lennel, *Histoire de Calais* II (Calais, 1911), 80 ff.

86. Knoop and Jones, "Overtime" (note 38); A. Deville, *Comptes des dépenses de la construction du château de Gaillon (1504)* (Paris, 1850), 133; B. Bucher, *Die alten Zunft- und Verkehrsordnungen der Stadt Krakau* (Vienna, 1889), 91.

87. Baulant (note 7), 465 note 2.

88. Complaints of the subjects: "Und damit wir in der Zeit der Sonnen irreten, hat uns der Herr von Rackel die Sonnenuhr von der Kirchen wegreißen und in Stücke zuhauen lassen. Und ließ der Herr unterweilen im Winter umb 9 Uhr anstatt 12 Uhr die Mittagsglocke laute, damit wir eher zu Hoffe fahren müßten, daß aniezo aber aus dem halben Wintertag auch ein ganzer Tag geworden, so daß wir Cossetten es weder mit den Handdiensten, noch weniger aber wir Bauren mit dem Gespahn es austehen können und dadurch ganz ruiniert werden." R. Lehmann, *Quellen zur Lage der Privatbauern in der Niederlausitz im Zeitalter des Absolutismus* (Berlin, 1957), 71.

89. T. Vernaleken, *Alpensagen* (Vienna, 1858), 319.

90. From W. Bosse, *Die Verhältnisse der Kammer als Domanialbehörde im Lande Stargard (1755–1806)* (Rostock, 1930), 111 f.; compare 38.

91. Engels, "Lage," in MEW 2, 398 ff.; Marx, "Rede über den Freihandel," MEW 4, 448; Marx, "Kapital," MEW 23, 391 ff.; E. P. Thompson, *The Making of the English Working Class* (London, 1968), 321, 337 f.

92. R. E. Prutz (Leipzig, 1851), vol. 1, 120.

93. Thompson, "Time," 85 f.; L. Allegato, *Socialismo e communismo in Puglia. Ricordi di un militante 1904–1924* (Rome, 1971), 46 (this reference furnished by V. Hunecke). *Factory Act* 1844, chap. 26: working time must be regulated through a public or publicly accessible clock, likewise *Factories Act* 1861; *Zürcherische Justizdirektion* 1858: working time in the cotton spinning mill in Engstringen must be regulated by the standard clock and may not be regulated by the arbitrarily set factory clock, in *Mitteilungen aus den Akten der Zürcherischen Fabrikkommission*, vol. 1 (Zurich, 1858), 286–288; *Argauisches Fabrikpolizeigesetz* 1862: working time must be regulated by the public clock. At the end of the nineteenth century, the factory regulations uniformly make the factory clock the binding time-giver.

94. In MEW 23, 448 f.

95. B. Hargreaves, *Recollections of Broad Oak* (Bowker, 1882), 42; quoted by G. Pearson, "Goten und Vandalen-Verbrechen in historischer Perspektive," *Kriminologisches Journal* 9 (1977): 289.

96. D. T. Rogers, *The Work Ethic in Industrial America 1850–1920* (Chicago and London, 1974), 153 ff.; graphic examples in B. F. Tolles, Jr., *Textile Mill*

Architecture in East Central New England: An Analysis of Pre–Civil War Design (1971), 223–253.

97. M. W. Flinn, ed., *The Law Book of Crowley Ironworks,* Surtees Society 167 (Durham and London, 1957), esp. "Order Number 103."

98. Control clocks: H. Kahlert, "Kontrolluhren des 19. Jahrhunderts," *AU* 7 (1984): 53 ff.; A. Burchall, "The Noctuary or Watchman's Clock," *AH* 15 (1985): 231 ff.; McKendrick (note 2); R. M. Currie, *Work Study,* 2nd ed. (London, 1963), 3–4; compare S. M. Macey, "Work Study before Taylor: An Examination of Certain Preconditions for Time and Motion Studies that Began in the Seventeenth Century," *Work Study and Management Services* 18 (1974), 530 ff.

99. R. A. Ferchault de Réaumur, *L'art de l'Épinglier . . . ;* German: *Der Nadler, oder die Verfertigung der Nadeln nebst Zusätzen von Herren du Hamel de Monceau und einigen aus Herrn Perronet . . . gezogenen Anmerkungen (übersetzt von J. H. G. von Justi),* in: *Schauplatz der Künste und Handwerke . . .* (Berlin, Stettin, and Leipzig, 1762), 191 ff., here 201, 212 ff., 227, 235.

100. Bernard Forest de Belidor, *Architectura Hydraulica, oder Die Kunst, die Gewässer des Meeres und der Flüsse zum Vorteil der Verteidigung . . . anzuwenden* (1739), part 2 (Augsburg, 1766), 21 ff.; the reference from H. Hämmerli, *Der Zeitakkord* (Berne, 1949), 42.

101. Municipal Archive Cologne, Französische Verwaltung (French Administration) No. 1336, 12.

102. R. Ehrenberg, "Krupp-Studien III," *Archiv für exakte Wirtschaftsforschung* vol. 3 (Jena, 1911), 88 f.

103. M. A. Bienefeld, *Working Hours in British Industry: An Economic History* (London, 1972); G. Cross, *A Quest for Time: The Reduction of Work in Britain and France, 1840–1940* (Berkeley, 1989); H. Pohl, ed., *Wirtschaftswachstum, Technologie und Arbeitszeit im internationalen Vergleich,* Zeitschrift für Unternehmensgeschichte, Beiheft 24 (Wiesbaden, 1983).

Chapter 10

1. See H. Kownatzki, "Geschichte des Begriffs und Begriff der Post," *APT* 51 (1923): 377–423.

2. Detailed discussion in V. Aschoff, *Geschichte der Nachrichtentechnik,* vol. 1.2, rev. ed. (Berlin, 1989), chaps. IV and VI.

3. O. Codogno, *Nuovo itinerario de poste per tutto il mondo* (Milan, 1608), 73 f., from J. Rübsam, "Zur Geschichte des internationalen Postwesens im 16. und 17. Jahrhundert," *Historisches Jahrbuch der Görres-Gesellschaft* 13 (1892): 72.

4. Marco Polo, *The Travels,* transl. by Ronald Latham (New York, 1958), 152–154.

5. B. Spuler, *Die Mongolen im Iran. Verwaltung und Kultur der Ilchanzeit (1220–1350),* 3rd ed. (Berlin, 1968), 422–426; P. Olbricht, *Das Postwesen in China unter der Mongolenherrschaft im 13. und 14. Jahrhundert,* Göttinger Asiatische Forschungen, vol. 1 (Wiesbaden, 1954).

6. H. Yule and H. Cordier, *Cathay and the Way Thither Being a Collection of*

Medieval Notices of China, new ed., Hakluyt Society vol. 33 (London, 1913; repr. 1967), 232.

7. Ibid., vol. 37 (1914; repr. 1967), 92.

8. Ed. by M. Letts, Haklyut Society, 2nd series, vol. 102, vol. II (1953; repr. 1967), 366 f.

9. B. Spuler (note 5), 425.

10. See "cursus publicus," in RE IV, cols. 1846 ff., and "Postwesen" in RE XXII, cols. 988 ff.; A. M. Ramsey, "The speed of the Roman Imperial Post," *Journal of Roman Studies* 15 (1925): 60–74.

11. J. Sauvaget, *La poste aux chevaux dans l'empire des Mamelouks* (Paris, 1941); M. Sabbagh and C. Löper, *Die Brieftaube schneller als der Blitz* (Strasbourg, 1897); compare the travel accounts of Johann Schiltberger of Munich (ca. 1420), ed. by K. F. Neumann (Munich, 1859; repr. 1976), 109 f.

12. See "Botenwesen" in Lexikon des Mittelalters II, cols. 484 ff.; E. Vaille, *Histoire Générale des Postes Françaises,* 2 vols. (Paris, 1947–1955), vol. 1, 149 ff.; a detailed description for the German-speaking region in E. Kießkalt, *Die Entstehung der deutschen Post bis zum Jahre 1932* (Erlangen, 1938); O. Lauffer, "Der laufende Bote im Nachrichtenwesen der früheren Jahrhunderte," *Beiträge zur deutschen Volkskunde* 1 (1954): 19–60; on the later development see S. Oettermann, *Läufer und Vorläufer. Zu einer Kulturgeschichte des Laufsports* (Frankfurt, 1984).

13. C. Löper, "Das Botenwesen und die Anfänge der Posteinrichtungen im Elsaß, insbesondere in der freien Reichsstadt Straβburg," *APT* 4 (1876): 197–204, 231–241. The payment of messengers by the mile is reported in the year 1345 for the ecclesiastical court in Constance. It is also found in a messenger statute from Constance from 1459 and in the Court Messenger Code of Wolfach from 1470; see Fuchs, "Zum Nachrichten und Verkehrswesen im Mittelalter am Oberrhein und am Bodensee," *APT* 14 (1886): 417–429; see also the code for running messengers of the city of Comar from the beginning of the sixteenth century, in "Urkunden über Botendienst und Postwesen im Elsaß," *APT* 14 (1886): 673 f.

14. ("temporibus debitis et statutis et per dietas continuas et directas"), A. Schaube, "Der Kurierdienst zwischen Italien und den Messen der Champagne," *APT* 24 (1896): 542–550, 571–581, 579.

15. F. Melis, "Intensità e regolarità nella diffusione dell'informazione economica generale nel Mediterraneo e in Occidente alla fine del Medioevo," in *Mélanges F. Braudel* (Paris, 1973), 389–424b.

16. F. Ludwig, *Reise- und Marschgeschwindigkeiten im 12. und 13. Jahrhundert* (Berlin, 1897); P. Thomas, "Délai de transmission de lettres françaises à destination de Lille pendant la fin du XIVe siècle," *Revue du Nord* 4 (1913): 89–122; M. N. Boyer, "A Day's Journey in Medieval France," *Speculum* 26 (1951): 597–608; E. Riedel, "Zur Geschichte der Reisegeschwindigkeiten," *Archiv für Postgeschichte in Bayern* 8 (1952–1954): 117–121; J. W. Nesbitt, "The Rate of March of Crusading Armies in Europe," *Traditio* 13 (1963): 168–181; R.-H. Bautier, "Recherches sur les routes de l'Europe médiévale," *Bull. phil. hist.* (1960), I:

102 f., 115 f.; Y. Renouard, "Routes, étapes et vitesses de marche de France à Rome au XIII au XIV siècles d'après les itinéraires d'Eudes Rigaud (1254) et de Barthélemy Bonis (1350)," in *Studi in Onore di Amintore Fanfani* III (Milan, 1962), 405–428; J. M. Cauchies, "Messageries et messages en Hainaut en XVe siècle," *Le Moyen Age* 82 (1972): 397 ff.; R. Elze, "Über die Leistungsfähigkeit von Gesandschaften und Boten im 11. Jahrhundert.—Aus der Vorgeschichte von Canossa 1075–1077," *Francia Beiheft* 9 (1980): 3–10; R. Schäffer, "Zur Geschwindigkeit des 'staatlichen' Nachrichtenverkehrs im Spätmittelalter," *Zeitschrift des historischen Vereins für die Steiermark* 76 (1985): 101–119; H. Hundsbichler, "Selbstzeugnisse spätmittelalterlicher Reisetätigkeit und historische Migrationsforschung," in G. Jaritz and A. Müller, eds., *Migration in der Feudalgesellschaft* (Frankfurt, 1988): 351–369, esp. 352, 359 f.

17. E. Melillo, *Le Poste nel Mezzogiorno d'Italia* (Naples, 1879), 64.

18. The text of the receipt has been edited by H. Bösch in *Mitteilungen aus dem Germanischen Nationalmuseum* I (1884–1886): 255 f.

19. W. von Stromer, *Die Nürnberger Handelsgesellschaft Gruber-Podmer-Stromer im 15. Jahrhundert,* Nürnberger Forschungen, vol. 7 (Nuremberg, 1963), 24 f., from K. O. Müller, *Welthandelsbräuche (1480–1540),* Deutsche Handelsakten des Mittelalters und der frühen Neuzeit V (Stuttgart, 1934), 188; the Augsburg Messenger Code of 1555 also designates just under six days for the route Venice-Nuremberg, *APT* 18 (1887): 198–200; however, P. Sardella, *Nouvelles et Spéculations à Venise au début du XVIe siècle* (Paris, 1949), 57, reconstructs, on the basis of Venetian sources, eight days as a quick and twenty-one days as the average delivery time for messages on this route.

20. In G. F. Pagini, *Della decima . . . , della moneta e della mercatura de' Fiorentini perfino al secolo XVI,* vol. 4 (Lisbon and Lucca, 1766), 103; compare Y. Renouard, "Information et Transmission des Nouvelles," in C. Samaran, ed., *L'histoire et ses Méthodes* (Paris, 1961), 112 f.

21. See P. Sardella, *Nouvelles* (note 19); for the earlier period see A. L. Udovitch, "Time, the Sea and Society: Duration of Commercial Voyages on the Southern Shores of the Mediterranean during the High Middle Ages," in *Navigazione mediterranea nell'alto medioevo,* Centro italiano di studi sull'alto medioevo XXV (Spoleto, 1978), 503–546.

22. R. Schäffer, "Nachrichtenverkehr," 104.

23. One of the rare studies is that of I. K. Steele, "Time, Communications and Society: The English Atlantic, 1702," *Journal of American Studies* 8 (1974): 1–21; scattered references also in Schäffer (note 16).

24. *Annali della Fabbrica* (chap. 6 note 45) I, 32 f.

25. F. Melis (note 15), 395 and note 23; on speed-dependent courier rates in Spain see also F. Ohmann, *Die Anfänge der deutschen Post und die Taxis* (Leipzig, 1909), 30.

26. Exceptions to this rule can be found already in antiquity, for example, in Cicero's oration for Sextus Roscius of Ameria (80 B.C.): the bringer of the news covered the distance from Rome to Ameria in a traveling carriage in ten night hours, "decem horis nocturnis . . . pervolavit," Cicero, *Pro Sexto Roscio*

Amerino 19, ed. with English transl. by J. H. Frese, Cicero vol. VI (London, 1930; 1984), 138.

27. The reference to the deviation in J. Rübsam, "Aus der Urzeit der modernen Post 1425–1562," *Historisches Jahrbuch der Görres-Gesellschaft* 21 (1900): 33 ff., here 39; the text from Gian Battista Ramusio, *Navigazioni et Viaggi,* vol. 1.2 (Venice, 1583), fol. 30 recto (repr. Amsterdam, 1968); on the textual tradition see L. F. Benedetto, *Marco Polo, Il Milione. Prima edizione integrale* (Florence, 1928), here 96; and the introduction to the English edition, A. C. Moule and P. Pelliot, *Marco Polo: The Description of the World,* vol. 1 (London, 1938).

28. B. Spuler (note 5), 425.

29. P. Ratchnevsky, *Un code des Yuan* (Paris, 1937), 281 ff.; O. Olbricht (note 5), 63 f., 79, 87.

30. That the arrival times of monastic messengers were already noted with hour indications in the eighth century is a misunderstanding that found its way into S. Maurice and K. Maurice, *Stundenangaben,* 156. The passage cited for this, E. Kießkalt (note 12), 32, speaks only in general terms of the recording of the arrival time. According to A. Ebner, *Die klösterlichen Gebetsverbrüderungen bis zum Ausgang des karolingischen Zeitalters* (Regensburg, 1890), the day was noted on the death certificate in order to prevent negligence on the part of the messengers. A death certificate from Admont from the late fifteenth century (1484–1485) contains a few scattered hour indications, but it gives them according to the canonical hours or monastic services; J. Wichner, ed., *Studien und Mitteilungen aus dem Benedictiner- und Cistercienserorden* 5 (1884): 64.

31. First references to the communication system of the German Order were given by J. Voigt in *Raumers Historische Taschenbuch,* vol. 1 (Leipzig, 1830), 219, who also published a piece (unfortunately undated) that was reproduced in Bilfinger, *Horen,* 209. His description was challenged—wrongly, as it later turned out—by P. Babendererde, in "Nachrichtendienst und Reiseverkehr des Deutschen Ordens um 1400," *Altpreußische Monatsschrift* 50 (1913): 189–244, esp. 243 f. Two other pieces from the Thorn Ratsarchiv (January 1 and 24, 1439), are printed in "Über die Entwicklung des Postwesen in Preußen zur Zeit des Deutschen Ordens and der Polnischen Oberhoheit," *APT* 10 (1882), 492. Reference to the great number of such notations in the Order's archive of letters since 1409, at the latest, in P. G. Thielen, "Die Rolle der Uhr im geistlichen und administrativen Alltagsleben der Deutschordenskonvente in Preußen," in E. Bahr, ed., *Studien zur Geschichte des Preußenlandes,* Festschrift E. Keyser (Marburg, 1963), 392–396; compare idem, *Die Verwaltung des Ordensstandes in Preußen vornehmlich im 15. Jahrhundert* (Cologne, 1965), 117–119. A description and cartographic analysis of the "express letters" was subsequently given by J. Jahnke and H. Zimmermann, "Die Postwege des Deutschen Ordens," *Historisch-geographischer Atlas des Preußenlandes,* ed. by H. Mortensen, G. Mortensen and R. Wenskus, vol. 1 (Wiesbaden, 1968). Unfortunately it does not say anything about the age of such notations.

32. J. Jahnke and H. Zimmermann, "Erläuterungen," ibid., 1.

33. P. G. Thielen, "Uhr" (note 31), 393.

34. See the letter of the commander (Komtur) Michael Küchenmeister to the Grand Master of the Order from April 29, 1418, in H. Koeppen, ed., *Die Berichte der Generalprokuratoren des Deutschen Ordens an der Kurie,* vol. 2, Veröffentlichungen der niedersächsischen Archivverwaltung Heft 13, (Göttingen, 1960) no. 253, 480 f. The letter was posted in Frankfurt an der Oder, but it bears praesentata notices only from Schlochau on.

35. See the views of: Balga, Brandenburg, Braunsberg, Christburg, Fischhausen, Frauenburg, Heiligenbeil, Insterburg, Kreuzburg, Königsberg, Ragnit, Rastenburg, Soldau, Tapiau, Tilsit, and Welau in *Von Danzig bis Riga. Ausstellungskatalog des Germanischen Nationalmuseums Nürnberg* (1982).

36. Zinner, *Räderuhren,* 32 f.; compare P. G. Thielen, "Uhr" (note 31), 394.

37. See M. Lindemann, *Nachrichtenübermittlung durch Kaufmannsbriefe. Briefzeitungen in der Korrespondenz Hildebrand Veckinghusens (1398–1428)* (Munich, 1978), 14 ff.

38. "Datum cavalario Brixie . . . hora decima octava." L. Osio, *Documenti diplomatici tratti dagli archivi Milanesi,* vol. 1 (Milan, 1864), 162 f.; vol. 2 (1869), 343; A. Sassi, *Cenni di Storia Postale del secolo XIV al XVIIII* (Estratti del Giornale "Il Francobollo") (Milan, 1897), 8 f., 1387: "per cavalarios postarum . . . et notentur hore"; 1389: "portentur velociter die noctuque"; C. Fedele and M. Gallenga, *Per un servizio di Nostro Signore. Strade, corriere e poste dei Papi dal Medieovo al 1870* (Modena, 1988), 25 ff.

39. A. Sassi (note 38).

40. C. Fedele and F. Gallenga (note 38), 25, 201.

41. For the history of the Taxis postal service see M. Dallmeier, *Quellen zur Geschichte des europäischen Postwesens 1501–1806,* part 1: Quellen- Literatur- Einleitung, part 2: Urkundenregesten, Thurn-und Taxis-Studien 9/I, II (Kallmünz, 1977) (cited as Dallmeier I and II), here 49 ff.; W. Behringer, *Turn und Taxis. Die Geschichte der Post und ihrer Unternehmen* (Munich, 1990).

42. Dallmeier II, nos. 41 and 188.

43. Behringer (note 41), 13 f.

44. See, for example, the nationalistic polemic of E. Kießkalt, *Die Post—ein Werk Kaiser Friedrichs III., nicht der Taxis. Die Aufdeckung einer Kultur- und Geschichtslüge* (Bamberg, 1926); E. Kießkalt's great work of 1938 (note 12) was more moderate in tone, though factually it pursued the same German-Prussian tendency.

45. From Behringer (note 41), 26 f.

46. The text in F. A. Isambert, *Recueil général des Anciens Lois Françaises,* book 5 (Paris, 1825), 487–492; G. Zeller, "Un faux du XVIIe siècle: l'édit de Louis XI sur la poste," *Revue Historique* 180 (1937): 286–292; L. Kalmus, *Weltgeschichte der Post* (Vienna, 1937), 316 ff., had also suspected a forgery, though he dated it to 1508–1509; compare E. Vaille, *Histoire* (note 12), vol. 2, 24–37; on the context see I. Mieck, *Die Entstehung des modernen Frankreich 1450–1619* (Stuttgart, 1982), 60 f.

47. The oldest extant hour pass from 1494 in an exhibit on the history of

transportation in Milan (April 1901); see J. Rübsam in *APT* 29 (1901), 482. Other passes in O. Redlich, "Vier Post-Stundenpässe aus den Jarhen 1496–1500," *Mitteilungen des Instituts für Österreichische Geschichtsforschung* 12 (1891): 494–504; J. Rübsam, "Das kaiserliche Postamt zu Mailand in der ersten Hälfte des XVI. Jahrhunderts unter Simon von Taxis," *APT* 29 (1901): 443–453; F. Ohmann (note 25), 130–147.

48. J. Rübsam, "Postamt," 452; reproduction of an hour pass with a drawing of a gallows in Behringer, 29.

49. F. Ohmann (note 25), 319 f.

50. Ibid., 133 ff.

51. Ibid., 326–329.

52. Dallmaier I, no. 2.

53. Ibid., no. 3.

54. In a Netherlandish-French postal agreement of December 15, 1660, the conveyance time of letters between Brussels and Paris was fixed at 40–42 hours (in winter 48–50); ibid., no. 325. On the actual conveyance times of imperial letters and delays caused by war see W. Bauer, "Die Taxische Post und Beförderung der Briefe Karls V. in den Jahren 1523 bis 1525," *MIÖG* 27 (1906): 436–459.

55. For example, in the "official regulations" of the posts in the Kingdom of Naples of September 18, 1559, in A. de Sariis, *Codice delle Leggi del regno di Napoli,* lib. V, tit. XX (Naples, 1794), from Melillo (note 17), 96 f.

56. Dallmeier I, 61 ff.

57. Dallmeier II, no. 124.

58. J. Rübsam, "Zur Geschichte des internationalen Postwesens im 16. und 17. Jahrhundert," *Historisches Jahrbuch der Görres-Gesellschaft* 13 (1892): 15–79, here 34 f.; *APT* 16 (1888): 212.

59. Dallmeier II, nos. 126–128, Rübsam (note 58), 43 f. and 75–77.

60. See the Imperial Postal Codes (Reichspostordnungen) of 1596 and 1698, Dallmeier II, nos. 106, 479, 480, and numerous individual statutes.

61. Dallmeier II, no 157; compare Behringer, 107 f.

62. "Relais- und Vorspann-Reglement 1703," ed. by C. O. Mylius, *Corpus Const. Magdeburgicarum novissimarum* (Magdeburg and Halle, 1714–1717), V, no. 110, 328; "Postordnung für den Kurs Berlin-Cleve 1710," *Corpus Constitutionum Magdeburgicarum.* V, § 13, 104; "Preußische Postordnung 1712," *Corpus Constitutionum Marchicarum* (chap. 8 note 91), IV, I, III, no. 97, cols. 1014, 1017; "Postordnung für Polen und Sachsen 1713," ed. by J. C. Lüning, *Codex Augusteus oder neuvermehrtes Corpus Juris Saxonici* (Leipzig, 1724, with three continuations to Dresden, 1824), II, cols. 1066 f.; "Ermahnung 1720:" II, cols. 1111 f.; "Oberpostamtsverordnung" (Sachsen) (1744), *Codex Augusteus,* cont. I, section I, book IV, chap. 6, cols. 1767–1768, 1777–1778.

63. *Corpus Const. March* (chap. 8 note 91), IV, I, III, no. 68, col. 916 f.; *Cod. Aug.* (note 62), II, II, IV, c. 6, cols. 1066 f., 1070 f.; H. Stephan, *Geschichte der preußischen Post* (Berlin, 1859; repr. Glashütten, 1976), 309.

64. *Cod. Augusteus* (note 62), II, cols. 1017, 1056.

65. *Corpus Const. March.* (chap. 8 note 91) IV, I, III, no. 167, col. 1103; see "Postordnung für Hessen-Kassel 1719," in H. Haass, "Das hessische Postwesen bis zu Beginn des 18. Jahrhunderts," *Zeitschrift des Vereins für hessische Geschichte und Landeskunde* 44 (1910): 65.

66. Ibid., no. 68, cols. 914 f., no. 124, cols. 1047 f.

67. P. S. Bagwell, *The Transport Revolution from 1770* (London, 1974); G. Arbellot, "Le grade mutation des routes de France au milieu du XVIIIe siècle," *Annales: Economies. Sociétés. Civilisations.* 28 (1973): 765–791; B. Lepetit, *Chemins de terre et voies d'eau — Réseaux de transports et organisation de l'espace en France 1740–1840* (Paris, 1984); W. Zorn, "Verdichtung und Beschleunigung des Verkehrs als Beitrag zur Entwicklung der 'modernen Welt,'" in R. Koselleck, ed., *Studien zum Beginn der modernen Welt* (Stuttgart, 1977), 115–134.

68. G. Arbellot, "Arthur Young et la circulation en France," *Revue d'histoire moderne et contemporaine* 28 (1981): 328–334.

69. A. H. Cole, "The Tempo of Mercantile Life in Colonial America," *Business History Review* 33 (1959): 277–299.

70. F. Braudel, *Civilization and Capitalism, 15th–18th Century,* vol. 1: *The Structures of Everyday Life,* transl. by Siân Reynolds (New York, 1979), 426–427.

71. H. Stephan (note 63), 791–791; W. Zorn (note 67), 121; P. Bagwell (note 67), 48.

72. P. Bagwell (note 67), 47, 53; Klaus Beyrer, "Das Reisesystem der Postkutsche. Verkehr im 18. und 19. Jahrhundert," in *"Zug der Zeit — Zeit der Züge." Deutsche Eisenbahnen 1835–1985,* 2 vols. (Berlin, 1985), vol. I, 39–59.

73. L. Börne, *Sämtliche Schriften,* ed. by I. Rippmann and P. Rippmann (Düsseldorf, 1964), 639–667.

74. From K. Sautter, "Die preußischen Schnellposten," *APT* 47 (1919): 451.

75. See Landes, *Revolution,* chap. II.

76. Enslin, "Bürgerliche Zeit."

77. K. A. von Kamptz, ed., *Annalen der preußischen inneren Staatsverwaltung* 9 (1825), no. 91, 415; 1.17.1833, no. 88, 144 and 4.17.1833, no. 80, 996.

78. The clocks of the mail coach guards supposedly already indicated Greenwich Mean Time: E. Zerubavel, "Standardization of Time," 6; see Howse, *Greenwich Time,* 83.

79. Nicholas Wood, *A Practical Treatise on Rail-Roads,* 2nd ed. (London, 1832), p XII; cited from W. Shivelbusch, *The Railway Journey: The Industrialization of Time and Space in the 19th Century* (New York, 1979), 8.

80. M. Riedel, "Vom Biedermeier zum Maschinenzeitalter," *Archiv für Kulturgeschichte* 43 (1961): 100–123, 112; Nuremberg and Fürth, from J. Jehle in *"Zug der Zeit"* (note 73) I, 69.

81. In addition to M. Riedel see D. Sternberger, *Panorama oder Ansichten vom 19. Jahrhundert,* 3rd. ed. (Hamburg, 1955), 50; see also the essays in *"Zug der Zeit — Zeit der Züge"* (note 73) by F. Sonnenberger, 24–37, M. Jehle, 68–93, D. Vorsteher, 404–431, and J. Tschoeke, 432–438.

82. L. Börne (note 72), 645.

83. *Quarterly Review* 63 (1839), cited from W. Schivelbusch (note 79), 34.

84. Streckert, "Stundenzonenzeit"; compare Zerubavel, "Standardization," and Howse, *Greenwich Time.*

85. In Adolf Glaßbrenner's comic play "Der Heiratsantrag," in *Buntes Berlin* (Berlin, 1843; repr. 1981), 29. The phrase "es ist höchste Eisenbahn" (lit., "it is high train") has become proverbial in German to mean "it is high time."

86. G. Schmoller, "Über den Einfluß der heutigen Verkehrsmittel," *Preußische Jahrbücher* 31 (1873): 424.

87. March 16, 1891, Helmuth Graf von Moltke, *Gesammelte Schriften und Denwürdigkeiten,* vol. 7 (Berlin, 1892), 38–43; Streckert, "Stundenzonenzeit," 509–511.

88. "Die Zukunft der öffentlichen Zeit-Angaben und Wetter-Anzeigen," *Reichsanzeiger,* 2, Beilage, No. 273 (Nov. 12, 1890); U. Merle, "Tempo! Tempo! Die Industrialisierung der Zeit im 19. Jahrhundert," in Jenzen, *Uhrzeiten,* 166–217.

BIBLIOGRAPHY

Autrand, Françoise. "Les Dates, la Mémoire et les Juges." In *Le Métier d'Historien au Moyen Age,* pp. 157–182. Edited by B. Guenée. Paris, 1977.

Bassermann-Jordan, Ernst von. *Alte Uhren und ihre Meister.* Leipzig, 1926; repr. 1979.

Battard. M. *Beffoirs, Halles, Hôtels de Ville dans le Nord de la France et la Belgique.* Arras, 1948.

Bedini, Sivlio A., and Francis R. Maddison. *Mechanical Universe: The Astrarium of Giovanni de Dondi.* Transactions of the American Philosophical Society, N. S., vol. 56, part 5. Philadelphia, 1966.

Beeson, C. F. C. *English Church Clocks 1280–1850: History and Classification.* London, 1971.

———. *Perpignan 1356: The Making of a Tower Clock and Bell for the King's Castle.* London, 1982.

Bergmann, Werner. *Innovationen im Quadrivium des 10. und 11. Jahrhunderts. Studien zur Einführung von Astrolab und Abakus im Lateinischen Mittelalter.* Sudhoffs Archiv, Beiheft 26. Stuttgart, 1985.

Bilfinger, Gustav. *Die mittelalterlichen Horen und die modernen Stunden. Ein Beitrag zur Kulturgeschichte.* Stuttgart, 1892; repr. 1969.

Borst, Arno. *Astrolab und Klosterreform an der Jahrtauendwende.* Sitzungsberichte der Heidelberger Akademie der Wissenschaften. Philosophisch-historische Klasse 1989/1. Heidelberg, 1989.

———. *Computus. Zeit und Zahl in der Geschichte Europas.* 1990.

Brusa, Guiseppe. *L'Arte dell'Orologeria in Europa — Sette secoli de orologi meccanici.* Busto Arsizio, 1978.

———. "Early Mechanical Horlogy in Italy." *AH* 18/5 (1990):484–514.

Cancellieri, Francesco. *Le due nove Campane di Campidoglio . . .* Rome, 1806.

Chapius, Alfred. *De Horologiis in Arte.* Lausanne, 1954.

Cipolla, Carlo M. *Clocks and Culture 1300–1700.* London, 1967.

Copenhaver, Brian P. "The Historiography of Discovery in the Renaissance: The Sources and Composition of Polydore Vergil's De Inventoribus Rerum I–III." *Journal of the Warburg and Courtauld Institutes* 41 (1978): 192–214.

Edwardes, Ernest L. *Weight-driven Chamber Clocks of the Middle Ages and Renaissance.* Altrincham, 1976.

Elias, Norbert. *Über die Zeit. Arbeiten zur Wissenssoziologie II.* Frankfurt, 1984.

Enslin, H. "Die Entwicklung der bürgerlichen Zeit seit 1800." *AU* 11 (1988/1): 60–75.

Fallet-Scheurer, Marius. "Die Zeitmessung im alten Basel." *Basler Zeitschrift für Geschichte und Altertumskunde* 15 (1916): 237–366.

Farré-Olivé, Eduard. "A Medieval Catalan Clepsydra and Carillon." *AH* 18 (1989): 371–380.

Franklin, Albert. *La vie privée d'autrefois. Arts et métiers, modes, moeurs, usages des Parisiens du XIIe au XVIIe siècle (4). La mesure du temps.* Paris, 1888.

Fraser, J. T., F. C. Haber, and G. H. Müller, eds. *The Study of Time II.* Berlin, 1975.

Gimpel, Jean. *The Medieval Machine: The Industrial Revolution of the Middle Ages.* London, 1977.

Grotefend, H. *Zeitrechnungen des deutschen Mittelalters und der Neuzeit.* 3 volumes. Hannover, 1891–1898; repr. 1970.

Gurevich, Aaron J. *Categories of Medieval Culture.* Transl. by G. L. Campbell. London, 1985.

d'Haenens, A. "Le clepsydre de Villers (1276). Comment on mesurait et vivait le temps dans une abbaye cistercienne au XIIIe siècle." In *Klösterliche Sachkultur des Spätmittelalters,* pp. 321–342. Vienna, 1980.

Hahnloser, Hans R. *Villard de Honnecourt. Kritische Gesamtausgabe des Bauhütenbuchs ms fr. 19093 der Pariser Nationalbibliothek.* 2nd, expanded edition. Graz, 1972.

Hammerstein, R. *Macht und Klang. Tönende Automaten als Realität und Fiktion in der alten und mittelalterlichen Welt.* Berne, 1986.

———. *The Book of Knowledge of Ingenious Mechanical Devices by Ibn al-Razzaz al-Jazari.* Dordrecht, 1974.

———. *On the Construction of Water-Clocks.* Turner & Devereux, Occasional Papers n. 4. 1976.

Hill, Donald R. *Arabic Water-Clocks.* Sources & Studies in the History of Arabic-Islamic Science, History of Technology Ser. 4. Aleppo, 1981.

Howgrave-Graham, R. P. "Some Clocks and Jacks, with Notes on the History of Horology." *Archeologia: Or Miscellaneous Tracts relating to Antiquity* 77 (1928): 257–312.

Howse, Derek. *Greenwich Time and the Discovery of the Longitude.* Oxford, 1980.

Jenzen, Igor A., ed. *Uhrzeiten. Die Geschichte der Uhr und ihres Gebrauches.* Frankfurt, 1989.

Jünger, Ernst. *Das Sanduhrbuch.* Frankfurt am Main, 1957.

Kurz, O. *European Clocks and Watches in the Near East.* Studies of the Warburg Institute, vol. 34. London, 1975.

Landes, David S. *Revolution in Time: Clocks and the Making of the Modern World.* London, 1983.

Leclercq, J. "Zeiterfahrung und Zeitbegriff im Spätmittelalter." In *Antiqui und Moderni. Traditionsbewußtsein und Fortschrittsbewußtsein im späten Mittelalter,* pp. 1–20. Edited by A. Zimmermann. Miscellanea Medievalia 9. Berlin, 1974.

Le Goff, Jacques. "Merchant's Time and Church's Time in the Middle Ages." In Jacques Le Goff, *Time, Work, and Culture in the Middle Ages,* pp. 29–42. Transl. by Arthur Goldhammer. Chicago: University of Chicago Press, 1980.

———. "Labor Time in the 'Crisis' of the Fourteenth Century: From Medieval Time to Modern Time." In Jacques Le Goff, *Time, Work, and Culture in the Middle Ages,* pp. 43–52. Transl. by Arthur Goldhammer. Chicago: University of Chicago Press, 1980.

Lehner, Franz. *Die mittelatleriche Tagesteilung in den österreichischen Ländern.* Quellenstudien aus dem historischen Seminar der Universität Innsbruck Heft 3. Innsbruck, 1911.

Leopold, J. H. *The Almanus Manuscript .* (Staats- und Stadtbibliothek Augsburg cod. in 2° no. 209). London, 1971.

Lippert, Elisabeth. *Glockenläuten als Rechtsbrauch.* Das Rechtswahrzeichen 3. Freiburg, 1939.

Lloyd, H. Allan. "Mechanical Timekeepers." In *History of Technology,* vol. 3, pp. 648–675. Edited by C. Singer et al. Oxford, 1957.

Luhmann, Niklas. "Die Knappheit der Zeit und die Vordringlichkeit des Befristeten." *Die Verwaltung* 1 (1968): 3–30.

———. "Weltzeit und Systemgeschichte." In P. C. Ludz, *Soziologie und Sozialgeschichte,* pp. 81–115. Opladen, 1972.

———. *Zweckbegriff und Systemrationalität. Über die Funktion von Zwecken in sozialen Systemen.* Frankfurt, 1973.

Maddison, F., B. Scot, and A. Kent. "An Early Medieval Water-Clock." *AH* 3 (1962): 348–353.

Marperger, Paul Jacob. *Horologiographica oder Beschreibung der Entheilung und Abmeßung der Zeit.* Dresden, 1723; repr. 1975.

Maurice, Klaus. *Die deutsche Räderuhr. Zur Kunst und Technik des mechanischen Zeitmessers im deutschen Sprachraum.* 2 vols. Munich, 1976; cited as Maurice I and II.

Maurice, Klaus, and Otto Mayr. *Die Welt als Uhr. Deutsche Uhren und Automaten 1550–1650.* Munich, 1980.

Maurice, S., and K. Maurice, "Stundenangaben im Gemeinwesen des 16. und 17. Jahrhunderts." In Maurice and Mayr, *Die Welt als Uhr,* pp. 146–158.

Mumford, Lewis. *Technics and Civilization.* New York, 1934.

Nahrstedt, Wolfgang. *Die Entstehung der Freizeit—Dargestellt am Beispiel Hamburgs. Ein Beitrag zur Strukturgeschichte und zur strukturgeschichtlichen Grundlegung der Freizeitpädagogik.* Göttingen, 1972.

Needham, Joseph. *Science and Civilization in China.* 6 volumes (in 15 parts) to date. Cambridge, 1954–.

———. *Heavenly Clockwork: The Great Astronomical Clocks of Medieval China.* 2nd ed., with suppl. by J. H. Combridge. Cambridge, 1986.
———. *The Hall of Heavenly Records: Korean Astronomical Instruments and Clocks, 1380–1780.* Cambridge, 1986.
North, John D. *Richard of Wallingford: An Edition of His Writings with Introductions. English Translation and Commentary.* 3 vols. Oxford, 1976; cited North I, II, III.
———. "Monasticism and the First Mechanical Clock." In *The Study of Time* II, pp. 381–393.
Patart, Christian. *Les cloches civiles de Namur, Fosses et Tournai au Bas Moyen âge.* Credit Communal de Belgique, Collection Histoire, Pro Civitate, ser. in-8, no. 44. Brussels, 1976.
Poulle, Emmanuel, *Un Constructeur d'instruments astronomiques au XVe siécle. Jean Fusoris.* Bibliothèque de l'École des Hautes Études, IV section, fasc. 318. Paris, 1963.
———. *Équatoires et Horlogerie planétaire du XIIIe au XVIe siècle.* 2 vols. Geneva, 1980.
Price, Derek J. de Solla. "Clockwork before the Clock." *Horological Journal* (1955): 810–814; (1956): 31–35.
———. "Gears from the Greeks: The Antikythera Mechanism—a Calender Computer from ca. 80 B.C. Transactions of the American Philosophical Society vol. 64, no. 7 (1974): 1–70.
Quinones, Ricardo J. *The Renaissance Discovery of Time.* Cambridge, Mass.: Harvard University Press, 1972.
Rigg, A. G. "Clocks, Dials and Other Terms." In *Middle English Studies,* pp. 255–274. Edited by D. E. Gray and E. G. Stanley. Oxford, 1983.
Robertson, J. Drumond. *The Evolution of Clockwork.* London, 1931; repr. 1972.
Schreiner, Klaus. "'Diversitas Temporum.' Zeiterfahurng und Epochengliederung im späten Mittelalter." Unpublished manuscript, Bielefeld (cited as Schreiner MS); shortened version in *Epochenschwelle und Epochenbewußtsein,* pp. 381–428. Edited by R. Herzog and R. Koselleck. Poetik und Hermeneutik XII. Munich, 1987.
Simoni, A. *Orologi Italiani dal Cinquecento all'Ottocento.* 2nd ed. Milan, 1980.
Sternagel, Peter. *Die Artes Mechanicae im Mittelalter. Begriffs- und Bedeutungsgeschichte bis zum Ende des 13. Jahrhunderts.* Kallmünz, 1966.
Streckert, W. "Die Stundenzonenzeit." *Jahrbuch für Nationalökonomie und Statistik* 59 (1892): 481–517.
Le Temps Chrétien de la fin du l'antiquité au moyen âge, IIe–XIIe siècles. Paris (C.N.R.S.), 1984.
Thompson, E. P. "Time, Work-Discipline and Industrial Capitalism." *Past and Present* 38 (1967): 56–97.
Thorndike, Lynn. "Invention of the mechanical clock about A.D. 1271." *Speculum* 16 (1941): 242–243.
Turner, Anthony J. "'The Accomplishment of Many Years': Three Notes towards a History of the Sand-glass." *Annals of Science* 39 (1982): 161–172.

———. *Water Clocks, Sand-Glasses, Fire Clocks.* The Time Museum. Catalogue of the Collection, vol. I. Time Measuring Instruments, pt. 3. Rockford, Ill., 1984.

———. *Astrolabes and Astrolabe Related Instruments.* The Time Museum. Catalogue of the Collection, vol. I. Time Measuring Instruments, pt. 1. Rockford, Ill., 1985.

Ungerer, A. *Les Horloges Astronomiques et Monumentales les plus remarquables de l'Antiquité jusqu'à nos jours.* Strasbourg, 1931.

Usher, Abbot P. *A History of Mechanical Inventions.* Rev. ed. New Haven, 1954.

Wendorff, Rudolf. *Zeit und Kultur—Geschichte des Zeitbewußtseins in Europa.* Opladen, 1980.

White, Lynn Jr. *Medieval Technology and Social Change.* Oxford, 1962.

———. "The Iconography of Temperantia and the Virtuousness of Technology." In *Action and Conviction in Early Modern Europe,* pp. 197–219. Edited by T. K. Rabb and J. Seigels. Princeton, 1969.

Wiedemann, Eilhard, and Fritz Hauser. *Über die Uhren im Bereich der islamischen Kultur.* Nova Acta Leopoldina, vol. 100, no. 5. Halle, 1915.

Zerubavel, Eviatar. "The Benedictine Ethic and the Modern Spirit of Scheduling: On Scheduling Social Life." *Sociological Inquiry* 50 (1980): 157–169.

———. "The Standardization of Time: A Sociohistorical Perspective." *American Journal of Sociology* 88/1 (1982): 1–23.

Zinner, Ernst. *Die ältesten Räderuhren und die modernen Sonnenuhren. Forschungen über den Ursprung der modernen Wissenschaft.* 28. Bericht der Naturforschenden Gesellschaft in Bamberg. Bamberg, 1939.

———. *Aus der Frühzeit der Räderuhr. Von der Gewichtsuhr zur Federzuguhr.* Munich, 1954.

INDEX

D

U

V